AF545055

JAVA with VISUAL J++

Dudley W. Gill
Gill Associates, Inc.
Monmouth University, Adjunct Professor
Raritan Valley Community College, Adjunct Professor

CRC Press
Boca Raton London New York Washington, D.C.

Library of Congress Cataloging-in-Publication Data

Catalog record is available from the Library of Congress.

Direct all inquiries to CRC Press LLC, 2000 Corporate Blvd., N.W., Boca Raton, Florida 33431.

International Standard Book Number 0-8493-2048-8
Printed in the United States of America 1 2 3 4 5 6 7 8 9 0
Printed on acid-free paper

About the Author

Dudley Gill is the principal of Gill Associates, Inc., a consulting organization specializing in training and application development involving Java, C++, and object technologies. He started programming with the first-generation languages, setting electrical levels in the early machines and wiring plug panels. He is currently involved in the development of distributed applications.

He holds B.S. and M.S. degrees in Electrical Engineering and a D.Sc. in Computer Science conferred by the George Washington University.

When not involved with information technology he enjoys the Central New Jersey and Southern California seashores. To communicate with the author, send email to dgill@dennardcomm.com

Acknowledgments

This book is dedicated to my wife VaLenda, my three children Dudley, Linda, and Marion, two mothers-in-law Agnes and Dorothy, brother and sister Patricia and Gregory, step-mother Trummell, aunt and uncle Doris and Lacey, all of whom have provided support; and in loving memory of my parents Marion and Dudley and wife Agnes, who enhanced and nourished the vision of what could be.

I am thankful for the support provided by the CRC Press family. Special thanks are due to Associate Editor Dawn Mesa, Project Editor Susan Fox, and Cover Designer Dawn Boyd.

Contents

Part 3 Abstract Window Toolkit

Part 4 Communications, Networking and Advanced Topics

Appendixes

List of Figures

List of Tables

Part 1

Language Fundamentals

1

Introduction to Java, Programs, Algorithms, and Integrated Development Environment

OBJECTIVES

After reading this chapter, you will be able to:

1. Understand Java's applicability to application development.
2. Understand the evolution of programming languages.
3. Understand the Integrated Development Environment.
4. Create a Java project using an Integrated Development Environment.

OVERVIEW

This chapter introduces the Java language and its applicability to application development. This chapter also introduces and develops an understanding of the Integrated Development Environment. This understanding will be applied in a hands-on exercise in which a project is created and a graphical user interface—including label, text field, and command button controls—is developed.

JAVA: A NEW LANGUAGE

James Gosling designed Java for use in embedded systems, that is computers used in appliances, machinery, automobiles, vending machines, and similar products. With the contribution of many others, Java has evolved into a mature programming language suitable for use in developing complex and sophisticated enterprise-wide applications.

The language is high-level, general-purpose, and object-oriented. Of particular interest to developers is the intent for Java to be implementation-independent, thereby facilitating porting programs from machine to machine or platform to platform.

Java is a relatively simple language when compared to languages such as C and C++. It incorporates many features of C and C++ as well as other languages. On the other hand, pointers (which are felt by many programmers to be a manifestation of the devil!) are absent in Java. Another advantage and simplifying factor is the inclusion of automatic memory management.

Java's relative simplicity and promise of machine-independence have captured the interest of programmers throughout the industry. This text is intended to provide an entry to the Java programming world.

PROGRAMS AND ALGORITHMS

A **program** is a set of instructions which when performed solve a problem. The steps required to bake a cake or service an automobile engine are examples of programs. Currently, complex and diverse problems such as managing airline reservations, stock trading, medical imaging and manufacturing processes have been expressed as programs. Mathematicians and computer scientists refer to procedures for solving problems as **algorithms** (after Al-Khowarizimi, a first century Arab mathematician.) Typically, algorithms involve the performing of repetitive operations many, many, many times. Such a procedure would overwhelm the most resourceful and determined person; however, computers have the ability to perform millions of operations per second and thus are able to successfully resolve these problems.

A procedure common to many programs is the ability to count backwards from a number by one and stopping at zero. For example, starting at ten the numbers counted are ten, nine, eight, seven, six, five, four, three, two, one, zero.

Procedures are easily represented by **flow charts**, graphical shapes representing the specific steps. Computational procedures are represented by rectangular shapes. Decisions are represented by diagonal shapes. Input and output operations are represented by trapezoidal shapes. Start and stop terminus are represented by rounded end rectangles. Arrows are used to show procedural flow. A flow chart illustrating the procedure of counting backward from ten to zero is illustrated in Figure 1.1.

The word "Count" is used to represent the current value in the counting process. The first step in the process is to set the current value of the count equal to ten. The next step is the enumeration of the count. This is followed by a decision. If the current value of the count is equal to zero, the procedure

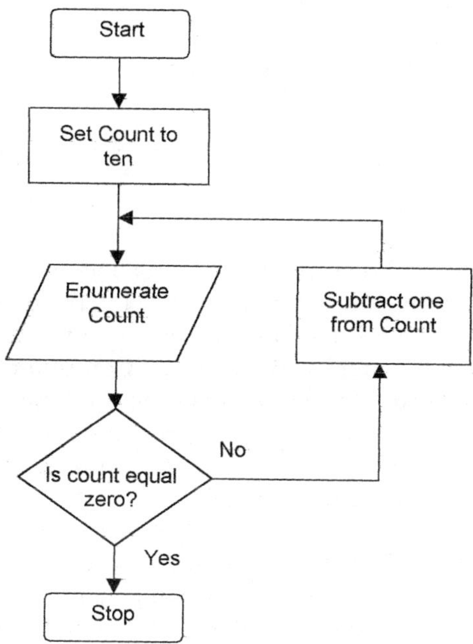

FIGURE 1.1
Flowchart of count from ten to zero by one.

ends. If the current value of the count is not equal to zero, the procedure continues with the subtraction of one from the current value of the count. The procedure continues with another enumeration of the count. The process of enumeration, testing and reduction of the current value by one continues until the current value becomes zero. The repetitive process of current count enumeration, testing for zero value and reduction of the current count value by one is referred to as a loop or iteration.

The procedure could also be expressed in a shorthand English language-like notation as shown in Figure 1.2.

```
Count = 0
Until Count = 0
    Output Count
    Count = Count - 1
Stop
```

FIGURE 1.2
Shorthand notation describing the procedure of counting backwards from ten to zero.

The shorthand notation shown in Figure 1.3 closely resembles the procedure as it would be expressed in the Java language.

The Java language has rules just as English or other spoken and written languages do. Java and other computer languages differ from most spoken and

```
Count = 10;
While ( Count != 0 ) {
    System.out.println (  Count );
    Count = Count - 1;
}
```

FIGURE 1.3
Java program code describing the procedure of counting backwards from ten to zero.

written languages in that the rules are exacting and errors in syntax are intolerable. Equally frustrating are errors in expressing the procedure, which generally leads to erroneous results. These errors are often difficult to detect and have led many a programmer to cry for mercy. Computers, unfortunately, are unfeeling and thus a programmer's lot is twofold. First, learn the language. Second, learn to clearly and accurately express the procedure.

Similarly, the flow chart expressing the procedure for counting backwards from ten to zero would probably be drawn by a Java programmer as shown in Figure 1.4. Note that the procedure has been generalized to enable input of the starting value. Now the same procedure can be used to count backwards from any number to zero by one. This form of generalization fosters the reuse of code in that it can be used over and over again in different circumstances. The reuse of code is an important factor in the reduction of application and system development costs.

Similar revisions can be incorporated to provide for the following:

- Input of the stopping value — it need not always be zero;
- Input of the step value — for example we often count by twos, fives and tens; and
- Input of the direction of count — we often need to count in a forward direction.

PROGRAMMING LANGUAGE EVOLUTION

The first computers were developed in the 1940s. These computers were programmed by wiring panels and setting switches. For all intents and purposes, the programs were written in machine language. Programs were expressed in terms of the binary values of the electronic devices comprising the machine. A typical instruction might appear as shown in Figure 1.5.

Although the original machines had a limited number of operations, present day machines have hundreds of machine level operations. Imagine trying to remember the binary number for a particular instruction. Even more difficult would be the specification of operand and result addresses in machines with 32- and 64-bit addresses. Programs written in this fashion are referred to as **first-generation language (1GL)** programs.

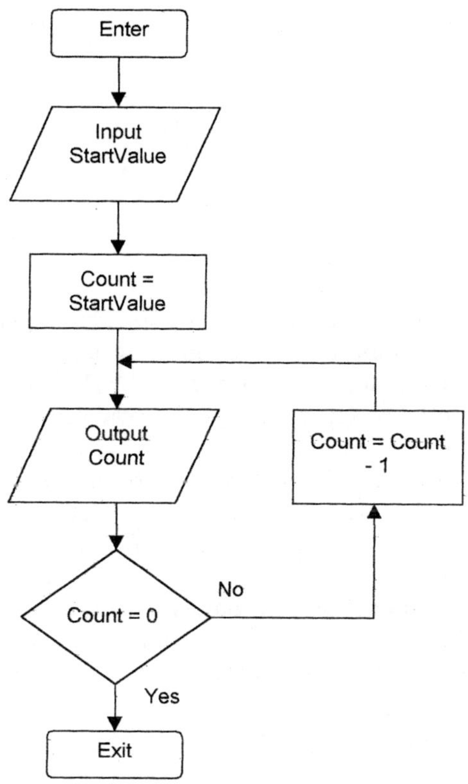

FIGURE 1.4
Java programmers flow chart describing the procedure of counting backwards from some starting value to zero.

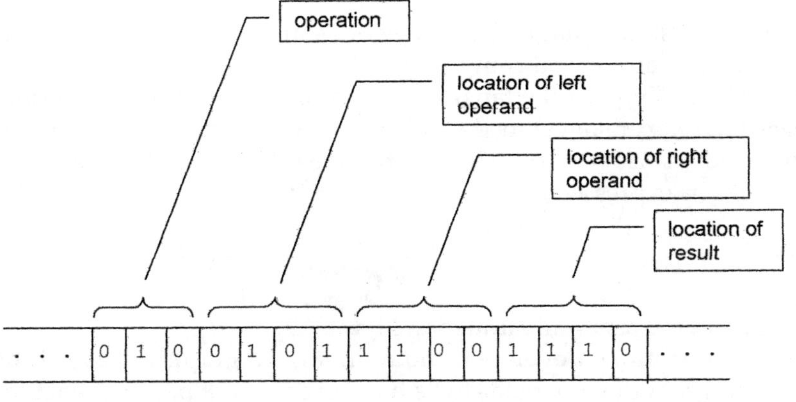

FIGURE 1.5
Typical machine language instruction.

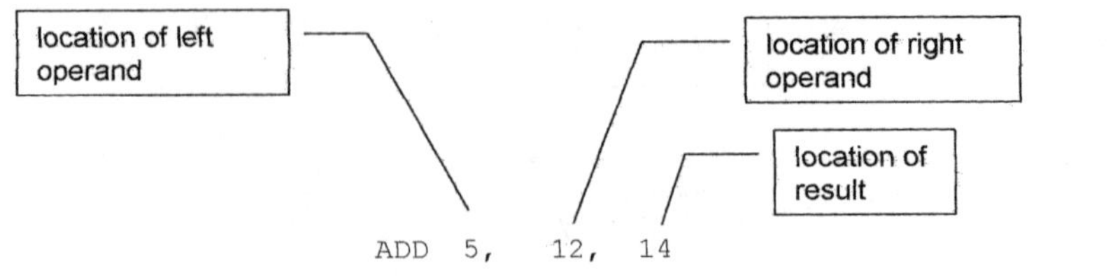

FIGURE 1.6
Typical assembly language instruction.

In the 1950s programs called assemblers were developed to convert assembler language instructions into machine language. These languages are referred to as **second-generation languages (2GL)**. The machine language instruction of Figure 1.5 might appear as the assembler language instruction shown in Figure 1.6.

In the 1950s the languages **FORTRAN** (Formula Translation) and **COBOL** (Common Business Oriented Language) were developed. FORTRAN was widely used by scientist and engineers. COBOL was, and still is, used to develop applications for business and commerce. These languages are referred to as **third-generation languages (3GL)**. Instructions in these languages more closely resemble natural language. For example, the FORTRAN expression converting degrees in Fahrenheit to Centigrade might appear as

```
CENT = 100 * (FAR - 32)/212
```

Similarly, the COBOL expression subtracting a FICA deduction from gross pay might appear as

```
SUB FICA FROM GROSSPAY
```

These English language-like instructions are converted to machine language by programs called **compilers**.

In the 1980s English-like languages, referred to as **fourth-generation languages (4GL),** were developed to access databases and produce reports. For example, an instruction to access a corporate database and list sales by territory might be expressed as

```
EXTRACT TERRITORY, SUM(SALES) ORDER BY TERRITORY
```

and result in the report illustrated in Figure 1.7.

Most applications developed today involve a **graphical user interface (GUI)**. Program development tools have been developed that enable these graphical user interfaces to be developed using Windows-based grab, drag and drop techniques. The program code developed by these languages is compiled into machine language instructions by 3GL and 4GL language

Sales Summary

Territory	Sales ($1000000)
Domestic	56,124
International — Europe	12,693
International — Asia	47,491

FIGURE 1.7
Report using data extracted from a database with a fourth-generation language instruction.

compilers. These new languages are referred to as **fifth-generation languages (5GL)**.

During the early days programs were written without attention to structure. These programs were written with what is often called "fall through" logic. These programs proved hard to develop as their complexity increased and they became very expensive to maintain over their lifetime. In the 1980s programs were developed using the well-defined control structures of sequence, decision and iteration and were said to be **structured programs**. These programs were easier to develop and less costly to maintain.

Today, most programs are developed with an object orientation. Programs of this nature are called **object-oriented programs (OOPs)**. The advantages of OOPs include the following:

- Ideally matches the results of object oriented analysis used by current analysts;
- Fosters development of highly reusable code, consider the buttons, textboxes and other graphical objects used in **graphical user interface (GUI)** development; and
- Facilitates team development of large complex systems as objects can be developed independently.

Java is an object-based language and thus ideally suited to object oriented programming. Java is supported by a variety of 5GL programming tools. As such, Java is quickly becoming the language of choice for program development!

INTEGRATED DEVELOPMENT ENVIRONMENT

Today, industry demands efficiency in the development of applications. Consequently, most application development is conducted within an integrated development environment. A typical **Integrated Development Environment (IDE)** includes the following:

- Visual graphical design editors for constructing Graphical User Interfaces

FIGURE 1.8
Typical Integrated Development Environment.

- Text Editor for writing the code required to implement the application
- A Project workplace that organizes the development process
- A Toolbox that contains the controls that can be used to develop the Graphical User Interfaces
- A Properties window that describes the properties associated with each item of the project
- Various toolbars containing icons that representing commonly used design actions

The IDE for a typical project is shown in Figure 1.8.

Learning to use the IDE can be a time-consuming task, as the components of the IDE are all linked together. Facility with the IDE enables efficiency and productivity. Most major application development is done in an IDE. IDEs are provided by many vendors. Visual J++ is emerging as a popular Java-oriented IDE. IDE features are demonstrated in the following with Visual J++.

The Hands-On Exercises in this text present the major components of an IDE. The features of major components, essential to the work at hand, are presented. The Exercises will provide hands-on experience with the components. Do not despair; facility with the components of the IDE comes with practice and by the time the course is finished, you will be an IDE savant. Most of the components will be revisited as additional features are presented.

Even then, all of the features will not have been uncovered. The major components of IDE, Visual J++, and others that will be utilized in this book are summarized below.

Project Explorer

The **Project Explorer window** lists all of the items in the project. The list is expanded or collapsed by clicking the plus (+) or minus (–) sign to the left of the project name. Typically, a project will consist of several items. Complex projects will include other projects. The Project Explorer window, shown in Figure 1.8, describes a project with a single item named Form1.

Selecting the View menu's Project Explorer command opens the Project Explorer window. Clicking the window's Close button closes the window.

Text Editor

The Text Editor provides for managing, editing, and printing source files. The features of the editor are similar to those of other Windows-based editors or word processors. Enhancements to the Visual J++ editor include the following:

- Writing code with statement completion;
- Dynamic syntax checking

When the Statement Completion option is enabled (Figure 1.9), Visual J++ continually performs a look-up and automatically displays the classes, members, and method parameters relevant to the code being written. The Statement Completion option is enabled in several steps. First click the Options… selection in the Tools menu to display the Options dialog box. Then in the leftmost window pane, expand the Java entry in the Per Language expanded list to display the relevant property page. Finally, click the Auto list members and Parameter information checkboxes to enable the Statement Completion options. Clearing the check boxes disables the Statement Completion options.

Visual J++ performs **Dynamic Syntax Checking** by automatically checking the syntax of code as it is being written. When a syntax error is detected, Visual J++ places a red squiggly line under the errant code. The red squiggly line indicates that the code as currently written contains syntactic errors. As the code is completed or revised, the squiggly lines may disappear, indicating the code is syntactically correct. An error message will be displayed when the cursor is rested over the red squiggly lines. Additionally, clicking the right mouse button while the cursor is resting over a red squiggly line will display a shortcut menu. Selecting the Error Help option in the menu opens online help to the entry relevant to the syntactic error represented by the red squiggly line (Figure 1.10).

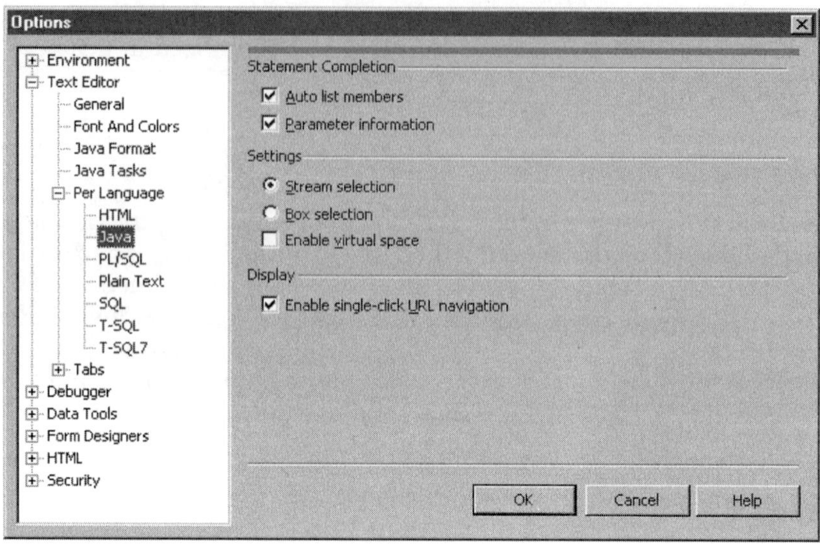

FIGURE 1.9
Enabling Statement Completion in the Options dialog box.

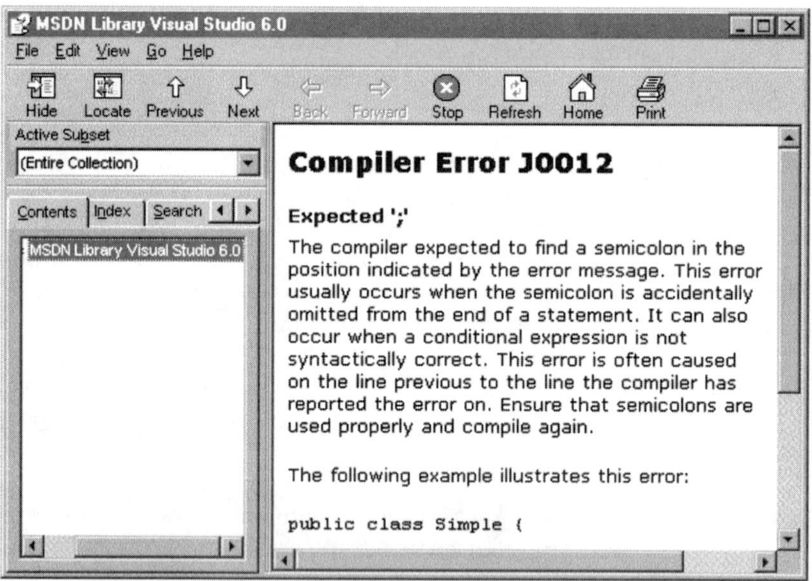

FIGURE 1.10
Online Help regarding a syntax error.

Capabilities for Advanced Programming

The Visual J++ IDE includes capabilities that are used by advanced programmers to develop applications utilizing the Microsoft Windows Foundation classes for Java.

- Properties window displays and provides for management of the attributes of a form or control. Typical properties include the backcolor of a form; the forecolor of a label; the visibility of a command button; the caption of a checkbox; or the font name, size or weight of the text in a textbox control. Selecting the View menu's Properties window command opens the Properties window. Clicking the window's Close button closes the window.
- Toolbox window displays controls that can be used to build an application. Typical controls include labels, command buttons, textboxes, and listboxes. The Toolbox can be customized to include special controls such as ActiveX controls. Selecting the View menu's Toolbox command opens the Toolbox window. Clicking the window's Close button closes the window.
- Form Designer enables the visual and dynamic creation of forms; hence the name Visual J++. With the Form Designer, the size and position of forms and controls are specified without the need to write Java code. Controls are positioned on the form by dragging selected Toolbox controls tools onto the form with the mouse. The form and controls are sized with the mouse. To view a form in the Form Designer, double-click the form's node in the Project Explorer window. Alternately, select the form in the Project Explorer window and click the View Designer icon on the Project Explorer's toolbar. The View Designer icon appears as a form with two textbox controls and labels.

This text focuses on the programming knowledge and techniques requisite to utilization of these tools to develop applications based on the Microsoft Windows Foundation classes for Java.

HANDS-ON EXERCISE 1: CREATING A JAVA APPLICATION

The application presents the user with a Welcome window. The window has a text field in which the user enters a name. The entered name is incorporated in a welcome message that is displayed when the user clicks the Display Greeting button. The user closes the application by clicking the Exit button.

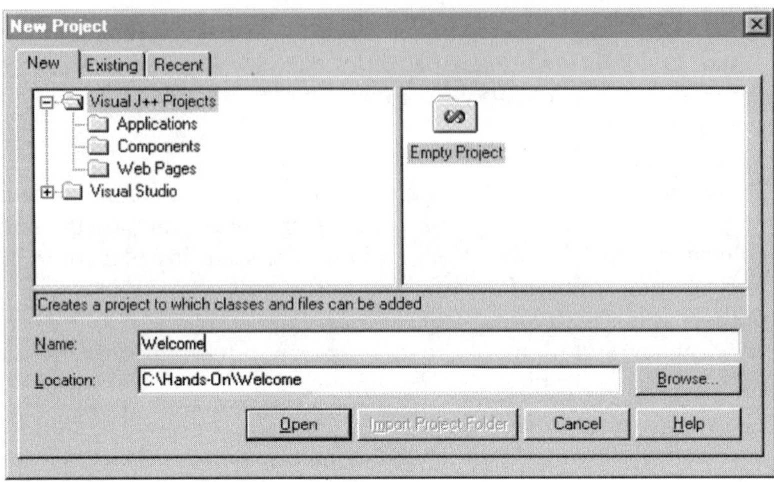

FIGURE HO-1.1
New Project window for Welcome application.

STEP 1: Create New Project

1. Start Visual J++.
2. Select the New Project … command in the File menu. The New Project dialog box is displayed with the New tab selected (Figure HO-1.1).
3. Template folders for selected types of applications are listed in the Options window, the leftmost pane. Expand and collapse the list by clicking + or –. Click the Visual J++ Projects folder.
4. The contents of each template folder are displayed in the right pane. Click the Empty Project.
5. Click the Browse… button and select the Location for the project's files.
6. Enter the project's name, **`Welcome`**, in the Name textbox by overtyping the default name `Project1`. Note the project name is added to the end of the project location.
7. Click the Open button to create the new project. (Clicking the Cancel button will abort the new projection creation.)

STEP 2: Examine the Project Explorer

The Project Explorer window (Figure HO-1.2) is opened automatically when a new project is created or a project is opened.

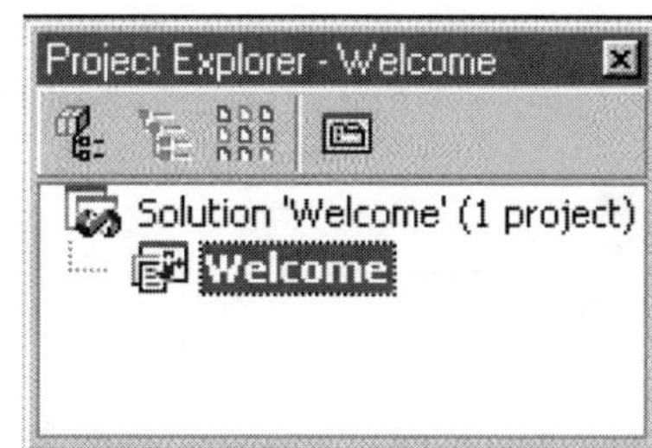

FIGURE HO-1.2
Initial Project Explorer window for Welcome application.

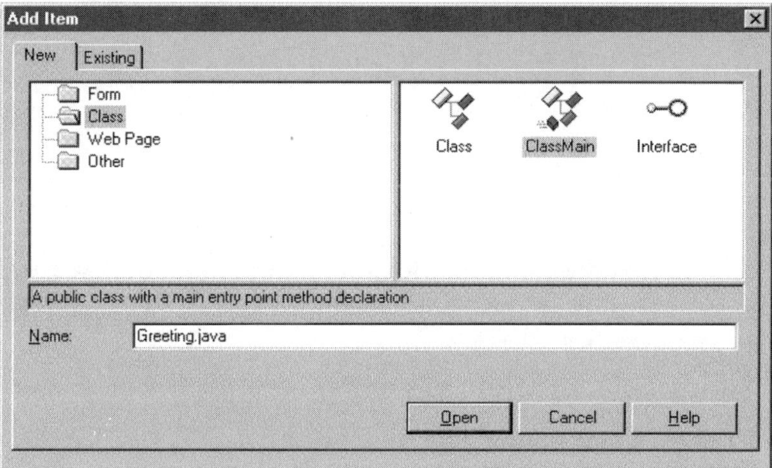

FIGURE HO-1.3
Add Item dialog box.

1. Close the Project Explorer.
2. Open the Project Explorer window clicking the Project Explorer option in the View menu.

STEP 3: Add Greeting Class

1. Select the Welcome project icon in the Project Explorer window.
2. Select the Add Class… option in the Project menu to display the Add Item dialog box (Figure HO-1.3). The left window pane displays the items that can be added. Note that the Class item is selected. The types of classes that can be added are shown in the right window pane.
3. Click the ClassMain icon.
4. Enter **`Greeting.java`** in the Name box, replacing the default name `Class1.java`. Note that the Java extension begins with a lower case `j`.

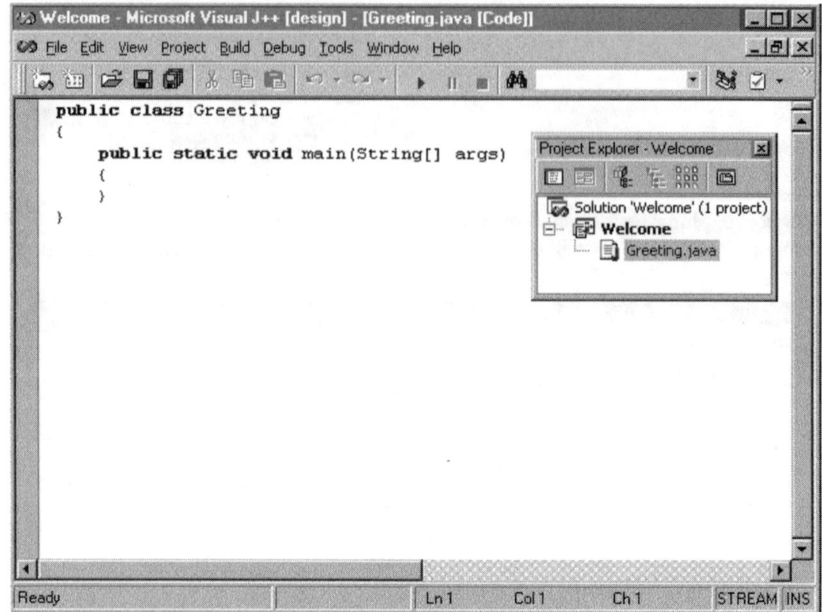

FIGURE HO-1.4
Text Editor open with Greeting.java file for editing.

5. Click the Open button.
6. The Text Editor opens with a skeletal class Greeting declaration. Note the declaration begins with the keyword `public` making the class accessible to other applications, the keyword `class`, which signals a class declaration and the name of the class `Greeting`. The skeletal class declaration includes a main method that is the entry point of all Java programs.
7. Note that the Project Explorer window remains open and that the Welcome project now contains a file `Greeting.java`. All Java classes are stored in a file with the same name as the class and with the `.java` extension.
8. The Text Editor is opened with the file highlighted in the Project Explorer by clicking the Code option in the View menu (Figure HO-1.4).

STEP 4: Complete the Greeting Class Declaration

Revise the skeletal to incorporate the following Java code shown in Figure HO-1.5.

> Alternately, copy the file Greeting.java from the disk included with the text.

```
/**
 * Enables access to classes declared in other packages
 */
import java.awt.*;
import java.awt.event.*;
/**
 * This class presents a welcome window including a text field
 * for entry of the user's name. When the Display Greeting
 * button is clicked, the user's name is included with a welcome
 * message that is displayed. Clicking the Exit button closes
 * the application.
 */
public class Greeting extends Frame
       implements WindowListener,
                              ActionListener
{
       /**
        * Creating the label, text field and button components
        */
       Label lName = new Label("Enter Name:");
       Label lGreeting = new Label(" " );
       Button bDisplayGreeting = new Button("Display Greeting");
       Button bExit = new Button("Exit");
       TextField tName = new TextField(" ");
       /**
        * Main entry point for the application
        */
       public static void main(String[] args)
       {
              Greeting greetingObject = new Greeting( );
       }

       /**
        * Positioning the label, text field and button
        * components in the frame window
        */
       Greeting( )
       {
              addWindowListener(this);

              setTitle("Welcome");
              setLayout( null );
              setSize(400,300);
              add( lName );
              lName.setBounds( 50, 60, 60, 20 );

              add( tName );
              tName.setBounds( 120, 60, 100, 20 );
              tName.setBackground(new Color(255, 255, 255) );

              add( lGreeting );
              lGreeting.setBounds( 50, 100, 300, 50 );
              lGreeting.setFont( newFont("Serif",Font.ITALIC,22)
);
              lGreeting.setForeground(new Color(255, 0, 0));
              lGreeting.setVisible(false);
              add( bDisplayGreeting );
              bDisplayGreeting.setBounds( 50, 200, 100, 50 );
              bDisplayGreeting.addActionListener(this);
              add( bExit );
              bExit.setBounds( 250, 200, 100, 50 );
              bExit.addActionListener(this);
              setBackground(new Color(0, 255, 255) );
```

```
            setVisible(true);
        }
        /**
         * Handling events associated with clicking the
         * Display Greeting and Exit buttons.
         */
        public void actionPerformed( ActionEvent ae )
        {
                if( ae.getSource( ) == bDisplayGreeting )
          {
                        lName.setVisible(false);
                        tName.setVisible(false);
                        lGreeting.setText( tName.getText( )+
", Welcome to Java!");
                        lGreeting.setVisible(true);
                }
                if( ae.getSource( ) == bExit )
          {
                        System.exit( 0 );
                }
        }

        public void windowClosing( WindowEvent we )
        {
                setVisible(false );
                dispose( );
                System.exit( 0 );
        }
        public void windowActivated( WindowEvent we ){}
        public void windowClosed( WindowEvent we ){}
        public void windowDeactivated( WindowEvent we ){}
        public void windowDeiconified( WindowEvent we ){}
        public void windowIconified( WindowEvent we ){}
        public void windowOpened( WindowEvent we ){}
}
```

FIGURE HO-1.5
Listing of Greeting.java

The code seems complex and to a newcomer unsettling. However, when the meaning and purpose of each line is understood the complexity will abate and the initiate will feel comfortable. The sole purpose of this text is to provide that enlightenment and feeling of comfort.

The code must be entered exactly as shown. Java is a case-sensitive language, and thus letters shown as lowercase must entered as lowercase. Similarly, letters shown as uppercase must be entered as uppercase.

> Again note that an alternative to manually entering the code is to copy the file `Greeting.java` from the disk included with the text.

STEP 5: Save the Project

Select Save Greeting.java in the File menu.

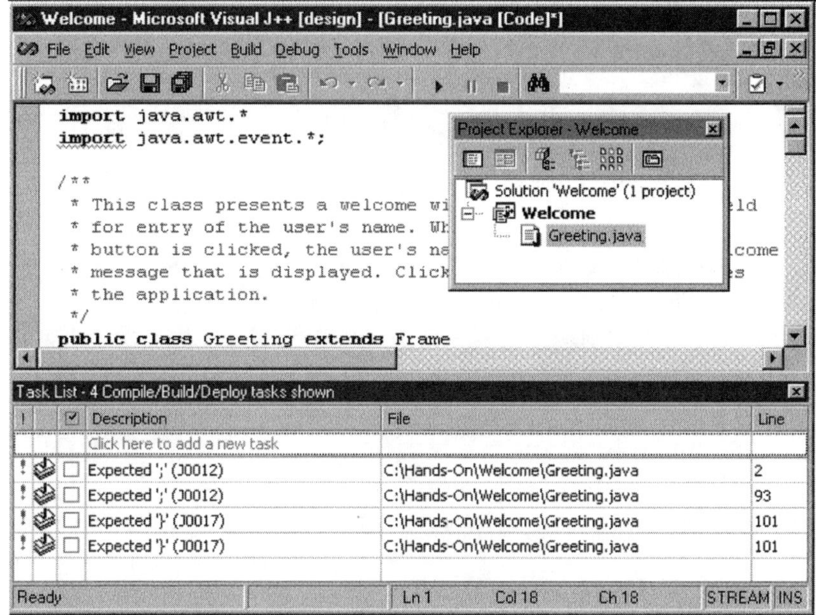

FIGURE HO-1.6
Output window showing program errors.

STEP 6: Build the Project

Select the Build command in the Build menu.

Typically, when a project is first built, many errors are detected. These errors are listed in an Output window that is automatically displayed (Figure HO-1.6). Sometimes the list can seem daunting!

For example, the errors listed due the omission of a single semicolon are shown in the figure. Programmers can spend a great deal of time correcting these errors if they are not careful to observe the syntactical rules of the language; observance of case, punctuation, spelling, and use of keywords. Note that clicking on the error will result in the cursor being positioned at the line on which the error was detected. Unfortunately, the error often actually occurs on prior lines and is not detected until a later line. It is to the later line that the cursor is directed; however, it does provide a starting point for chasing down the error. The error messages will often seem cryptic. However, their meaning will become apparent as knowledge of the Java language increases. Again, the purpose of the book is to provide that knowledge.

With regards to correcting errors in the first Hands-On Exercise, observe that every newcomer to a language has experienced frustration in debugging first programs. Welcome to an elite group!

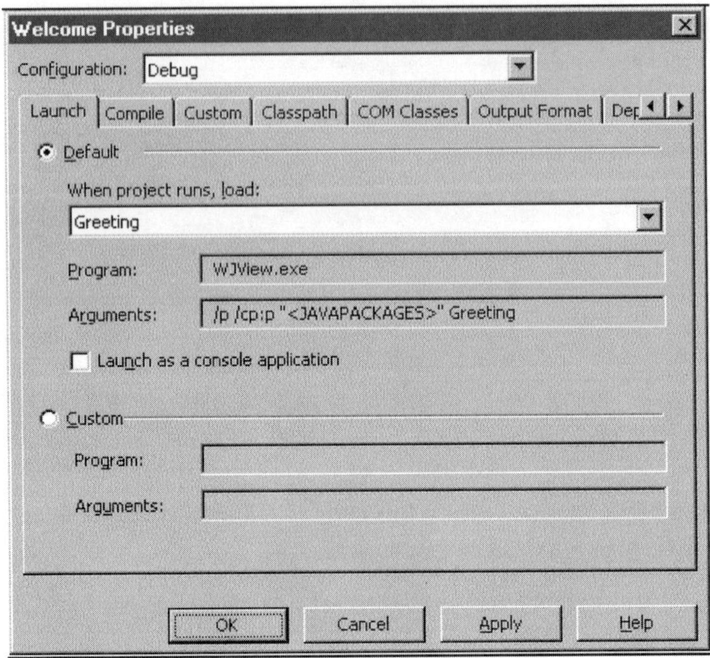

FIGURE HO-1.7
Welcome project Properties Window.

STEP 7: Run the Project

1. Select the Start Without Debugging command in the Debug menu. The first time the application is run, the project Properties window will be displayed (Figure HO-1.7).
2. The default project properties are acceptable. In future applications, we will sometimes click the Launch as Console Application checkbox to enable the application's output to be viewed as if on a console. The current application is intended for the Windows environment.
3. Note also that the application is in a Debug configuration.
4. Click the OK button to close the Properties window.
5. Enter a name in the text field labeled Enter Name (Figure HO 1.8). In Figure HO-1.9, the name VaLenda has been entered in the text field. Note that text fields are often called textboxes by Windows programmers.
6. Note also that the greeting Welcome to Java! is invisible. The greeting is supposed to become visible when the Display Greeting command button is clicked.

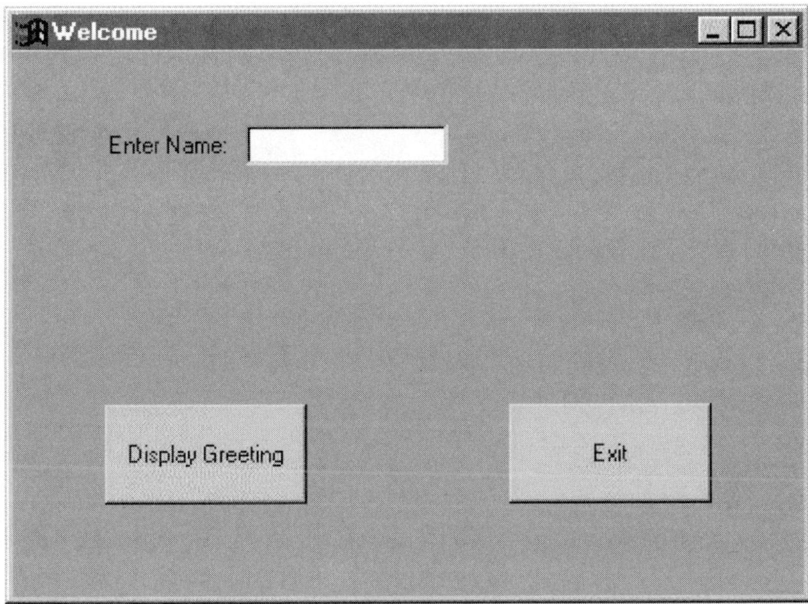

FIGURE HO-1.8
Initial Welcome application window.

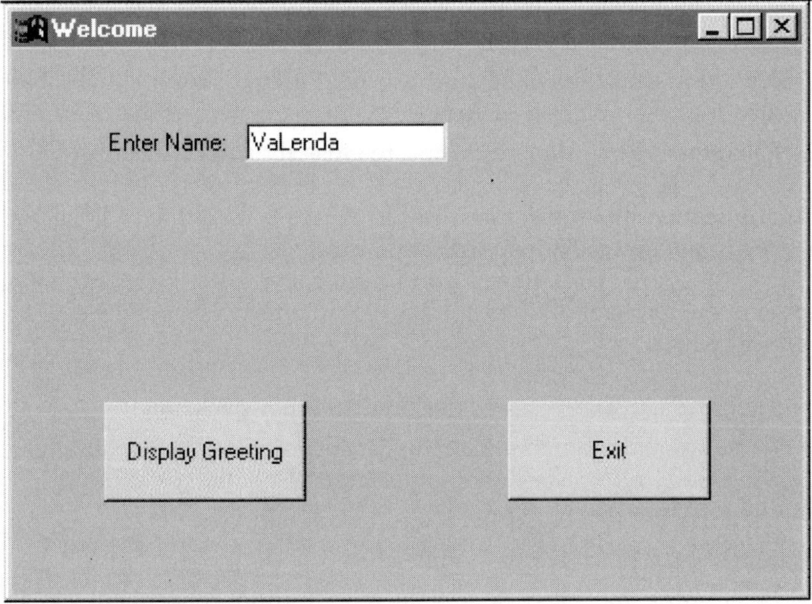

FIGURE HO-1.9
Welcome window with name entered in text field.

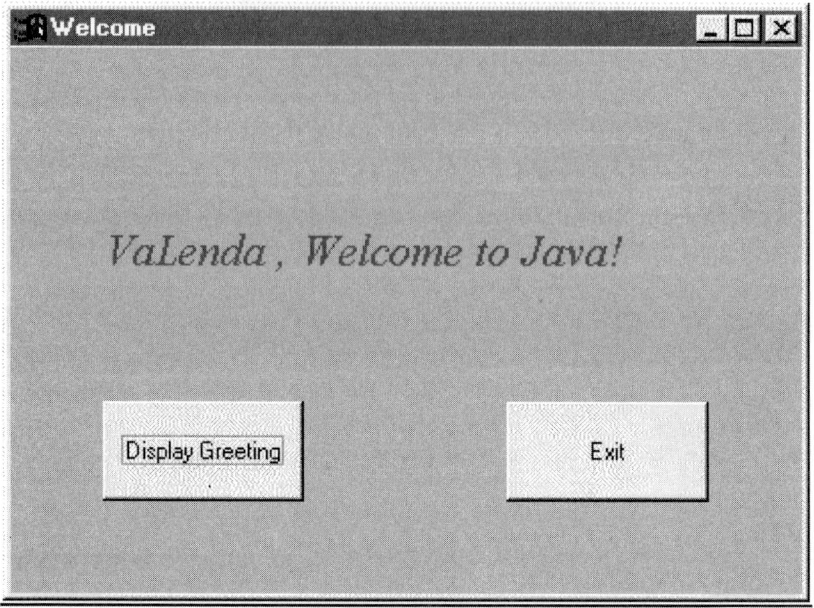

FIGURE HO-1.10
Welcome window with greeting.

7. Click the Display Greeting command button. The greeting is displayed (Figure HO-1.10).
8. Note that the Name label and text field are no longer shown. Note also that the greeting includes the name entered in the text field.
9. Click the Exit command button to close the application.

This application illustrates many of the elements common to Java applications used in business, engineering, and science.

Review the source code in the Code Editor and note the major elements comprising the application identified by comments (Comments begin with a `/*` and end with a `*/`):

- Enabling access to classes declared in other packages.
- Declaration of the class Greeting incorporating the capabilities of a Frame window class that is declared elsewhere by others.
- Creating the label, text field, and button objects.
- Main entry point for the application.
- Positioning the label, text field, and button components in the frame window as well as setting the properties of each of these components. For example, the greeting is displayed in serif font, italic style, and 22 point height.

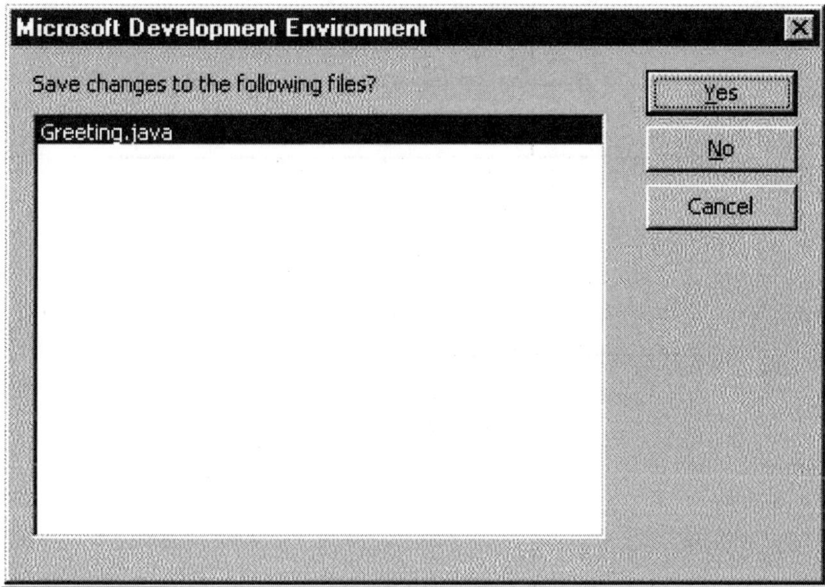

FIGURE HO-1.11
Visual Studio prompt regarding saving files.

- Handling events associated with clicking the Display Greeting and Exit buttons. For example, the greeting is made visible when the Display Greeting button is clicked. Similarly, the application is closed when the Exit button is clicked.

STEP 8: Close the Project and Exit Visual J++

1. Click the Close button, the X, at the top right corner of the Visual J++ window. The project closes. Alternately, select the Exit command in the File menu. The Visual Studio may prompt the saving of files (Figure HO-1.11).
2. Click the Yes command button. Visual J++ closes.

DEVELOPING APPLICATIONS WITH JAVA

Java is an object-oriented language that can be used to develop applications for business, engineering, and science. Consider, for example, the graphical user interface shown in Figure 1.11. Every business has need for a customer application with the type of interface shown. The graphical interface provides for the entry and display of the following customer characteristics:

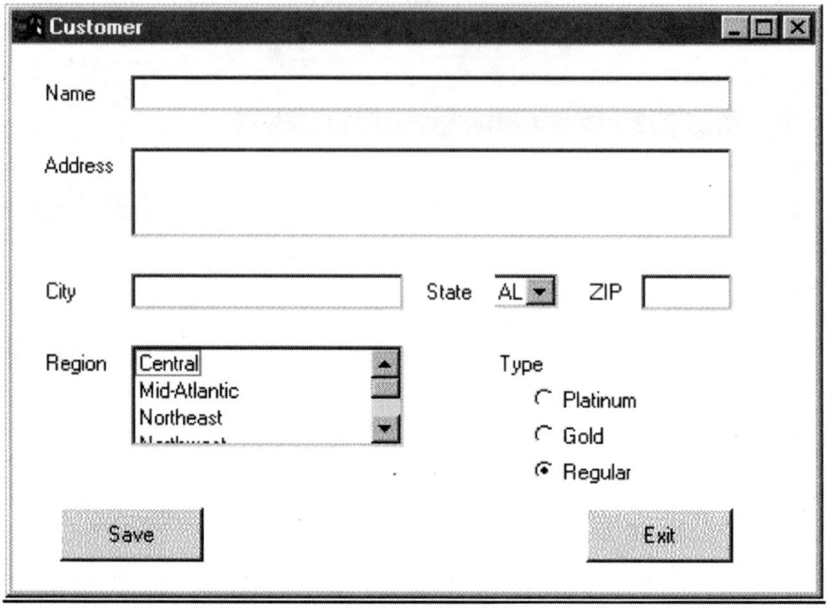

FIGURE 1.11
Typical business application interface

- Name
- Address
- City
- State that is selected from a combo-box drop-down list
- Sales region selected from a list
- Type selected from a radio button group.

The interface also includes buttons for saving entered data as well as closing the application. Generally, the application will also include menus enabling database information retrieval. Similar interfaces are required in science and engineering.

Java provides the capability of developing these interfaces. Chapters 1 through 5 present language fundamentals:

1. Introduction to Java including the Visual J++ Integrated Development Environment.
2. Language preliminaries including primitive data types and their use.
3. Class data type including methods declaration and use.
4. Object-oriented programming properties of encapsulation, polymorphism, and inheritance and the Java language implementation.
5. Operators and their use in formulating and evaluating expressions.

These chapters are intended to provide a firm foundation in the fundamental elements of the language, particularly data types, methods, and object-oriented programming concepts and use.

Chapters 6 through 9 present programming fundamentals:

6. Control structures and their use in implementing algorithms required in application development.
7. Data handling techniques and structures required to manipulate tables and lists.
8. Java's approach to handling exceptional conditions.
9. Java's package-based organization and the contents of selected packages.

These elements establish the ability to program an algorithm in Java. Included are an understanding of the techniques of looping and decision making in problem solving.

Chapters 10 through 13 examine the structure and use of the Abstract Window Toolkit:

10. Development of windows and event-handling model.
11. Graphic fundamentals including the color model, fonts, and rendering of lines and figures.
12. Basic Graphical User Interface components: buttons, labels, and text components.
13. Additional Graphical User Interface components: list, scrollbars, checkboxes, checkbox groups, and choice components.

Windows-based interfaces are the primary user interface today. These topics examine in detail the components of the Abstract Window Toolkit that can be used to build these interfaces.

Finally, Chapters 14 through 16 present Java's communication, networking, and threads.

14. Stream input and output model and implementation.
15. Programming for network communications and network browsers.
16. Threads and their use.

At this point, the reader should be able to build Java based applications.

SUMMARY

1. Java is a new language that incorporates many aspects of C and C++ while excluding the unwieldy and often confusing aspects of C and C++.
2. Visual J++ contains an Integrated Development Environment that facilitates application development.
3. Java is a new language that can be used to develop applications for business, engineering, and science.
4. The File menu commands are used to manage the project's files and folders
5. Creation, development, and execution of the application Welcome demonstrated Java and the Visual J++ environment.

KEY WORDS AND CONCEPTS

`.java` (extension)
`.vjp` (extension)
algorithm
class file
Button Control
COBOL
fifth-generation language (5GL)
first-generation language (1GL)
flow chart
FORTRAN
fourth-generation language (4AGL)
graphical user interface (GUI)
Integrated Development Environment (IDE)
Java
Label Control
object-oriented programs (OOPs)
program
Project Explorer window
project file
Project menu
second-generation language (2GL)
Statement Completion
Text Editor
Textfield Control
third-generation language (3GL)
View menu

REVIEW QUESTIONS AND EXERCISES

1. What menus are pulled down to:
   a. Save a project in a file?

   b. Open the Project Explorer window?
   c. Open a file in the Text Editor?
2. Name key elements of an Integrated Development Environment.
3. List the steps and operations used to create a Java application.
4. What is the purpose of the main method in a Java program?
5. When is the `new` keyword used in a Java program?
6. Methods for handling events enables the program to respond to ______________.
7. Give examples of statements that enable a program to access the contents of other files.
8. What file extension is required for a Java class file?
9. Mark the following true or false.
   a. (T/F) Dynamic Syntax Checking detects syntax errors.
   b. (T/F) Statement Completion cannot be disabled.
   c. (T/F) The Start command is found in the Project menu.
   d. (T/F) Syntax errors detected are listed in the `Output` file at the completion of program building.
   e. (T/F) Java does not include all of the features of C and C++.
10. Describe typical applications that can be developed with Java.

PRACTICE WITH VISUAL J++

Modify the Welcome application to provide the following:

1. Display the greeting in 36 point type.
2. Leave the Name label and textfield visible when the Display Greeting button is clicked.
3. Change the size and location of the buttons. (*Hint:* set Bounds method.)

2

Java Language Preliminaries

OBJECTIVES

After reading this chapter, you will be able to:

1. Use comments to promote program documentation and readability.
2. Understand the Java variables and primitive data types.
3. Understand and use identifiers.
4. Use an Integrated Development Environment on-line Help capability

OVERVIEW

This chapter introduces syntax, lexicographic, and naming conventions used in the Java language. Included are a description of the character set as well as the various methods for expressing comments. The chapter also introduces variables and primitive data types, including their size, value range, and use. The chapter will also present the on-line Help included in an Integrated Development Environment.

LEXICAL STRUCTURE

This section describes the lexical structure of Java; that is, the words and vocabulary of a Java program, such as the program illustrated in Figure 2.1a. This program displays the message "Hello Java World" on the monitor (Figure 2.1b).

```
// File: INTRO.JAVA
class Intro
{
      public static void main(String args[])
      {
            System.out.println("Hello Java World");
      }
}
```

FIGURE 2.1A
Traditional first Java program.

```
Hello Java World
```

FIGURE 2.1B
Output of traditional first Java program.

Included in this section are the characters used to form the words, rules used to form words, conventions for forming comments, and rules for forming identifiers.

Unicode

The characters available to a language are known as the language's **character set. Unicode** is a 16-bit character set capable of representing characters worldwide including accented and ideographic characters. Unicode is a superset of the ASCII character set. The first 128 Unicode characters are ASCII characters. Translations provide for expressing Unicode characters as a sequence of **ASCII characters.** Consequently, Java programs can be expressed as ASCII characters only. A table listing the ASCII characters and their numerical values is presented in Appendix A.

Comments

Java provides for three types of comments: single-line, traditional, and documentation. **Comments** are not translated into the code that is executed by the computer.

The double slashes (//) mark the beginning of a **single-line comment:** The comment begins immediately to the left of the slashes and terminates at the end of the line.

```
// Hello, I am a Java application
```

The end of a single-line comment is signaled by the line termination character.

The `/* .... */` **traditional comment** format can also be used. The comment begins immediately to the left of the characters `/*` and ends immediately to the right of the `*/` characters. For example:

```
/*  Hello, I am a Java application  */
```

Documentation comments are processed separately and used to generate documentation. The `/**` and `*/` characters are used to begin and end documentation comments.

```
/**  Displays: Hello, I am a Java application
*/
```

Comments can be included anywhere the program and are very useful in documenting the program. Comments cannot be nested:

```
//       /* This is a nested comment */
```

In the above case, the `/*` and `*/` characters do not have a special meaning, that is the beginning and end of a comment. Similarly, the `//` characters do not have a special meaning in comments beginning with `/**` and `/*`.

Identifiers

An **identifier** is a name given to a variable, function, or data type. More generally, an identifier is a name given to any user-defined object. The rules for forming an identifier are the following:

1. Can be an unlimited sequence
2. Contain only the Java characters, including the Java letters `A` to `.Z` and `a` to `z`, the Java digits `0` to `9`, the underscore character (_), and the dollar sign (`$`)
3. Must start with a Java letter or underscore
4. Must not be a keyword
5. Must not be the `true` or `false`, the Boolean literals
6. Must not be `null`, that is, the null literal.

Examples of valid identifiers are:

```
no
accounts_payable
buffer_index
i
I
ACCOUNT_NO
Index
_firstname
```

Invalid identifiers and the reason why they are invalid are given in Table 2.1.

TABLE 2.1

Invalid identifiers

Identifier	Explanation
Acct_#	is not allowed
0_or_1	must start with a letter or underscore
ACCOUNT-NO	dash is not allowed (the dash and the underscore are different characters)
END OF FILE	imbedded spaces (space is not allowed)

Good programming practice dictates the use of identifiers that accurately represent the variable, function, or data type. For example, it is easier to use the character sequence `an` for account number; however, if the procedure must be read by another person or by yourself several years and thousands of procedures latter, the identifier `account_no` will make more sense. Some programmers avoid and others use the underscore sparingly, as the blinking prompt and the underscore are often confused. Most programmers follow a pattern in forming identifiers. Often, large application development departments required all identifiers to conform to a standard. Typically, the Java classes begin with a upper-case letter. Variables are often formed by lower-case letters representing the data type followed by capitalized nouns describing the property represented. For example, `iAccountNo` identifies an int type variable representing an account number. Additionally, method identifiers are generally a lower-case verb followed by nouns representing the object of the method. For example, `getAccountNo` identifies a method that retrieves the account number.

Note that Java is case-sensitive; therefore;

```
UPPER
Upper
upper
```

are all different identifiers.

Keywords

Keywords are reserved words that have special meaning to the compiler, can only be used as intended, and cannot be used as identifiers. Java keywords are listed in Table 2.2.

> Note that the keywords `const` and `goto` are not used in the Java language. Their inclusion enables the easy compiler identification of inappropriate code based on C++ type control statements.

Note also that there are many Java compilers in existence. In addition to processing the Java keywords, many of these compilers have added key-

TABLE 2.2

Java keywords

abstract	instanceof
boolean	int
break	interface
byte	long
case	native
catch	new
char	package
class	private
const	protected
continue	public
default	return
do	short
double	static
else	super
extends	switch
final	synchronized
finally	this
float	throw
for	throws
goto	try
if	void
implements	volatile
import	while

words and features intended to make them more productive and easier to use. As such, a compiler can have keywords unique and meaningful to it and its and supporting software but not necessarily having meaning to other compilers and supporting software.

Java is a **case-sensitive** language; that is, lower-case letters and upper-case letters are different. Thus `short` and `SHORT` are different. In fact, `short` is a keyword and `SHORT` is not a keyword.

PRIMITIVE DATA TYPES AND VALUES

All data, variables, literals, and expressions have a type, referred to as a **data type,** that prescribes the values that can be assumed. In fact, Java is said to be a **strongly typed** language in that the types are known at compile time. Java data types are divided into two categories: primitive types and reference types. The primitive data types are also known as intrinsic data types. The amount of memory, in terms of bytes, used to implement a type is referred to as the type's **size.**

Primitive types are defined by the Java language and named by keywords. There are numeric and Boolean primitive data types. The numeric types are further divided into integral types and floating-point types.

Boolean Data Types

The **`boolean` type** represents a logical quantity and has two values represented by the literal values `true` and `false`.

Integral Data Types

The **integral types** are **`byte`**, **`short`**, **`int`**, and **`long`**. The values represented by the integral types are predefined as 8-bit, 16-bit, 32-bit, and 64-bit signed two's-complement numbers. Note that the numbers are treated as defined regardless of the size implemented on a particular type of hardware. Thus, a short may be stored by a computer as a 32-bit number, but only the low order 16-bit part will be used in operations performed on the Java machine. This feature enhances portability.

A **`char`** type is also an integral type that is implemented as a 16-bit unsigned integer value representing a Unicode character.

Floating-Point Data Types

The **floating-point types** are `float` and `double`. The values are single-precision 32-bit and double-precision 64-bit floating-point numbers.

Values

The range of values and default values for the primitive data types are summarized in Table 2.3.

TABLE 2.3

Values assumed by primitive data types

Type	Size (bits)	Value range	Default value
`boolean`	8	`true, false`	`false`
`byte`	8	-128 to 127	
`short`	16	-32,768 to 32,767	
`int`	32	-2,147,483,648 to 2,147,483,647	
`long`	64	-9,223,372,036,854,775,808 to 9,223,372,036,854,775,807	
`char`	16	\u0000' to '\uffff'	`'\u0000'`
`float`	32		`0.0F`
`double`	64		`0.0D`

LITERALS

Specific values of the primitive data types are represented in the source code by **literals.**

Integral

Integral literals can be expressed as decimal (base 10), hexadecimal (base 16) or octal (base 8) source code representations (Table 2.4). The decimal representation is generally used. The hexadecimal representation is sometimes used in source code and is widely used to display the values of nonprintable characters used in data transfer and communications.

TABLE 2.4

Source code representations of selected integer data type values

Source code representation	Value
`0`	Decimal value zero
`09`	Decimal value nine
`0x0`	Hexadecimal value zero
`0X9`	Hexadecimal value nine
`0xB`	Hexadecimal value eleven
`00`	Octal value zero
`011`	Octal value nine
`013`	Octal value eleven

Literal values of type long are suffixed with the character `L` or `l`. Note that the lower case letter `l` cannot be distinguished from the number `1` in some type faces; consequently, the upper case `L` is generally used to signify a long data type literal. Examples of long data type literals are listed in Table 2.5.

TABLE 2.5

Source code representations of selected long data type values

Source code representation	Value
`0L`	Decimal value zero
`0x0L`	Hexadecimal value zero
`0XfL`	Hexadecimal value fifteen
`00L`	Octal value zero
`013L`	Octal value eleven

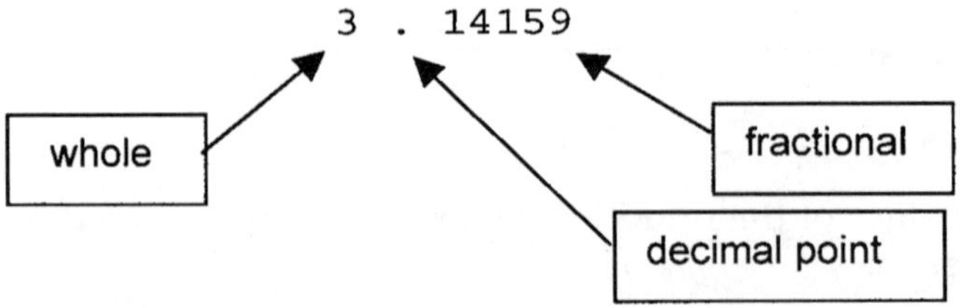

FIGURE 2.2
Parts of a floating-point number.

Floating Point

Floating point or real number literals are expressed in three parts: a whole part followed by a decimal point which is followed by a fractional part (Figure 2.2).

Floating point literals are double by default. The characters `d` and `D` can be appended to indicate the double data type. Exponents are represented by appending `e`*nn* or `E`*nn* where *nn* is the value of the exponent. Source code representations of selected double values are shown in Table 2.6. The characters `F` or `f` can be appended to make the real literal a float data type (Table 2.7).

TABLE 2.6
Source code representations of selected double data type values

Source code representation	Value
`7.11`	7.11
`7.11d`	7.11
`7.11D`	7.11
`7.11e0`	7.11
`0.00711E3`	7.11
`71.1de-1`	7.11

TABLE 2.7
Source code representations of selected float data type values

Source code representation	Value
`7.11f`	7.11
`7.11F`	7.11

Boolean

`Boolean`-type values are represented in source code statements by the literal `true` and `false`.

Character

Character literals are enclosed in single quotes ('), (Table 2.8).

TABLE 2-8

Source code representations of selected character literals

Source code representation	Value
`' '`	space
`'a'`	lower case letter a
`'7'`	character seven

The Unicode characters are denoted as '`\x`*nn*' where *nn* is the Unicode value assigned to the character. For example, '`\x20`' is the source code representation of the Unicode character space.

Nonprintable character literals also have source code representations, as shown in Table 2.9.

TABLE 2.9

Source code representations of selected nonprintable character literals

Nonprintable character literal	Source code representation
Backslash	'\\'
Backspace	'\b'
Carriage return	'\r'
Double quote	'\"'
Form feed	'\f'
Horizontal tab	'\t'
New line	'\n'
Single quote	'\''

String

String literals are any number of characters enclosed within double quotes (") and ("); for example, `"Hello, Java world!"` String literals are implemented in the class String (which will be covered Chapter 9.)

Null

The **`null`** value is special and will be discussed along with classes and objects. Remember that it does not have a known numeric value. The source code representation is `null`.

HANDS-ON EXERCISE 2: GETTING ON-LINE HELP

Help regarding the Java language and the Visual J++ IDE are available on-line. Selecting the Content, Index, or Search options in the Help menu launches the **Visual J++ Documentation** dialog box. The left window pane consist of four tabs corresponding to the various Help modes (Table 2.10).

TABLE 2.10

Visual J++ on-line documentation search modes

Tab	Explanation
Content	Displays books on-line and their contents in a hierarchal list.
Index	A list of subjects contained in the documentation.
Search	Facility for entering search words and the results of the search.
Favorites	A list of favorite help topics compiled by the user.

Selection of a topic in the Content, Index, Search or Favorites window pane results in the display of documentation regarding the topic in the right window pane. Selection and display of the On-line Book topic Visual J++ Start Page topic in the left pane and its display in the right pane are illustrated in Figure HO-2.1.

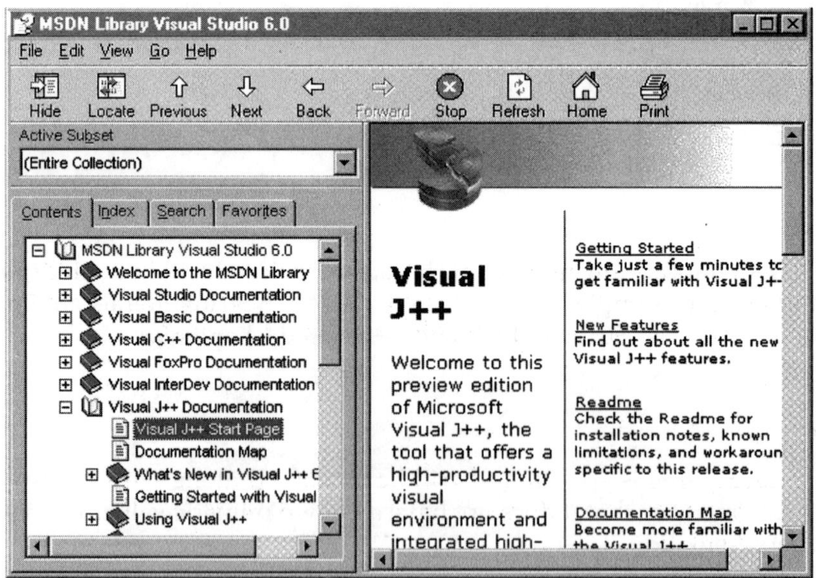

FIGURE HO-2.1
On-line help opened to Visual J++ Start Page.

The *Using Visual J++* book contains two books:

- *User's Guide*—contains documentation regarding use of Visual J++
- *Programmer's Guide*—contains documentation regarding the Windows Foundation Class library, programming Dynamic HTML, building ActiveX controls, and building COM objects.

The reference book also contains several books that are of special interest to beginning as well as experienced application developers:

- *Java Language Specification*
- *Reserved Words*
- *Windows Foundation Classes for Java*

The Index of Help Topics is complete and enables accessing the Help documentation based on specification of a keyword. The Search facility enables accessing of Help documentation based on the specification of words that are used in a search.

Rest assured, answers regarding Java language and Visual J++ IDE questions that you might have are contained in the Help documentation. As you use the Help documentation, you will become increasingly familiar with the contents and gain greater facility with its operation and use. Experienced developers will say that once you have learned the language and gained facility with the development environment, you will be able to find the answer to any question that you might have. Unfortunately, this feeling is developed only after much use of the Help documentation.

Learn to use the Help Documentation.

Step 1: Launch Visual J++

1. Start Visual J++.
2. The New Project dialog box will appear if its Don't Show this Dialog in the Future checkbox has not been clicked. If it appears, close the New Project dialog box.

Step 2: Books On-line—Contents—Looking for Keyword `null`

1. Click the Contents option in the Help menu. Books in the MSDN Library Visual Studio 6.0 are displayed.
2. Scroll through the list and click the plus (+) sign to the left of the Visual J++ Documentation book. *Visual J++ Documentation* expands to show all of the on-line books regarding Java and Visual J++.

3. Scroll through the expanded list and click the plus sign to the left of the Visual J++ Reference book. *Visual J++ Reference* expands to show all of the on-line books regarding Visual J++.
4. Scroll through the expanded list and click the plus sign to the left of the *Reserved Words* book. The book expands to show all of the on-line books regarding reserved words.
5. Scroll through the expanded list and click the plus sign to the left of the *Reserved Words (M-Z)* book. Visual J++ Reference expands to pages in the book corresponding to the reserved words.
6. Scroll through the pages and select the page titled `null`. On-line help regarding the reserved word `null` is displayed in the right window pane, Figure HO-2.2.

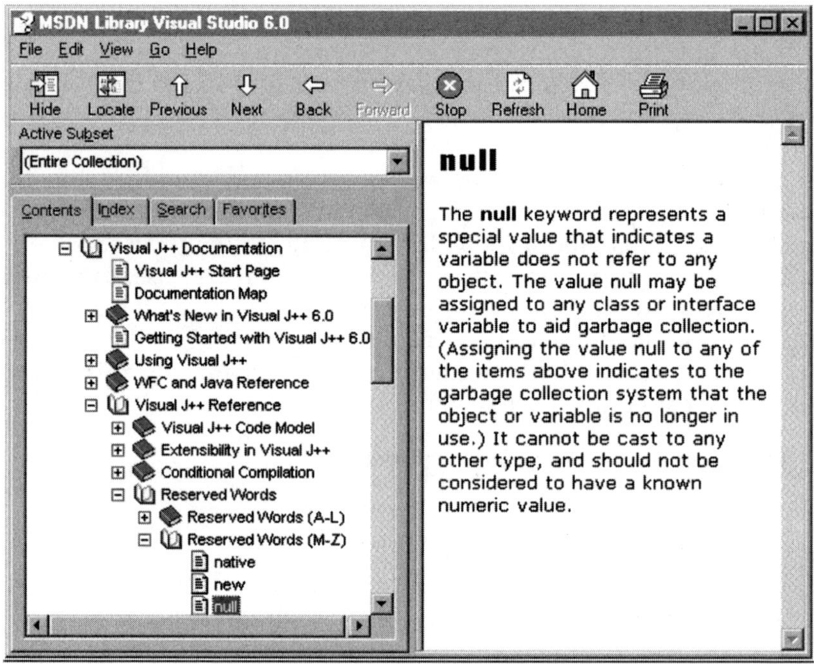

FIGURE HO-2.2
On-line Help window about the reserved word `null`.

Step 3: Index—Looking for double data type documentation

1. Click the **Index** tab in the Help dialog box. The left pane changes to reflect the index search mode. Note that the cursor is position in the text field under the label—Type in the keyword to find. Note also the index of subjects displayed in the lower-left pane.

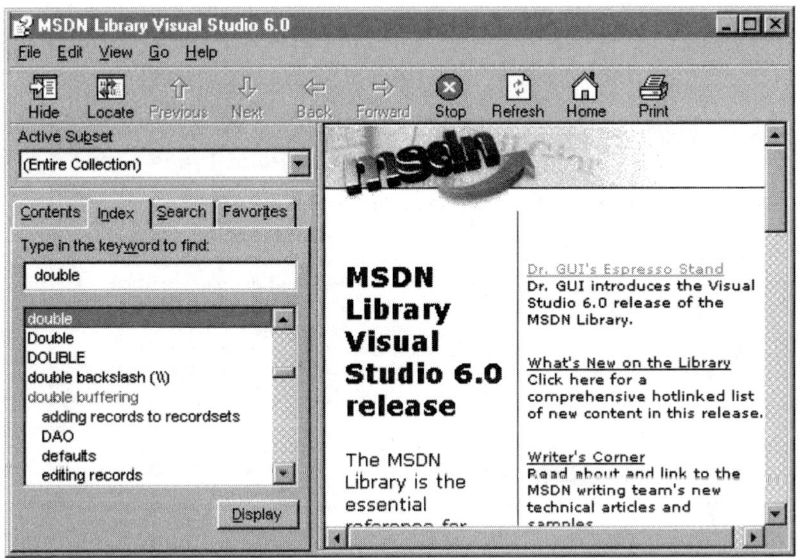

FIGURE HO-2.3
Beginning a search of the index for `double`.

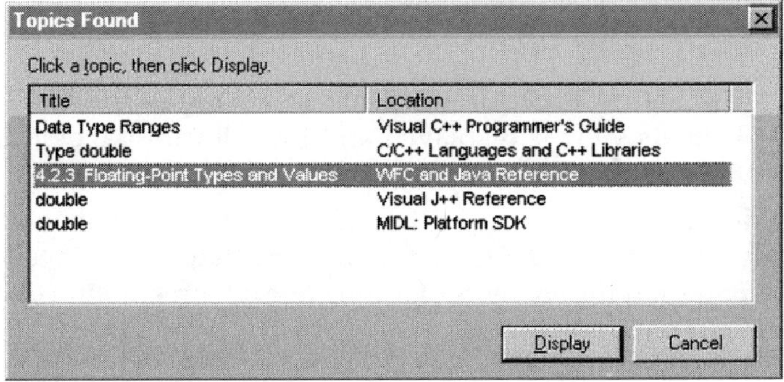

FIGURE HO-2.4
Topics found in the documentation related to the index entry `double`.

2. Enter `double`. The index is automatically positioned to show and highlight the index word best matching the entered word, Figure HO-2.3.

3. Click the Display button. A list of documentation topics related to the keyword is displayed (Figure HO-2.4). All references to the topic in the on-line documentation are listed. In the present case five references are listed.
4. Select the most likely topic.
5. Click the Display button. Documentation regarding the `double` keyword is displayed in the right pane (Figure HO-2.5).

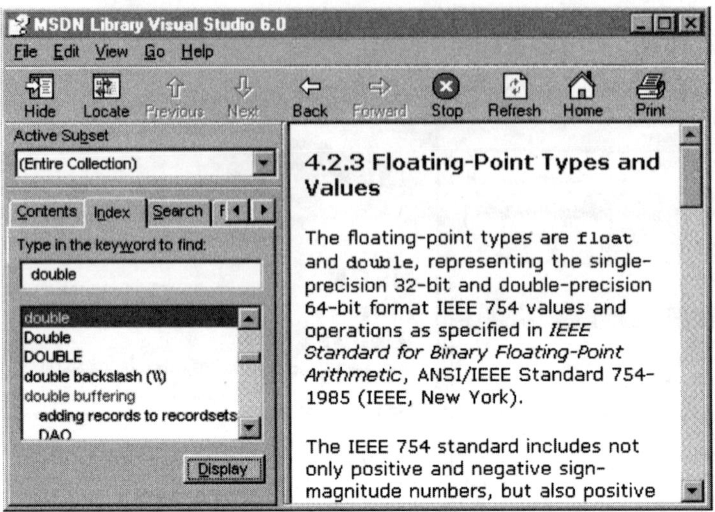

FIGURE HO-2.5
Documentation about the double type.

Step 4: Synchronizing the right window documentation and the contents of the Books On-line

1. Click the Locate Topics in Contents command in the View menu. The Contents view is displayed and the relevant page of the documentation is highlighted (Figure HO-2.6).
2. Clicking the Locate icon in the Help window's Toolbar will also synchronize the contents of the documentation displayed in the right pane and the contents of the on-line documentation displayed in the left pane.

Step 5: Printing Help Documentation

1. Click the Print command in the File menu. The Print dialog box is displayed.
2. Select the desired print properties in the Print dialog box.
3. Click the OK button. The contents of the Help Documentation right window-pane are printed and the Print dialog box is closed.

Step 6: Close the Help Documentation Dialog Box

1. Select the Exit command in the Help Documentation's dialog box File menu.
2. Exit the Visual J++ session.

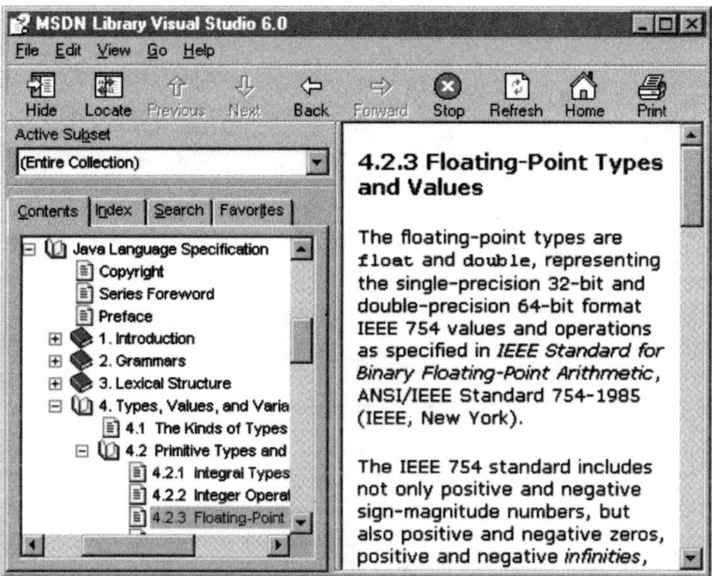

FIGURE HO-2.6
The location of the Help topic displayed in the documentation table of contents.

VARIABLES—REPRESENTING DATA IN MEMORY

A **variable** is a region in memory that can be referenced and its contents accessed. Variable names are identifiers that reference a memory location. The referenced memory location can consist of one or more bytes. Each byte referenced contains an 8-bit binary number. The number of bytes referenced is dependent upon the size of the data type. For example, one byte is referenced for char data types, whereas four bytes are referenced for a float data type. The numeric value of the bits in the bytes referenced determines the value of the datum. Consider for example, the identifier `data`, which refers to a memory location that contains the binary number 01100000. The binary number can mean the decimal number 48 or it can mean the character zero. The specific meaning of the binary number—that is, the way in which Java interprets the meaning of the number—is determined by the type of data stored in the memory location. If the type of data is `byte`, the stored data is the number 48. On the other hand, if the type of data is `char`, the data is the character 0. In practice, the representation of data in memory is much more complex, however, the general concept is one hundred percent accurate. The message is that *the meaning of the number stored in a memory location is determined by the type of data associated with the memory location.*

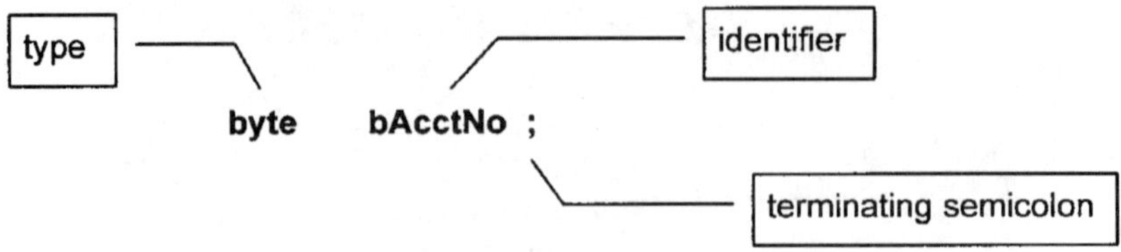

FIGURE 2.3
Variable declaration.

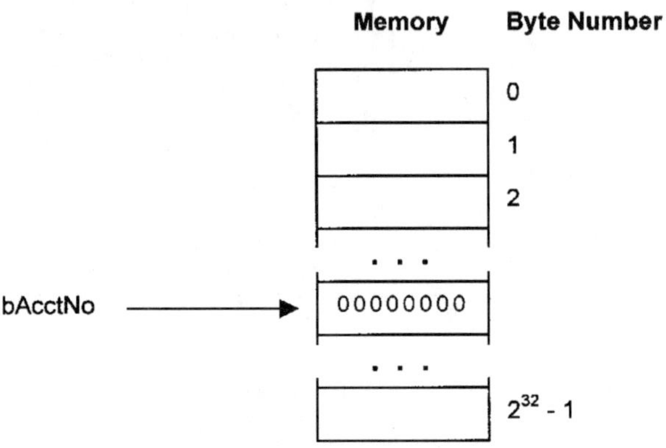

FIGURE 2.4
Accessing the memory location referenced by `bAcctNo`.

Declaring Primitive Type Variables

Primitive data type variables are declared with a **declaration statement.** Declaration of a `byte` type that is accessed with the identifier `bAcctNo` is shown in Figure 2.3. The variable `bAcctNo` could be used to programmatically represent a business customer's account number.

Note that the declaration begins with a type, is followed with the identifier used to access the memory location, and is terminated with a semicolon. Access to the memory location is illustrated in Figure 2.4. Memory can be envisioned as a stack of blocks where each block is a byte. The first byte is numbered 0, the next 1, and so on until the last which is numbered $2^{32} - 1$. These numbers are referred to as the address of the byte. When a variable is declared memory corresponding to the type is allocated. In this case one byte is allocated as the type `byte`. Note that the pattern of ones and zeros, that is the binary number, in the allocated byte corresponds to the `byte` value of 0, which is the `byte` type default. Any byte can be addressed programmatically by specifying its address. The identifier, in this case `bAcctNo` can be viewed as an alias for the address of the byte to which it refers.

> Memory is allocated in a primitive type declaration. The identifier named in the declaration assumes the value stored in the memory location. The actual value assumed is based on the type declared.

More than one variable can be specified in a declaration. For example, the declaration

```
double currentBalance, highBalance, cred-
itLimit;
```

declares three type double variables that represent the customers current balance, previous high balance, and credit limit.

A primitive variable can be initialized in its declaration. For example, the statement

```
double currentBalance = 0, highBalance,
          creditLimit = 1000;
```

initializes the customer's current balance to 0 and the credit limit to 1000. Note that the customer's previous high balance is initialized to the type `double` default of 0.

Similarly,

```
int CreditRating = 5;
double CreditLimit = 1000*CreditRating;
```

the field `CreditRating` is initialized to value 5 and the field `CreditLimit` is initialized to the value of the expression 1000*`CreditRating`, which is 5000.

Using Primitive Data Types

The primitive data types can be used to hold values that represent the characteristics of objects encountered in business, engineering and science. Consider first the representation of the account of a banking customer whose identification number is 8421 and whose account balance is $952.48. This representation is made with the declaration of the two variables:

```
long  ID = 8421;
double  AcctBal = 952.48;
```

Similarly, a product in a manufacturing operation with product number 7521, costing $95,421.69, and with 35 currently on hand might be represented with the following variable declarations:

```
long  ProdNo = 7521;
double  UnitCost = 95421.69;
long  QytOnHand = 35;
```

Primitive type variables can be used to represent the characteristics of "real world" entities.

MORE LEXICAL ELEMENTS

The Java language also includes operators, punctuation, provision for white space, line termination, and the ability to represent the Unicode characters.

Java **operators,** used to develop expressions, are one of the following:

```
=      !=     |      &=
>      &&     ^      |=
<      ||     %      ^=
!      ++     <<     %=
~      --     >>     <<=
?      +      >>>    >>=
:      -      +=     >>>=
==     *      -=
<=     /      *=
>=     &      /=
```

Characters used for **punctuation** are the following:

```
(      }      ;
)      {      ,
{      }      .
```

Terminating characters mark the end of source code lines. These characters are the carriage return (CR) and line fault (LF) or the carriage return and line fault (CR LF) characters together. Pressing the ENTER key while entering source code automatically inserts these characters at the end of a source code line. Typically, they are not displayed or printed.

Collectively, the space character, the horizontal tab, form feed characters, and line terminators are referred to as **white space.**

SUMMARY

1. The Unicode character set is implemented.
2. Identifiers are formed from the ASCII character subset of the Unicode character set.

3. Comments can be used to document programs and improve readability.
4. Java includes words reserved for special use only that are known as keywords.
5. Primitive data types provide for representing both integral and floating-type data.
6. Each of the primitive data has a size and a range of values that can be assumed.
7. Variables are used to represent data in memory.
8. Variables and associated memory are established in a declaration statement.
9. Specific values are represented by literals.
10. Characters for line termination, operators for forming expressions, and characters for punctuation are also included in the language.
11. Visual J++ includes an on-line Help capability.

KEY WORDS AND CONCEPTS

ASCII characters
`boolean`
`byte`
case-sensitive
`char`
character set
comment
data type
declaration statement
documentation comment
`double`
`false`
`float`
floating-point
`int`
integral type
Java keywords
Identifier
literals
`long`
`null`
operators
primitive types
punctuation character
`short`
single-line comment
size
strongly typed
terminating character
traditional comment
`true`
Unicode
Visual J++ Documentation
variable
white space

REVIEW QUESTIONS AND EXERCISES

1. Which of the following are valid keywords?
   a. `string`
   b. `javapp`

   c. `int`
   d. `INT`
   e. `void`
   f. `true`
   g. `null`
   h. `synchronized`
   i. `new`
   j. `default`

2. Which of the following comments are valid?
   a. `/* A traditional comment */`
   b. `// /* A nested comment */`
   c. `/** another comment */`
   d. `/** // Another nested comment */`
   e. `/* /* Still another nested comment */ */`

3. Which of the following are valid identifiers?
   a. `string`
   b. `student_no`
   c. `id`
   d. ID
   e. #_per_gm
   f. `true`
   g. `null`
   h. `synchronized`
   i. `MyName`
   j. `01Aisle`

4. Which of the following are valid operators or separators?
   a. `**`
   b. `->`
   c. `;`
   d. `//`
   e. `>>>=`
   f. `|`
   g. `~`
   h. `%`
   i. `+=`

5. Declare variables to represent a student's identification number, major that is a three digit numeric value, total number of credits,

grade point average, year first enrolled, and expected year of graduation.

6. Declare variables to represent an airline ticket in a file. Assume that all origination and destination points can be represented by a unique five-digit number. Include in the representation an origination code, destination code, departure time, arrival time, price for first, business, and economy seats.
7. Declare variables to represent a bank loan or mortgage. Include in the representation an integral customer identification code, amount loaned, loan origination date, term, monthly payment, current balance, amount due to date, and amount paid to date.
8. Declare variables to represent an employee health benefits record. (*Hint*: assume the employee and various health plans are represented by integral codes.)
9. Declare variables to represent the daily high, low, and close for a stock. Assume the stock is represented by a unique integral number rather than the usual stock symbols.
10. What assumptions are required to represent a chemical composition in terms of primitive data types? Is such a representation possible?

PRACTICE WITH VISUAL J++

1. Locate documentation regarding the null keyword using the Help Documentation search facility.
2. Add the null documentation topic to the list of topics in the Help Documentation's Favorites list.

3

Class Data Type

OBJECTIVES

After reading this chapter, you will be able to:

1. Understand the class data type and its use in modeling "real world" objects.
2. Model object attributes with primitive data type fields.
3. Manipulate the class fields members with member methods.
4. Use the class data type in applications.
5. Understand constructors and their use in the creation of object instances.
6. Create classes using reference type fields.
7. Differentiate between passing-by-value and passing-by-method method calls.
8. Demonstrate use of the call data type in building applications.

OVERVIEW

Objects are a useful tool in understanding and modeling "real world" processes and systems. For example, business systems have customer and salesperson objects, manufacturing systems have inventory objects, and educational systems have student and class objects. Objects have attributes that describe the characteristics of a specific instance of the object. For example, a customer object would typically have attributes describing the customer's unique identification code, name, address, current balance, and credit limit. Additionally, objects include methods for manipulating the values of the attributes. For example, a customer object would include methods for setting the customer identification code, adjusting the current balance to reflect payments and charges, and determining business risk using the customers current limit attribute.

The Java class data type implements the object concept. The Java class includes fields that model the object's attributes as well as methods for manipulating the fields.

CLASS DECLARATION

The Java class data type is an effective object model. Contents of a file containing a class modeling a business customer are shown in Figure 3.1.

The class declaration specifies a reference data type. The declaration consists of an optional class modifier, `class` keyword, an identifier naming the class, and a body in which the class' fields and methods are defined.

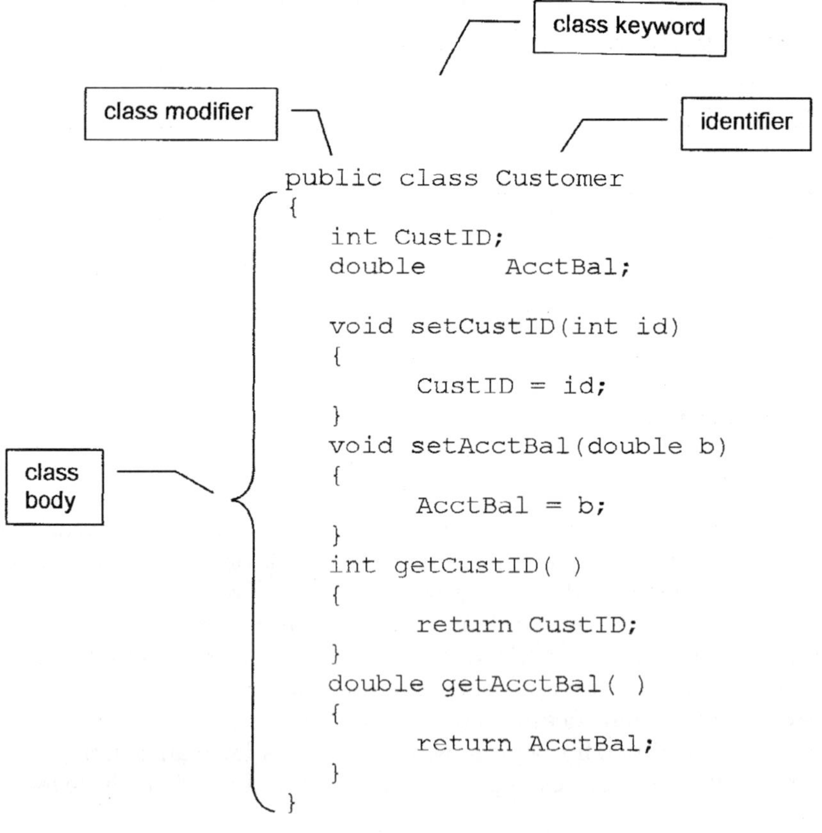

FIGURE 3.1
Class modeling a business customer.

The identifier in the declaration names the class. The keyword `class` is followed by the class name, in this case `Customer`. The name of the class is available to other files in the same directory or folder, which we will come to know as a **package,** in which the class is declared. That is, the class can be referenced by other classes declared in the same package.

The class body follows the class name and begins and ends with braces. The body of the class definition contains members describing the class' attributes and methods.

Java class definitions are not terminated with a semicolon.

A Java class is defined in a file with the same name as the class and the extension `.java`. The class `Customer` is saved in a file named `Customer.java`.

Comments provide a useful method for documenting a program. Without extensive documentation, Java programs are almost unintelligible—even to the person that authored the program as recently as yesterday! Therefore it is good programming practice to document all programs.

It is common practice to state the file name in which the class is declared as the first line of a program, as shown in Figure 3.2.

The class body contains the member fields and member methods declarations. **Member fields** represent the attributes of the object modeled. In the class `Customer`, the attributes represented are customer identification number and account balance. These fields are declared with the statements

```
int CustID;
double AcctBal;
```

Member methods provide for manipulating the values of the member fields, for example, setting and retrieving or getting the values of the field members. Member methods declared in the `Customer` class are the following:

```
setCustID( )
setAcctBal( )
getCustID( )
getAcctBal( )
```

A class that successfully models an entity must not only have attributes that completely describe the entity but also member functions that will enable the external access to the data members corresponding to the entity's attributes.

```
// Customer.java

public class Customer
{
        int CustID;
        double     AcctBal;

        void setCustID(int id)
        {
                CustID = id;
        }
        void setAcctBal(double b)
        {
                AcctBal = b;
        }
        int getCustID( )
        {
                return CustID;
        }
        double getAcctBal( )
        {
                return AcctBal;
        }
}
```

file name

FIGURE 3.2
Include the file name as a comment.

FIELD MEMBERS

Member fields, data members or **class variables** represent the properties of the object modeled by the class. The terms "member fields," "data members," and "class variables" are often used interchangeably.

Field Declaration

The member fields are declared in the class body. The field `CustID`, modeling the customer's identification, is declared in Figure 3.3.

A field declaration consists of a type and identifier. Two fields in a class cannot have the same name. The scope of a field name is the entire body of the class in which it is declared.

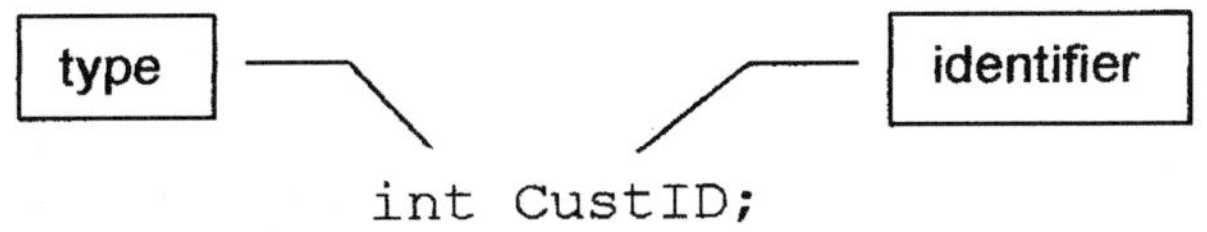

FIGURE 3.3
Member field declaration.

More than one field can be specified in a declaration. For example, the declaration

```
Double currentBalance, previousBalance, cred-
itLimit;
```

declares three type double variables that represent the customers current balance, previous balance, and credit limit.

Field Initialization

A field can be initialized in its declaration. For example, the statement

```
Double currentBalance = 0, previousBalance,
       creditLimit = 1000;
```

initializes the current balance to 0 and the credit limit to 1000. Similarly, in

```
int CreditRating = 5;
double CreditLimit = 1000*CreditRating;
```

the field `CreditRating` is initialized to value 5 and the field `CreditLimit` is initialized to the value of the expression `1000*CreditRating`, which is 5000.

Field Modifiers: `static` and `final`

Field modifiers can describe the character of the field. The field modifier is optional and precedes the type specification in the declaration. Examples and explanations of the `static` and `final` field modifiers are presented below.

A variable is defined as static with the keyword `static`. For example, the variable `Discount` is declared static:

```
static int Discount;
```

A variable declared static is said to be a **static variable.** Variables not declared static can be referred to as **non-static variables.** Static variables differ from non-static variables in that every instance of a class shares a static variable whereas each instance of the class has its own copy of its non-static variable. This situation is represented conceptually for the class Customer in Figure 3.4

```
// Customer.java

public class Customer
{
      int CustID;
      double AcctBal;
      static int Discount = 40;
      . . .
}
```


FIGURE 3.4
Class including a static variable.

An application using the Customer class, declared in Figure 3.4, is shown in Figure 3.5. Two instances of the class `Customer` are created: `custx` and `custy`. The static variable `Discount` is initialized to the value 40 when an instance of the class is created. Values are also assigned to the non-static variables in the portion of the `UsingCustomer` application that is not shown. Eventually, the static variable `Discount` is assigned the value 50 with the statement

```
custx.Discount = 50;
```

```
public class UsingCustomer
{
      public static void main(String[] args)
      {
            Customer custx = new Customer();
            Customer custy = new Customer();
            . . .
            custx.Discount = 50;
            Customer.Discount++;
            . . .
      }
}
```

FIGURE 3.5
Application illustrating accessing a static variable.

In this case, the variable `Discount` is accessed as a member of the `custx` instance of the `Customer` class. Because the variable is static, all instances of the `Customer` class share the same value. Multiple instances of a class sharing the same value of a static variable is shown conceptually in Figure 3.6.

Static variables can also be accessed as class variables. For example, the static variable `Discount` in the `Customer` class is accessed with the statement

```
Customer.Discount++;
```

When the statement is executed, the value of `Discount` is incremented to 51. This value of the variable is shared by all instances of the class.

A variable declared `final` must be initialized at compilation and cannot be assigned a value during program execution. Declaring a field that cannot

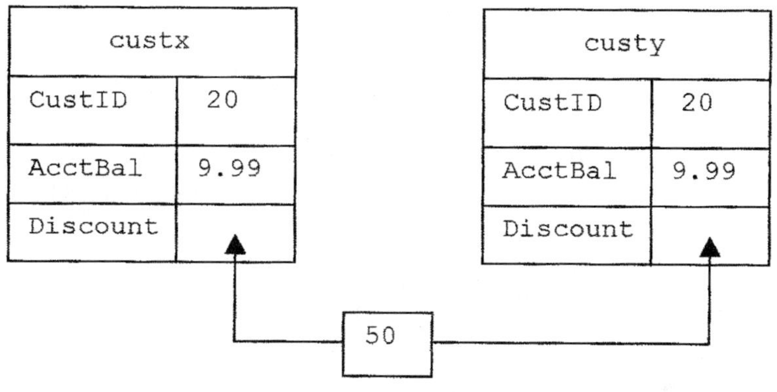

FIGURE 3.6
Conceptual sharing of the value of a static variable by multiple instances of a class.

change as `final` is a useful documentation tool. For example, the field CreditLimit is declared final by the statement

```
final double CreditLimit = 10000;
```

and is initialized to the value 10,000. This value cannot be changed in the program. For example, the statement

```
CreditLimit = 10000;
```

is invalid. Note that both static and non-static fields can be declared final.

METHOD MEMBERS

A **method** is a group of statements that perform a computation. Methods are referred to as "functions" in some languages. A method is **executed, invoked,** or **called** by stating its name. When invoked, control passes to the method named. When the invoked method is completed, control is returned to the system or program that invoked the method.

Methods provide for the following:

- Breaking a program into independent functional components, commonly called "modules"
- Elimination of repeated code for frequently performed tasks
- Debugging of small independent modules

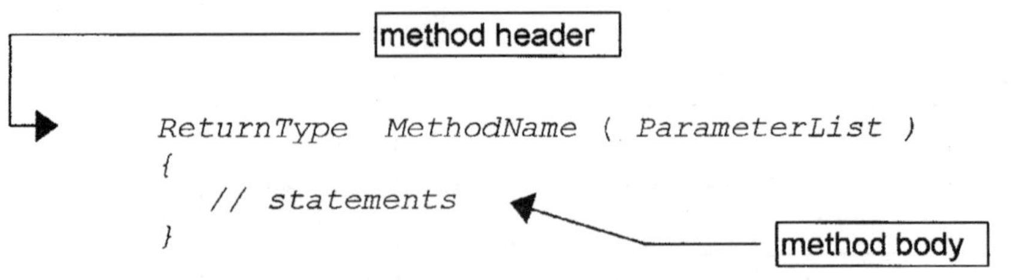

FIGURE 3.7
Method declaration syntax.

Declaring Methods

A method must be **declared.** A method's declaration specifies the actions performed by the method. A method declaration consists of two parts: a method header and a method body. The formal syntax for declaring a method is shown in Figure 3.7.

The **method header,** or simply **header,** specifies the data type of the value the method returns, the name of the method, and lists parameters the method requires. The **parameter list** is a comma-separated list of data type declarations each followed by an identifier enclosed in parentheses. The identifiers, also referred to as formal parameters, act as place holders for the values that are passed when the method is invoked in a program. All methods can accept arguments. The way in which the arguments are processed to produce a result is specified in the method body. The **method body** follows the parameter list and is a block of code that begins with an opening brace **{** and ends with a closing brace **}**. The block contains statements that specify the functionality of the method. In addition, definitions for all data items used in the computation that are not specified in the header are specified in the body. An example of a method definition is shown in Figure 3.8.

Note that the **method name,** or **method identifier,** `setCust`, is specified in the header. A method can return or not return a result. The result's data type is specified in the header. The `void` data type is specified as the result data type for methods that do not return a result. The method `setCust` does not return a result, consequently `void` is specified as the return type. The header also specifies an argument list. The argument list for the method `setCust` is the following:

```
(int id, double ab)
```

The method receives two arguments: the first is a type `int` argument named `id` and the second is type `double` and named `ab`. Generally, the arguments will be of different types, reflecting the data represented.

The body of the method's definition follows the header and is enclosed in braces. Statements describing the action performed when the method is called, or invoked, are contained in the method's body. For example, consider

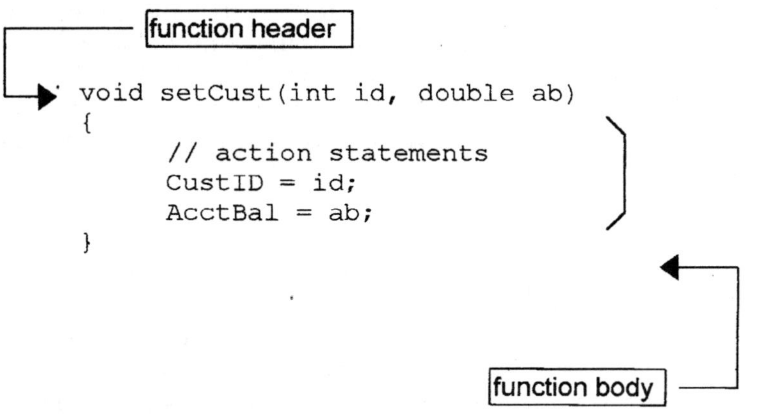

FIGURE 3.8
Declaration of the method `setCust`.

the method `setCust` that is used to assign values to the customer identification and account balance variables. In this case, the statements

```
CustID = id;
AcctBal = ab;
```

assign the value of the parameters `id` and `ab` the fields `CustID` and `AcctBal`.

Note that the method ends at the closing brace, and control is returned to the method's invoker.

Returning Values

A **`return`** statement returns control to the method's invoker. When a return statement is encountered in the method, control is returned to the method's invoker. The `return` keyword can be used anywhere within a function. The return statement need not return a value. The return statement ends the function no matter where it appears. A return statement is not needed if the function does not return a value—a function ends automatically at its closing brace. An example of a return statement is shown in Figure 3.9.

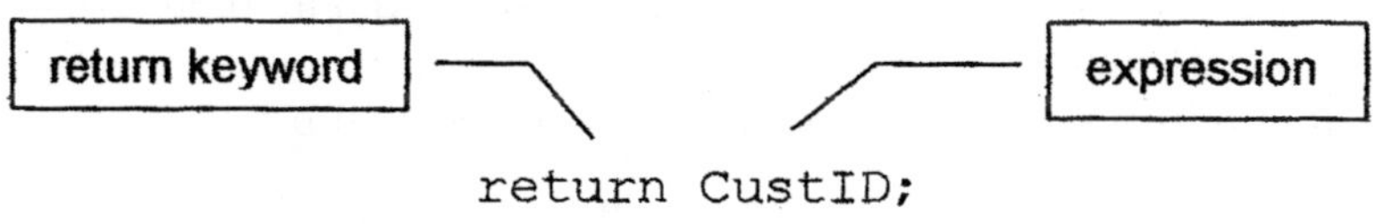

FIGURE 3.9
Returning the value of the variable CustID.

If the return statement contains an expression, then the expression is evaluated and returned to the invoker. In the example the value of the customer identification variable is returned to the calling method. The return statement only returns one value, which is the value of the expression. The complexity of the expression is irrelevant as the most complex expression evaluates to a single value. For example,

```
return ( Math.sqrt( 16 ) * 2 + 3 );
```

returns the value 11. `Math.sqrt()` is a Java method that computes the square root of a specified value.)

A function can return or not return a value. The type of value returned is specified in the header. *The return value must be the same data type as the method.* For example, when a method's result type is defined as `int` the return value must be data type `int`. Similarly, if the method's return data type is defined as `char` the returned value must be data type `char`.

A return statement can be placed anywhere in the method and immediately ends the method. In addition, a method can have more than one return statement. For example, consider the following code to determine if an error has occurred:

```
If(errorFlag == 1)
    return 16;
else
    return 0;
```

Assume the `errorFlag` is set to value `true` when an error occurs. (The code will be examined in detail in Chapter 6.) If an error has occurred, the method returns the value of 16—a value traditionally reserved for disastrous errors. When the error flag is not set, a value of 0 is returned—the value traditionally used to indicate successful execution or execution without error.

The `void` data type is used as the return value data type for functions that do not return a value to the calling function. Consider the method that is to set the value of the customer identification code:

```
void setCustID(int id)
{
    CustID = id;
}
```

In this case a value is not returned and hence the void data type is the appropriate return data type.

Note that return statement's expression is optional; thus

```
return;
```

is valid.

Parameters

Whether a value passed between a calling program and a called program is called an "argument" or a "parameter" depends on the point of view. In the calling program, the values passed to the method are called **arguments.** In the called program, that is the method itself, the values passed are referred to as **parameters.** In either case, the values are the same.

Parameters are specified as a comma-separated list in which each parameter consists of a type and an identifier. For example, consider the method `setCust` which assigns values to the customer identification and account balance variables.

```
void setCust(int id, double ab)
{
    CustID = id;
    AcctBal = ab;
}
```

In this case, values of the parameters `id` and `ab` are assigned to the fields `CustID` and `AcctBal`.

Note that the parameter list can be empty, as in the case of the method `getCustID`:

```
int getCustID( )
{
    return CustID;
}
```

Two parameters cannot have the same name. The following code illustrates an invalid parameter list:

```
(int id, double ab, int id)
```

as the parameter `id` is listed twice.

The name and type of the parameters is called the method's **signature.** Two methods cannot have the same signature.

> The value passed to a method is referred to as an argument in the calling statement program and as a parameter in the called statement.

USING THE CLASS DATA TYPE

An application using the Customer class is shown in Figure 3.10. The application `UsingCustomer` creates an instance of the Customer class, assigns values to the customer identification and account balance variables of the instance, and retrieves the values of these variables.

The method `main` is a special method that serves as the entry point to a Java application programs. Every J++ application contains one method called

```
public class UsingCustomer
{
      public static void main(String[] args)
      {
            int newCustId = 21;
            double newCustBal = 99.99;
            int outBufID;
            double outBufBal;

            Customer cust = new Customer();

            cust.setCustID(newCustId);
            cust.setAcctBal(newCustBal);
            outBufID = cust.getCustID();
            outBufBal = cust.getAcctBal();
      }
}
```


FIGURE 3.10
Application using the Customer class data type.

`main`. There can only be one method `main` in a program. The method `main` must have a return type of `void`. As stated, the `main` method is the method performed when the application is first executed. Other methods can be called by the main method. These methods can in turn call other methods.

The variables `newCustID` and `newAcctBal` represent buffers holding customer identification and account balance values that are typically input by a user or retrieved from a database or file. Note that these variables have been initialized. Similarly, the `outBufID` and `outBufBal` represent buffers holding values that will be output to a user or written to a database or file.

Creating a Class Instance

An instance of the Customer class is created with the new expression (Figure 3.11).

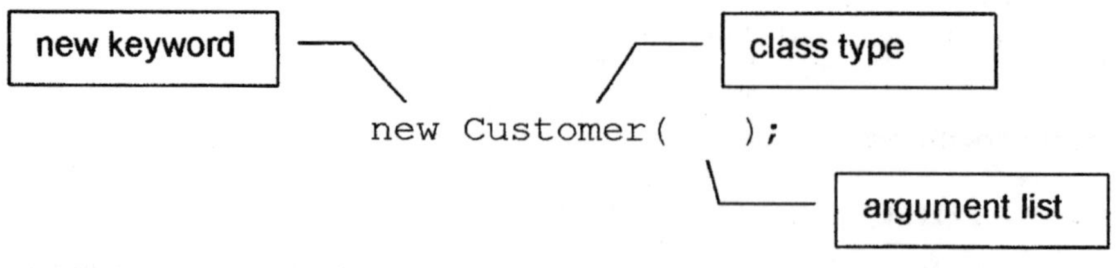

FIGURE 3.11
Creation of a `Customer` class instance.

The class type, in this case the Customer class, must be a class that is not abstract. The argument list is used to select a constructor declared in the class with a matching signature. (Constructors are methods executed automatically during the creation of an object and are explained later in the chapter.) Creation of the class instance proceeds in several steps.

1. Memory is allocated for the instance of the class. Memory is set aside for all the fields declared in the class. Each field is initialized to its default value.
2. The appropriate constructor is called. Note that the call is automatic and is performed immediately after the memory for the class instance is allocated and the member fields initialized.
3. The value of the class instance creation expression is a reference to the newly created object. This reference value can be assigned to a reference to the type of object created as in the statement

```
Customer cust = new Customer(    );
```

Note that the reference to the newly created `Customer` data type object is assigned to a `Customer` data type reference named `cust`.

In this case, a default constructor `Customer( )` is called and executed. This constructor does not perform any action, but this will not be the case as we proceed to more complicated classes modeling "real world" conditions.

Accessing Instance Members

A field in an instance of a class is accessed with an expression consisting of the identifier referencing the instance of the object followed by the "." token followed by the field's identifier. For example, the field `CustID` in the `cust` instance of the `Customer` class is accessed with the expression

```
cust.CustID
```

Similarly, a method in an instance of a class is accessed with an expression consisting of the identifier referencing the instance of the object followed by the "." token followed by the method's identifier. For example, the method `setCust` in the `cust` instance of the `Customer` class is accessed with the expression

```
cust.setCust( . . . )
```

Manipulating Members

After an instance of the object is created, member methods of the class can be invoked to manipulate the member fields of the class instance. For example, fields in the instance of the class can be accessed. To illustrate, the field

CustID in the instance cust of the Customer class is accessed with the expression shown in Figure 3.12. In this case the field member CustID of the cust instance of the Customer class is assigned the value 99.

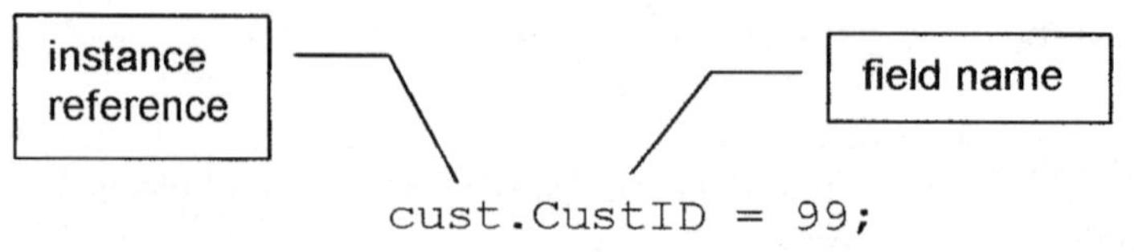

FIGURE 3.12
Accessing the setCustID field for the Customer class instance cust.

Similarly, the member method setCust of the class Customer instance cust is invoked with the statement shown in Figure 3.13.

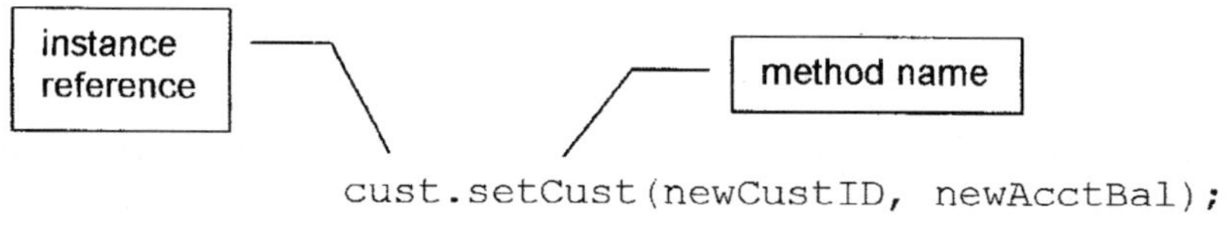

FIGURE 3.13
Accessing the setCust method for the Customer class instance cust.

In both cases, access is for a specific instance of the Customer class, in this case cust. Note that the name of the instance is followed by a "." token which is followed by the name of the member accessed.

CONSTRUCTORS

Constructors are class members that are invoked every time an instance of the class is created. Because they are invoked at instance creation time they are often used for initialization of the class members.

Every time an instance of a class is created a constructor is called.

Constructor Declaration

Constructor declarations are similar to method declarations, except that they do not have a return type. An example of a constructor declaration is presented in Figure 3.14.

Constructors have the same name as the class in which they are declared. Note that unlike member methods, constructor declarations do not have a return type. In the example, the constructor is named Customer, the same as

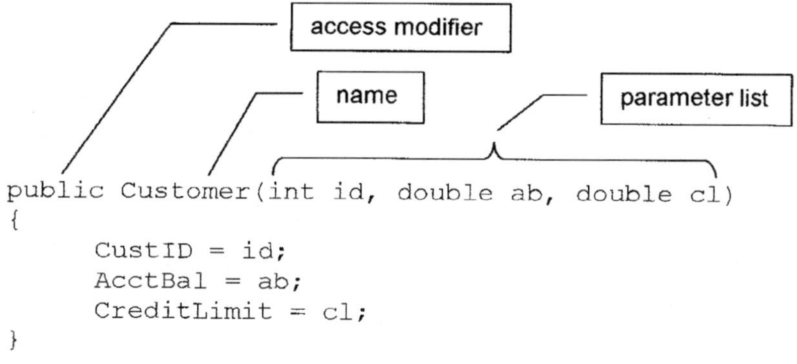

FIGURE 3.14
Constructor declaration for the class Customer.

the class. The parameter list in constructor declarations is specified and used in the same manner as the parameter list in member method declarations.

Constructors are invoked by the class instance creation expression. For example,

```
Customer cust = new Customer(10, -75.30,
5000.00);
```

Note that the parameters specified in the class name match the parameters specified in the constructor. An exact match of parameters specified in the new instance expression with the parameters list in the constructor declaration is required. The constructor is invoked automatically after the fields in the class have been initialized.

Consider the class `Customer` shown in Figure 3.15. As stated above, an instance of the class is created with the statement

```
Customer cust = new Customer(10, -75.30,
5000.00);
```

First, memory for the fields declared in the class is allocated for the instance. Then, the fields are initialized as specified in the declaration. In this case the field `CreditLimit` is initialized to the value `1000.00`. Finally, the constructor for the class is executed. In this case, the constructor parameter list specifies values, 10, -75.30,and 5000.00, which are assigned to the fields `CustID`, `AcctBal` and `CreditLimit`. Note that the initial value `1000.00` assigned to the variable `CreditLimit` is changed when the constructor is called.

Step in instance creation are:
1. Memory Allocation
2. Field Initialization
3. Constructor Execution

```
// Customer.java

class Customer
{
        Customer(int id, double ab, double cl)
        {
                CustID = id;
                AcctBal = ab;
                CreditLimit = cl;
        }

        protected double CreditLimit = 1000.00;
        protected int CustID;
        protected double AcctBal;

        public void setCust(int id, double ab)
        {
                CustID = id;
                AcctBal = ab;
        }
        public void setCustID(int id)
        {
                CustID = id;
                return ;
        }
        public void setAcctBal(double b)
        {
                AcctBal = b;
        }
        public void setCreditLimit(double cl)
        {
                CreditLimit = cl;
        }
        int getCustID( )
        {
                return CustID;
        }
        double getAcctBal( )
        {
                return AcctBal;
        }
        double getCreditLimit( )
        {
                return CreditLimit;
        }
}
```

FIGURE 3.15
`Customer` class declaration with constructor.

Return statements can be included in the constructor body but must not include an expression. Remember that a constructor declaration does not include a return type and does not return a value.

Note that when a constructor is declared for the class, the parameter list following the class name must match the parameter list in the constructor declaration. In other words, the parameter list specified in the new instance creation must match the constructor's signature. For example, the new instance creation statement

```
Customer cust = new Customer();
```

is invalid as the parameter list, said to be empty, does not match the signature declared in the class shown in Figure 3.15.

Constructor access modifiers are `public`, `protected`, and `private`. Only one of the modifiers can be specified in the declaration.

Default Constructor

A constructor with no parameters is said to be a **default constructor.** If a class contains no constructors, a default constructor is automatically provided. If the class is declared public the default constructor has public access; otherwise, the default constructor has default access. For example, the declaration

```
public class Customer
{
}
```

is equivalent to the declaration

```
public class Customer
{
    public Customer()
    {
    }
}
```

Even though a default constructor might have an empty parameter list, the body need not be empty. An example of a default constructor that is used to initialize a member field is shown in Figure 3.16 in which the `CreditLimit` variable is assigned the value `10,000.00`.

```
public class Customer
{
    public Customer()
    {
        CreditLimit = 10000.00;
    }
}
```

FIGURE 3.16
Default constructor with nonempty body.

REFERENCE TYPE CLASS MEMBERS

Reference types can be declared as fields in a class. Consider, first, the class SalesPerson that models a salesperson with fields to represent the salesperson's sales region and identification as well as methods to manipulate the field values shown in Figure 3.17.

```
public class SalesPerson
{
     private int RegID;
     private int SlsPerID;

     public SalesPerson(int r, int sp)
     {
          RegID = r;
          SlsPerID = sp;
     }
     public void setSalesPerson(int r,int sp)
     {
          RegID = r;
          SlsPerID = sp;
     }
     public int getRegId()
     {
          return RegID;
     }
     public int getSlsPerID()
     {
          return SlsPerID;
     }
}
```

FIGURE 3.17
SalesPerson class declaration.

A field of the type SalesPerson can be included in the Customer class to represent the salesperson assigned to the customer. Consider the Customer class as shown in Figure 3.18, which includes a SalesPerson field SlsPer.

```
// Customer.java

class Customer
{
     final int CreditRating = 10;
     double CreditLimit = 1000*CreditRating;
     protected int CustID;
     protected double AcctBal;
     public SalesPerson SlsPer;
```

```
    Customer(int id, double ab, double cl, SalesPerson sp)
    {
        CustID = id;
        AcctBal = ab;
        CreditLimit = cl;
        SlsPer = sp;
    }
    public void assignNewSalesPerson( SalesPerson spObj )
    {
        spObj.setSalesPerson(44, 22);
    }
    Customer()
    {
    }
    public void setCust(int id, double ab)
    {
        CustID = id;
        AcctBal = ab;
    }
    public void setCustID(int id)
    {
        CustID = id;
        return ;
    }
    public void setAcctBal(double b)
    {
        AcctBal = b;
    }
    public void setCreditLimit(double cl)
    {
        CreditLimit = cl;
    }
    int getCustID( )
    {
        return CustID;
    }
    double getAcctBal( )
    {
        return AcctBal;
    }
    double getCreditLimit( )
    {
        return CreditLimit;
    }
}
```

FIGURE 3.18
Customer class with a `SalesPerson` type field.

The initial values of the fields after the cust object is created are shown in Table 3.1.

TABLE 3.1

Initial values of Customer object after instantiation.

Field	Value	Type
CreditRating	10	final int
CreditLimit	1000.0	double
CustID	0	protected int
AcctBal	0.0	protected double
SlsPer	null	public SalesPerson

The `SalesPerson` field is initialized to value `null`. The default value assigned to all reference data types is `null`. The difference between the default values assigned to reference data types and primitive data types is illustrated by the default value assigned to the `int` data type `CustID`. Note that the primitive data type field `CustID` is initialized to its default value `0`. In addition, note that `SlsPer` is simply a reference to a `SalesPerson` object.

> The default value assigned to all reference data types is null.

Creation of the `SalesPerson` object can be done in a constructor. Consider the revised `Customer` class with a constructor that creates a `SalesPerson` object and assigns values to its fields, as shown in Figure 3.19.

```
// Customer.java
class Customer
{
     final int CreditRating = 10;
     double CreditLimit = 1000*CreditRating;
     protected int CustID;
     protected double AcctBal;
     public SalesPerson SlsPer;

     Customer(int id, double ab, double cl, int r, int sp)
     {
          CustID = id;
          AcctBal = ab;
          CreditLimit = cl;
          SlsPer = new SalesPerson(r, sp);
     }
     public void assignNewSalesPerson( SalesPerson spObj )
     {
          spObj.setSalesPerson(44, 22);
     }
     Customer()
```

```
        {
        }
        public void setCust(int id, double ab)
        {
             CustID = id;
             AcctBal = ab;
        }
        public void setCustID(int id)
        {
             CustID = id;
             return ;
        }
        public void setAcctBal(double b)
        {
             AcctBal = b;
        }
        public void setCreditLimit(double cl)
        {
              CreditLimit = cl;
        }
        int getCustID( )
        {
              return CustID;
        }
        double getAcctBal( )
      {
           return AcctBal;
      }
      double getCreditLimit( )
      {
           return CreditLimit;
      }
   }
```

FIGURE 3.19
Creation of `SalesPerson` object in `Customer` class construction

The constructor contains the statement

```
SlsPer = new SalesPerson(r, sp);
```

that creates a new instance of `SalesPerson` and assigns the reference to the `SalesPerson` type field `SlsPer`. Following memory allocation, the new instance creation calls the `SalesPerson` constructor with a parameter list representing the salesperson's region identification code and personal identification code. These values are assigned to the fields `RegID` and `SlsPerID` in the newly created `SalesPerson` object.

The salesperson region identification code can be retrieved with the statement

```
SlsPer.getRegID( );
```

Similarly, the salesperson's identification code can be retrieved with the statement

```
SlsPer.getSlsPerID( );
```

Class field members can be reference types.

REFERENCE TYPE METHOD PARAMETERS

Reference types can be passed as method parameters. Consider the method `reassignSalesPerson` declared in the class `UsingCustomer` shown in Figure 3.20. The method enables the reassignment of a salesperson. This is common to business operation, as salespersons are a traveling lot resulting in the need to regularly reassign a new salesperson to a customer. The method has a single parameter which is a `SalesPerson` object representing the salesperson to be assigned. The method invokes the `SalesPerson` class method `setSalesPerson`, which takes two parameters. The first parameter is assigned to the invoking `SalesPerson` object's `RegID` and the second is assigned to the invoking `SalesPerson` object's `SlsPerID`. In this case, the `SalesPerson` object returned is the `SalesPerson` represented by the `Customer` class object `cust`'s field `SlsPer` returned by the expression

```
cust.getSalesPerson()
```

The method `getSalesPerson` returns the `Customer` class' `SalesPerson` object reference `SlsPer`.

```
// UsingCustomer.java
 class UsingCustomer
{
     public static void main(String[] args)
     {
          Customer cust = new Customer(77,77.77,7777.77,5,99);
          UsingCustomer uc = new UsingCustomer();
          uc.reassignSalesPerson(cust.getSalesPerson());
     }
     private void reassignSalesPerson(SalesPerson sp)
     {
          sp.setSalesPerson(21,12);
     }
}
```

FIGURE 3.20
Passing a reference type parameter in the method `reassignSalesPerson`.

The method `reassignSalesPerson` invokes the `setSalesPerson` method for the instance of `SalesPerson` passed as a parameter. In the case illustrated, the `RegID` and `SlsPerID` variables are set to the values `21` and `12`, respectively.

Now, if we were to look at the `RegID` and `SlsPerID` values of the object referenced by `SlsPer` of the `Customer` instance `cust`, we will find that they have the values `21` and `12`, respectively. Note that the values of the reference object passed have been altered. Compare this with primitive variables in which the passed variable itself is not changed but only the copy used in the method, and the variable is said to be **passed-by value.** In the case of reference types, however, the reference type passed is altered, and the reference type is said to be **passed-by reference.**

HANDS-ON EXERCISE 3: PASSING VALUE IN A METHOD CALL

Methods are a fundamental building block. This exercise examines the passing of values in method calls.

Debugging is the process of finding and correcting errors, both logical and syntactical, in an application. The Integrated Development Environment contains a debugger application that enables a developer to step through an application program, examine and modify date values as well as examine conditions. The debugger provides for the real-time examination of values during a method call.

A key debugging tool is the ability to **single-step** through a program's statements. In this mode, execution of each program statement is initiated by user input. The statement that will be executed is referred to as the **current program statement.** The current program statement is indicated by a yellow arrow in the left margin of the code window. When a function call is encountered, the first statement of the function becomes the current program statement. Tools enabling debugging include **breakpoints.** A breakpoint can be toggled on and off at a line of code by clicking in the left margin opposite the statement. another tool is the **Debug menu,** which contains commands that enable stepping through a program, execution to the cursor insertion, execution to a breakpoint, as well as setting and disabling breakpoints. Table 3.2 lists key items that may appear in the debug menu.

A **Debug toolbar** contains icons corresponding to key commands in the Debug menu.

When the program execution is stopped, windows enable the examination of the values program variable and registers. Windows useful in debugging include Autos, Immediate, Locals and Watch windows, which are summarized in Table 3.3. These windows are toggled on and off using the options displayed when the Debug Windows command I View menu is selected.

TABLE 3.2

Key debug menu items

Command	Action
`Start`	Initiates program execution in the debug mode.
`Start Without Debugging`	Initiates program execution is a normal mode—that is, not debugging mode.
`Break`	Halts program at its current location.
`End`	Completes execution.
`Continue`	Execute code from the current statement until a breakpoint, program exit, or user input point is reached.
`Restart`	Restarts program execution at the first program statement.
`Step Into`	Single-steps program execution—that is, current program statement is executed.
`Step Over`	Single-steps through program execution executing called functions without stepping.
`Step Out`	Executes, without stepping, the current function and stops at the statement immediately following the statement's call.
`Run to Cursor`	Executes the program code, without single-stepping, stopping at the insertion point.

TABLE 3.3

Windows useful during debugging

Window	Description
`Autos`	Displays variables and values within the scope of the current line of program execution.
`Locals`	Displays variables and values in the current procedure.
`Watch`	Displays values within the scope of the current line of execution within the scope of the current line of execution within the current procedure.
`Immediate`	Enables changing values and evaluating expressions.

Step 1: Create New Project

1. Start Visual J++
2. Select the New Project… command in the File menu (Figure HO-3.1).
3. Select the Visual J++ Projects folder in the Options window, the leftmost pane. The contents of the folder are displayed in the right pane.

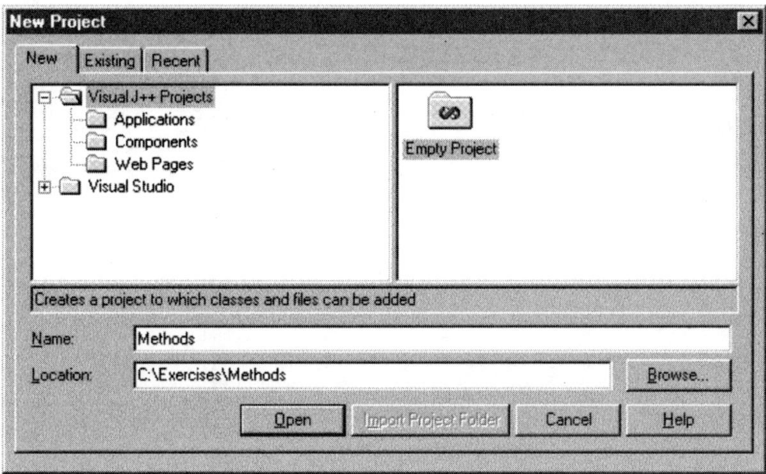

FIGURE HO-3.1
New Project window.

4. Click the Empty Projects folder, which creates an empty project to which files can be added.
5. Click the Browse… button and select the location for the project's file.
6. Enter the project's name, `Methods`, in the Name textbox.
7. Click the Open button to create the new project.

Step 2: Create the File `Method.java`.

1. Select the add Class… command in the Project menu. The Add Item dialog box is displayed with the Class folder selected.
2. Select the Class Main option in the right pane, which creates a public class with a main method entry point.
3. Enter **`Method.java`** in the Name textbox.
4. Click the Open button.
5. The `Methods.java` class is opened in the Code Editor (Figure HO-3.2).
6. Enter the following code:

```
public class Methods
{
    public static void main(String[] args)
    {
        int returnedValue;
        int a = 1;
```

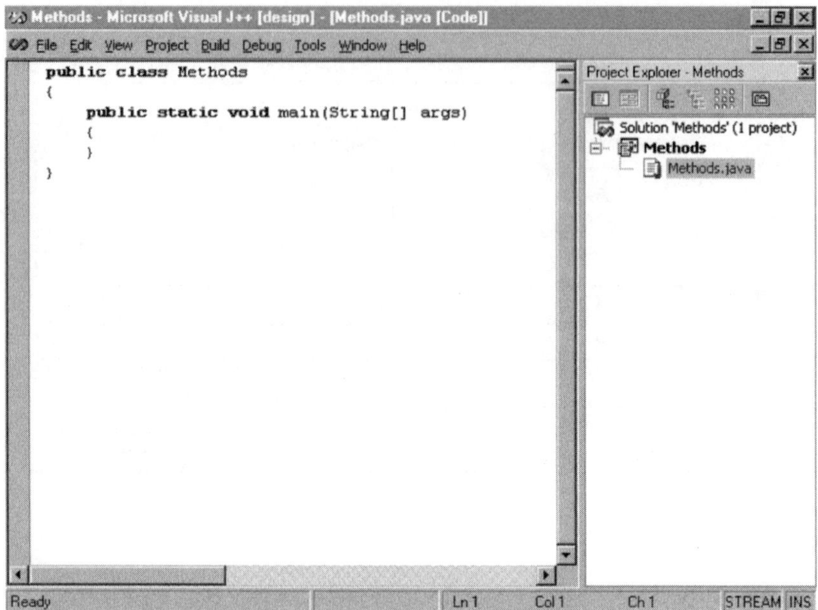

FIGURE HO-3.2
`Methods.java` class.

```
            int b = 2;
            Methods m = new Methods();
            returnedValue = m.swap (a, b);
        }
        Methods()
        {
        }
        public int swap (int x, int y )
        {
            int temp = x;
            x = y;
            y = temp;
            return x + y;
        }
}
```

7. Select the Build command in the Build menu to build the project. Correct any syntax errors and complete the build process.

Step 3: Basic Debugger Operations

1. Set the cursor on line 6, `int a = 1;` , and click the mouse button to set the insertion point. The insertion is represented by a blinking bar.

2. In the Debug menu, select the Run to Cursor command. The first time the program is executed a dialog box regarding Properties is displayed. Click the OK button to accept the default settings. Note that Launch as Console Application is unchecked. Note also that the Properties dialog box is opened by Select in the Method Properties command at the bottom of the Project menu.
3. Program execution begins and proceeds to the cursor. The current statement is marked by a yellow arrow in the left margin of the Code window.
4. Open the Locals window by selecting the Locals option in the Debug Windows in the Project menu. The locals window is displayed (Figure HO-3.3).

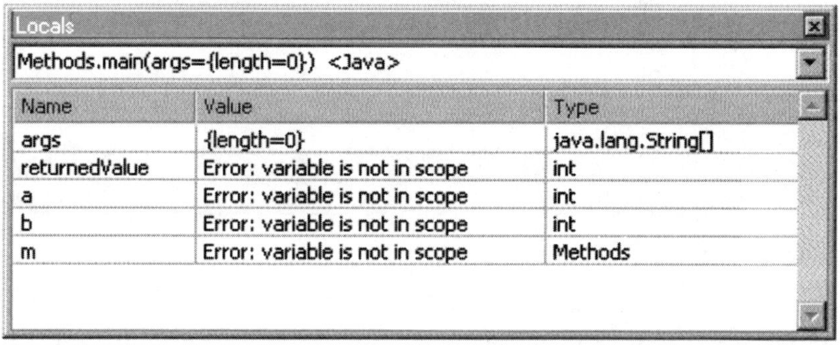

Locals

Methods.main(args={length=0}) <Java>

Name	Value	Type
args	{length=0}	java.lang.String[]
returnedValue	Error: variable is not in scope	int
a	Error: variable is not in scope	int
b	Error: variable is not in scope	int
m	Error: variable is not in scope	Methods

FIGURE HO-3.3
Locals window.

- Why is an error message displayed for the Value of the local variables?

  ____

5. Select the Step Into command in the Debug menu. The current statement is executed. The yellow arrow moves to the next statement. Note the change in the Locals window.
   - What changes in the Locals window? Why? ______________

     ____
6. Select the End command in the Debug menu.

Step 4: Using Breakpoints

1. Click the cursor in the left gray margin at line ten, `opposite returnedValue = m.swap(a, b);` to set a breakpoint (Figure HO-3.4).
2. Select the Start command in the Debug menu.

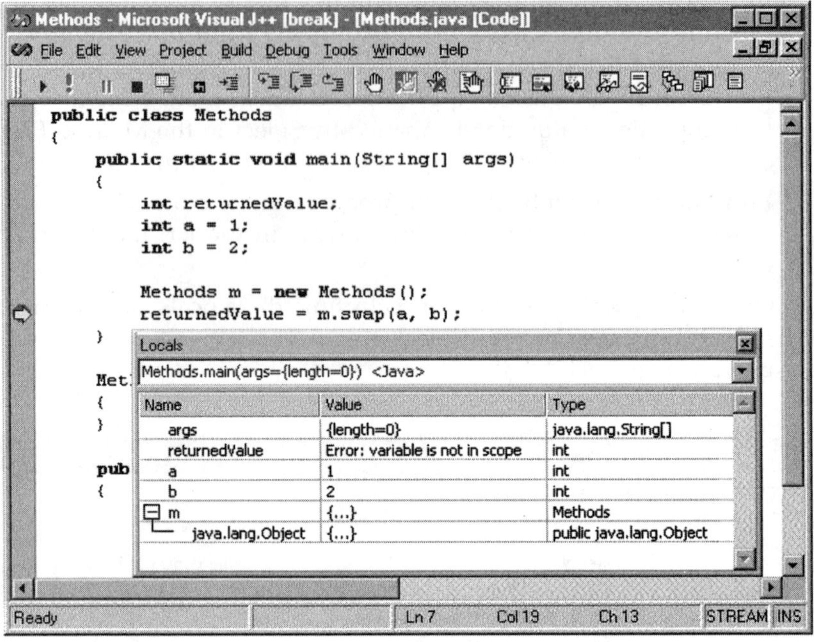

FIGURE HO-3.4
Setting a breakpoint.

3. Program execution is initiated and continues normally until the breakpoint is reached. Note that the current statement, indicated by the yellow arrow, is the method call `m.swap(a, b);`
4. Select the Step Into command in the Debug menu to execute the current program statement. The method is invoked and the current program statement becomes the first statement in the invoked method.
   - What variables are reflected in the Locals window?

     ____

5. Select the Step Into command in the Debug menu three times.
   - Are the passed values swapped? ____________________
6. Step through the called method and return to the main procedure.
   - Have the values for `a` and `b` been swapped? Why? __________
   - Explain the difference between passed-by value and passed-by reference!

     ____

     ____

     ____

     ____

Step 5: Evaluating Expressions and Changing Values

1. Select the Clear All Breakpoints in the Debug menu to clear all set breakpoints.
2. Set a breakpoint at line 22, opposite the statement `return x + y`.
3. Select the Start command in the Debug menu. Program execution stops at the breakpoint.
4. Open the Immediate window.
5. Type `x - y` in the immediate window (Figure HO-3.5). Press the ENTER key.
   - Is the expression `x - y` evaluated in the Immediate window? __________
6. Type `x = 6` in the Immediate window (Figure HO-3.5). Press the ENTER key.
   - Is the value of x changed? __________
7. Select the End command in the Debug menu.

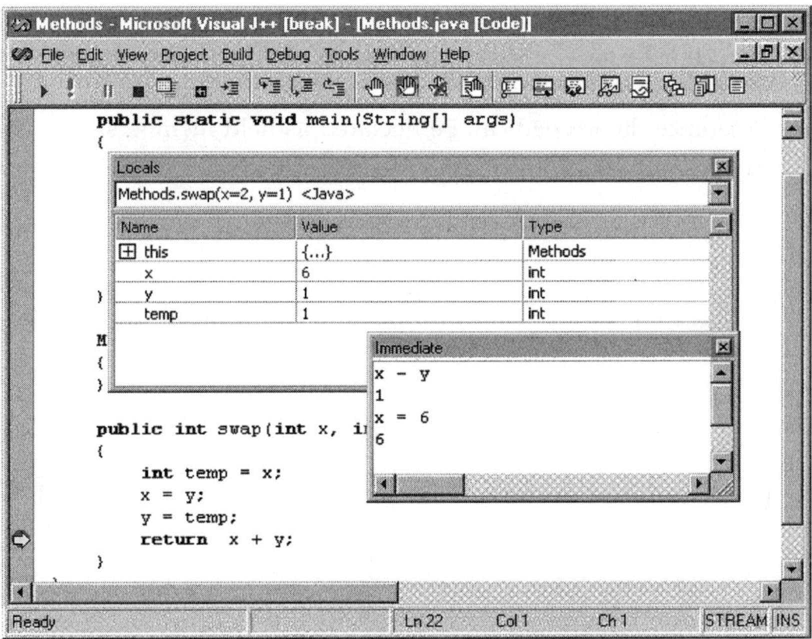

FIGURE HO-3.5
Immediate window.

Step 6: Passing Objects

1. Modify the program to pass objects as parameters.
   - Are objects passed by reference? __________

SUMMARY

1. Objects are modeled by the Java class data type.
2. Fields declared in the class represent object attributes.
3. Methods declared in the class manipulate the fields.
4. Members, both fields and methods, are accessible throughout the class in which they are declared.
5. Members declared private are not accessible externally.
6. Memory is allocated for objects when they are created with the new instance expression.
7. Special members called constructors are called automatically during object creation.
8. Reference data typed can be declared as field members.
9. Reference type method arguments are passed-by reference.

KEY WORDS AND CONCEPTS

.java
argument
call
class
class variable
constructor
data member
default constructors
fields
final
header
invoke
main
method
member field
method header
method identifier
method name
non-static variable
parameters
parameter list
passed-by reference
passed-by value
signature
static
static variable
return
void

REVIEW QUESTIONS AND EXERCISES

1. Are the following statements valid? If not, why?

```
final double CreditLimit;
. . .
CreditLimit = 21;
```

2. Is the following declaration valid? If not, why?

```
final static CreditLimit = 10000.00;
```

3. (True/False) A class must contain both fields and methods.
4. (True/False) Only primitive data types can be used as class field members.
5. Are private members accessible externally?
6. (True/False) The default value for reference data type objects is `null`.
7. Is access to a member with default access possible outside the package in which the member's type is declared?
8. (True/False) The method header specifies only the arguments.

9. (True/False) The method declaration statement is terminated with a semicolon?
10. (True/False) Every method must have at least one return statement.
11. (True/False) A method that returns a value must have a return statement.
12. (True/False) A method's return statement expression cannot contain a call to another method.
13. What is the difference between an argument and a parameter?
14. (True/False) Reference data types used as method arguments are passed-by value.
15. (True/False) Constructors are automatically called when reference data types are created.
16. (True/False) Both the parameter list and body of a default constructor must be empty.
17. (True/False) The default constructor for the `Customer` class object `cust` is invoked by the statement

cust.Customer( );

PRACTICE WITH VISUAL J++

1. Develop and demonstrate a class called Loan that contains the variables principal, rate, and term as well as methods to manipulate their values. *Challenge*: develop a method to compute monthly payments.
2. Develop and demonstrate a class to model a rectangle. Include methods to return the rectangle's perimeter and area.

4

Object-oriented Programming

OBJECTIVES

After reading this chapter, you will be able to:

1. Understand the concepts of objected-oriented programming (OOP) and its use in application development.
2. Use access modifiers to enable encapsulation of Java objects.
3. Apply Java's inheritance properties to develop new classes incorporating the fields and methods of existing classes.
4. Use the same name for methods with similar functionality, thereby gaining polymorphic benefits.
5. Reference current and superclass members.
6. Understand and use interfaces and abstract classes.
7. Demonstrate the use of OOP in developing applications.

OVERVIEW

This chapter introduces the concepts of object-oriented programming, generally known as OOP. When access to the member fields and methods are qualified with access modifiers, the Java class fulfills the OOP concept of encapsulation. Overloading of class methods and constructors fulfills the OOP concept of polymorphism. In addition, Java provides for declaring new classes incorporating the fields and methods of existing classes and thus fulfills the OOP concept of inheritance. Examples are used to demonstrate Java's ability to develop OOP-based applications.

OBJECT-ORIENTED PROGRAMMING CONCEPT

Traditionally, programs have input data, transformed the data, and output the transformed data. The data abstracts the business entities such as a customer, product, order, and sales. The design and programming process was based on an understanding of the data generally described by a file and record descriptions or form or table. The design also was based on an understanding of the transformations applied to the data. In addition, the design included an understanding of the procedures required to handle the data: for example, access the database or files for data input or output.

Object- oriented programming (OOP) is the development of programs, systems, and applications utilizing the concepts of encapsulation, polymorphism, and inheritance. These three factors are referred to as the object-oriented paradigm. **Object-oriented modeling (OOM)** is currently used to abstract and model applications. OOP languages provide a natural conversion of these models to computer programs.

Conceptually an **object** is autonomous; that is, it is complete within itself. It has the capacity to act on its internal members and communicate with the outside world through "well mannered" interfaces (Figure 4.1).

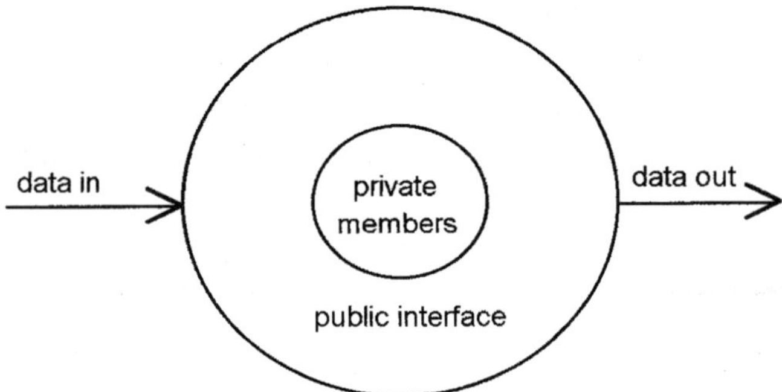

FIGURE 4.1
Access to private data members is through public interface members.

Typically, the private members are the attributes that describe the object. The public interface consists of methods for manipulating the private or internal members. **Encapsulation** refers to the property of combining the attributes and methods of an object in a single package. Moreover, the concept provides the ability to restrict access to the private members except through the public interface.

Polymorphism is the use of the same name or symbol to represent technically different though conceptually similar actions. This aspect of object-oriented programming provides the ability to express conceptually similar operations with the same name and makes the intent of the code easily understood. The benefit of this is a reduction in the need to examine an application's code in detail to understand what is accomplished. This factor becomes very important in the development of large applications where many analysts and programmers are independently developing modules that must be used by others.

Inheritance is the property of using existing objects to create new objects that inherit the property of the existing object. This property is deemed important in the modeling of real-world relationships that exhibit inherited properties. For example, in modeling human relation characteristics of a firm, a manager could be assumed to inherit all of the properties of an employee. The result of this property is the simplification of design in that well-defined and debugged objects can be used in creating new objects.

Object-orient programming languages provide a natural vehicle for realizing object-oriented models of business applications. The benefits include the following:

1. The OOPs process promotes an environment that fosters the development of *reusable code.* Lower-level objects can be developed and tested to the point of being "bullet proof." These objects can be used with confidence in the development of higher-level objects. This benefit results from the completeness associated with encapsulation. The result is a reduction in the time and cost of applications development.
2. The OOPs process promotes code that is maintainable. This is extremely important, as experience has shown that the programs used in business applications undergo more hours in maintenance than they took to develop originally. This benefit results from the modularity and independence of encapsulation as well as the ease of understanding resulting from polymorphism.
3. The OOPs process promotes code that is reliable. The development and thorough testing of lower-level elements provides an environment in which the developers of higher-level objects can utilize the lower-level objects with confidence, thereby promoting an environment in which they are free to concentrate on the development and thorough testing of the higher-level object. This benefit results from the modularity of encapsulation and the ability to create new objects using existing objects through inheritance.

Note that OOP is not a panacea—objects must be well crafted to be effective, must be well documented to be understood, and their existence widely known in order to be used. In addition, management must make inheritance and code reuse a design imperative—that is, they must stamp out the "not coded here" attitude that so easily defeats code reuse and inheritance.

ENCAPSULATION

The concept of encapsulation speaks to completeness and controlled access. The class `Customer`, shown in Figure 4.2, fulfills the "completeness" aspect of a customer object model in that it contains both the attributes of a customer and the functionality for manipulating the attributes.

```
class Customer
{
        int CustID;
        double AcctBal;

        void setCust(int id, double ab)
        {
                CustID = id;
                AcctBal = ab;
        }
        int getCustID( )
        {
                return CustID;
        }
        double getAcctBal( )
        {
                return AcctBal;
        }
}
```

FIGURE 4.2
The `Customer` class is complete.

However, the class `Customer` as defined does not limit access, in that all members, both fields and methods, are accessible outside the body of the class declaration. For example, in the application `UnlimitedAccess`, shown in Figure 4.3, the customer identification and account balance fields are accessed with the statements

```
cust.CustID = 711;
cust.AcctBal = 10.01;
```

```
class UnlimitedAccess
{
        public static void main(String[] args)
        {
                Customer cust = new Customer();

                cust.CustID = 711;
                cust.AcctBal = 10.01;
        }
}
```

FIGURE 4.3
Accessing both `Customer` fields and methods.

This access is contrary to the limitation of access though "well-mannered methods." Note that the method `setCust` qualifies as a well mannered method. If access to the attributes of customer identification and account balance were through the well mannered methods the class `Customer` would satisfy fully the requirements of the OOP encapsulation requirement. Java provides for the specification of access to class members. This section presents Java's approach to access control.

If access to a class or class member is permitted, the accessed class or class member is said to be **accessible.** The following sections presents access control; that is, the ability to permit or limit access to classes and class members.

Class Access

Classes without an access modifier are accessible to other classes in the same package. (Packages are examined in Chapter 9 and for all intent and purposes are the directories or folders used to organize Java files.) A class with a `public` access modifier is accessible to any Java code that has access to the package. The access modifier precedes the `class` keyword in the class declaration. The class `Customer`, shown in Figure 4.4, is declared `public` and is accessible outside the package in which it is declared. Such classes are said to have public access.

Access to a class declared without the public access modifier is limited to the package in which it is declared. Such a class is said to have **package access.**

Member Access

A class member is accessible only if the class is accessible and the member is declared accessible. Member access modifiers are summarized in Table 4.1.

TABLE 4.1

Member access modifiers

Modifier	Explanation
`public`	Access is permitted to all code that has access to the package in which the member is declared.
`private`	Access is limited to the class in which the member is declared.

The access modifier precedes the data type of the field or return type of the member. An example of a filed member declared private is shown in Figure 4.5.

Consider the class customer shown in Figure 4.6. Note that the class `Customer` has been declared public and thus is accessible to any code that has access to the package in which `Customer` is declared. The fields `CustID` and `AcctBal` are declared private and thus are accessible only to code within the class. The methods `setCust`, `getCustID`, and `getAcctBal` are declared public and thus are accessible wherever the type is accessible.

access modifier

```
public class Customer
{
        int CustID;
        double     AcctBal;

        void setCust(int id, double ab)
        {
                CustID = id;

                AcctBal = ab;
        }
        int getCustID( )
        {
                return CustID;
        }
        double getAcctBal( )
        {
                return AcctBal;
        }
}
```

FIGURE 4.4
Public class access modifiers make the class accessible to code in other packages.

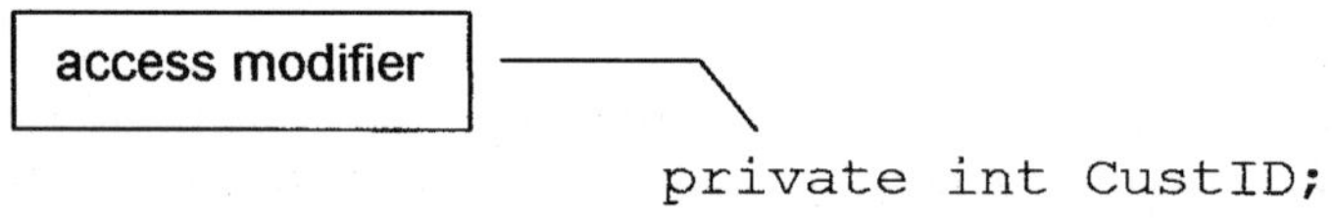

FIGURE 4.5
Member access modifiers precede the data type in member declaration statements.

In the formulation of the class `Customer` shown in Figure 4.6, the filed members, representing the attributes of the object modeled by the class, are declared private and thus are not accessible outside the class. Access to these fields are through the methods defined in the class which are declared public. As such, the class data type with fields declared to have public access and the methods declared to have private access fulfill the OOP requirement of encapsulation. The implications of this access are illustrated in Figure 4.7.

The statement

```
cust.CustID = 711;
```

is in the class `LimitedAccess` and not in the class `Customer`. As such, the statement attempts to assign a value to a private member outside the class in which the member is declared and thus is invalid. On the other hand, the statement

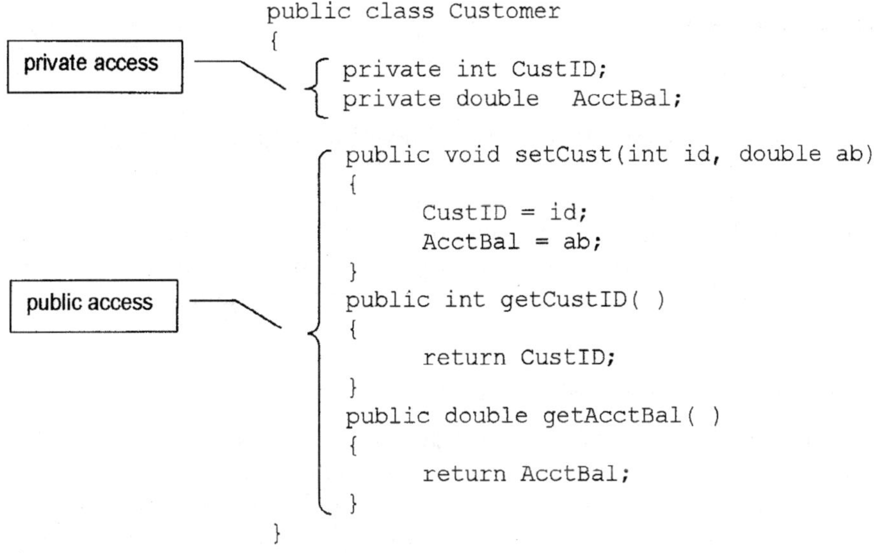


```
public class Customer
{
    private int CustID;
    private double  AcctBal;

    public void setCust(int id, double ab)
    {
         CustID = id;
         AcctBal = ab;
    }
    public int getCustID( )
    {
         return CustID;
    }
    public double getAcctBal( )
    {
         return AcctBal;
    }
}
```

FIGURE 4.6
`Customer` class declaration with access modifiers.

```
class LimitedAccess
{
      public static void main(String[] args)
      {
            Customer cust = new Customer();

            cust.CustID = 711;

            cust.setCust = (7.11, 10.01);
      }
}
```

invalid access

valid access

FIGURE 4.7
Illustrating valid and invalid access.

```
cust.setCust = (7.11, 10.01);
```

invokes a method that is declared public and thus is accessible outside the class in which it is declared.

If a member is not declared public or private, it is said to have **default access,** and access is possible only from within the package in which the type is declared.

INHERITANCE

Inheritance is the ability of a class to incorporate the properties, including fields and methods, of an existing class in a new class. In object-oriented design it is not uncommon to have strings of classes each inheriting from the other. Inheritance provides the means for building complex classes from simpler classes that have some of the attributes and functionality but not all that is required. This section presents Java's implementation of inheritance.

Declaring a Derived Class

Java classes can inherit from other classes with the `extends` keyword. That is, using the `extends` keyword a new class, called the **subclass** or **derived class,** is created that inherits the fields and methods of an existing class, called the **superclass** or **base class.** The existing class must be accessible.

For example, the class Customer models the business customer. Consider now a company that has both domestic and international customers. Companies operating in foreign country have attributes that are not present in this country. For example, businesses in some countries are subject to a value-added tax (VAT). Consider also the difficulty that might be encountered in trying to make a time sensitive business contact without knowledge that a particular business operated in a region that observed a siesta custom. On the other hand, few foreign businesses are subject to a state sales tax. Note, however, that the domestic and foreign businesses both require the attributes and methods contained in the `Customer` class. Using the `extends` keyword, domestic and foreign classes can be declared that inherit the fields and methods of the `Customer` class. A class modeling a domestic customer is shown in Figure 4.8.

The class `DomesticCustomer` extends the class `Customer`. Thus, even though they are not explicitly stated in the declaration, the class `DomesticCustomer` inherits the members of class `Customer`. In additional, a field modeling the existence or nonexistence of a state sales tax is declared. Methods to manipulate the `SalesTax` field are also included. The application shown in Figure 4.9 illustrates use of the class `DomesticCustomer`.

A `DomesticCustomer`-type customer object is created with the statement

```
DomesticCustomer domCust = new DomesticCustomer();
```

The class `DomesticCustomer` includes the fields and has access to the methods of its subclass, or base class, `Customer`. Thus, the customer identification and account balance variables are set by invoking the `Customer` class `setCust` method.

```
domCust.setCust(69, 69.69);
```

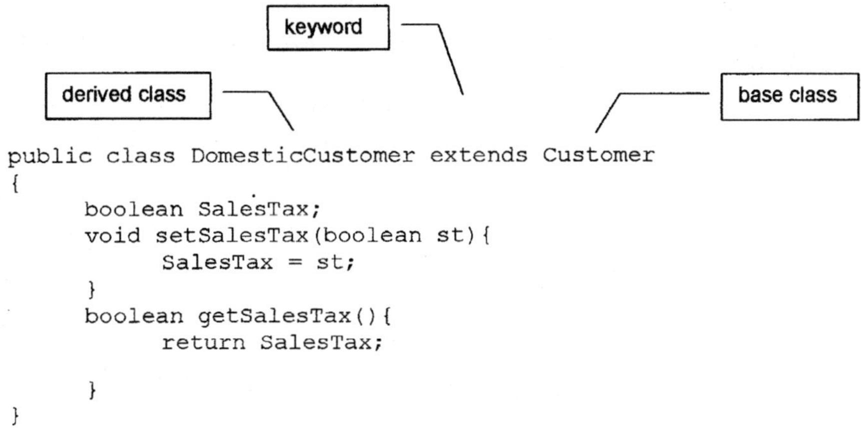


```
public class DomesticCustomer extends Customer
{
      boolean SalesTax;
      void setSalesTax(boolean st){
            SalesTax = st;
      }
      boolean getSalesTax(){
            return SalesTax;

      }
}
```

FIGURE 4.0
DomesticCustomer class inherits the Customer class fields and methods.

```
// UsingCustomer.java
 class UsingCustomer
{
      public static void main(String[] args)
      {
            int outID;
            double outBal;
            boolean outSalesTax;

            DomesticCustomer domCust = new DomesticCustomer();

            domCust.setCust(69, 69.69);
            domCust.setSalesTax(true);
            outID = domCust.getCustID();
            outBal = domCust.getAcctBal();
            outSalesTax = domCust.getSalesTax();
      }
}
```

FIGURE 4.9
Using the DomesticCustomer class.

Similarly, the customer identification and account balance are retrieved with the getCustID and getAcctBal methods.

```
outID = domCust.getCustID();
outBal = domCust.getAcctBal();
```

> Member methods of the base class can be called directly, no special identification is required, as they are members of the derived class.

The value of the DomesticCustomer field SalesTax is set by invoking the setSalesTax method. Retrieval of the SalesTax value is accomplished by invoking the getSalesTax method.

A class modeling foreign customers is similarly declared and shown in Figure 4.10.

```
  // Foreigncustomer.java
 class ForeignCustomer extends Customer
{
        double VAT;
        boolean SiestaCustom;
        void setVAT(double vat){
             VAT = vat;
        }
        void setSiestaCustom(boolean sc){
             SiestaCustom = sc;
        }
        double getVAT(){
             return VAT;
        }
        boolean getSiestaCustom(){
             return SiestaCustom;
        }
}
```

FIGURE 4.10
ForeignCustomer class inherits fields and methods of the Customer class.

One way to view the relationship between the base class and the derived class is to consider the derived class as a specialization of the base class. The derived class takes the information provided by the base class—that is, customer identification and account balance—and adds to it information specific to the derived class. Because of this reliance on the base class, the base class must be "bullet proof." For example, imagine the embarrassment and problems that can be caused when it is uncovered that one of the methods of a class that is used as the base class of a class that is used as a base class that in turn is used as the base class of a derived class does not work!

Class Hierarchies

The class being declared is said to be a subclass of the class being extended. The class extended is said to be a superclass of the class being declared. Class hierarchies, illustrating the relationship between superclasses and subclasses, can be graphically displayed as shown in Figure 4.11. In this graph, the arrow points to the subclass or derived class. The classes DomesticCustomer and ForeignCustomer are subclasses of the class Customer. Similarly, the class Customer is said to be a superclass of the classes DomesticCustomer and ForeignCustomer. Note that a class can be a subclass and a superclass at the same time.

Note that another style of graphing class hierarchies exists in which the arrow points to the superclass. Consequently, care must always be exercised to ascertain the convention used in the presentation when using a particular

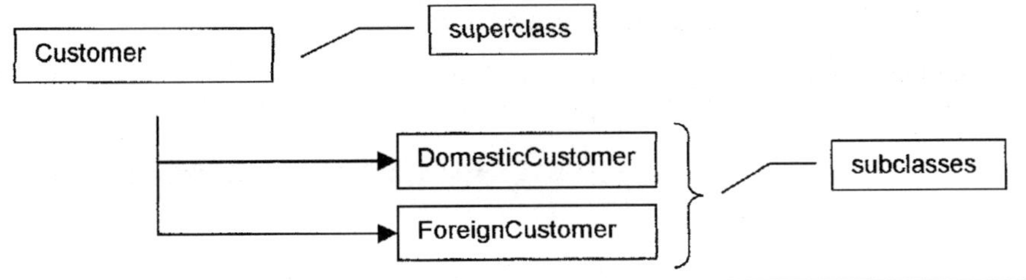

FIGURE 4.11
Class hierarchy chart showing the relationship between subclasses and superclasses.

graph to analyze a class hierarchy. In this text, the arrow always points to the subclass.

Protected Access Modifier

A class member declared with the `protected` access modifier is accessible from within the package containing the class in which the member is declared or within a subclass of the class in which the protected member is declared.

Consider the project organization shown in Figure 4.12. The `Customer` class and its subclasses are declared in the `customers` package. The application using the `Customer` class and its subclasses is declared in the `application` package.

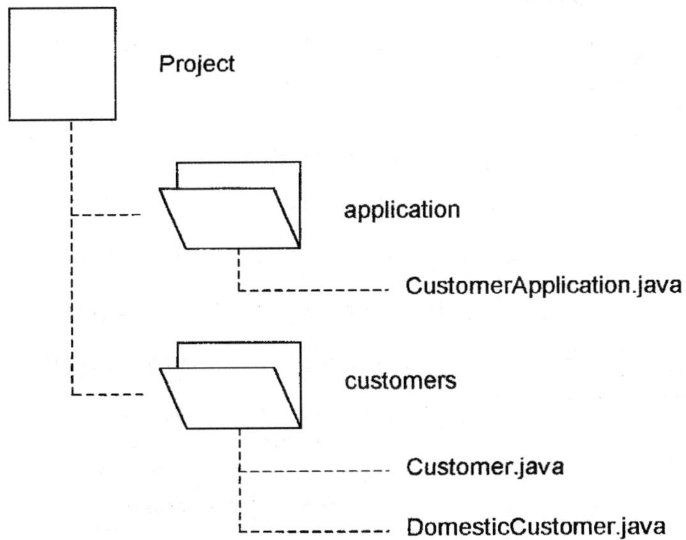

FIGURE 4.12
Project organized with `application` and `customers` packages.

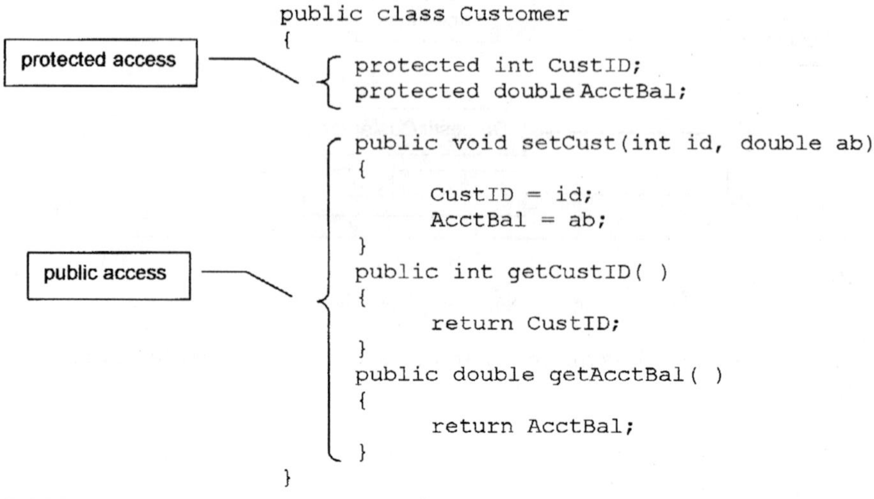


```
public class Customer
{
    protected int CustID;
    protected double AcctBal;

    public void setCust(int id, double ab)
    {
        CustID = id;
        AcctBal = ab;
    }
    public int getCustID( )
    {
        return CustID;
    }
    public double getAcctBal( )
    {
        return AcctBal;
    }
}
```

FIGURE 4.13
`Customer` class declared with protected members.

For example, consider the version of the `Customer` class declared in Figure 4.13. The fields `CustID` and `AcctBal` are declared with the protected access modifier.

Use of the `Customer` class is demonstrated in the `CustomerApplication` class declared in the `application` package, Figure 4.14. Note the statement

```
import customers.*;
```

which is needed to enable access to the `Customer` class that is declared in the `customers` package.

```
package application;

import customers.*;

public class CustomerApplication
{
    public static void main(String[] args)
    {
        Customer cust = new Customer();

        cust.CustID = 711;
        cust.setCust( 711, 99.99 );
    }
}
```

invalid access

valid access

FIGURE 4.14
Application illustrating valid and invalid access statements.

An instance, `cust`, of the `Customer` class is created. The attempt to access `CustID` with the statement

```
cust.CustID = 711;
```

is invalid, as `CustID` is declared `protected`. For instances of the `Customer` class a member must be declared public if it is to be accessed in code outside the package in which its class is declared. Note that the access of the method `setCust` with the statement

```
cust.setCust( 711, 99.99 );
```

is valid, as the method is declared with `public`. For instances of the class in which it is declared, `protected` access is similar to `private` access.

Typically, subclass of `Customer` would be declared in the `customers` package. However, sometimes, as when the subclass is specific to the application, it might be declared in the application package such as the `DomesticCustomer` class shown in Figure 4.15.

```
package application;

import customers.*;

public class DomesticCustomer extends Customer
{
      public boolean SalesTaxFg;

      public void setDomCust(int id, double ab,
            boolean stf)
      {
            CustID = id;
            AcctBal = ab;
            SalesTaxFg = stf;
      }
}
```

enables access to package customers

FIGURE 4.15
Subclass extending a superclass in a different package.

The statement

```
import customers.*;
```

enables access to classes declared in the customers package. Because `CustID` and `AcctBal` are declared `protected`, subclasses of `Customer` have access to them. For all intent and purposes, protected members appear to subclasses as public members.

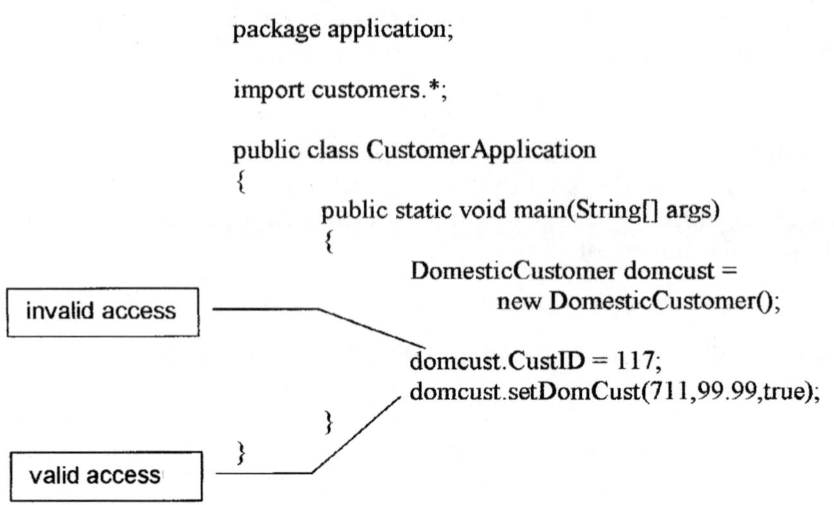

FIGURE 4.16
Application illustrating access to members of `Customer` class protected members.

A revised version of the `Customerapplication`, using the `DomesticCustomer` class, is shown in Figure 4.16.

The attempt to access the `protected` subclass member CustID with the statement

```
domcust.CustID = 117;
```

is invalid. Access to protected inherited members are, for all intents and purposes, the same as subclass `private` members.

POLYMORPHISM

Java provides for using the same name for methods thereby enabling methods with similar functionality to be similarly named. In addition, fields representing similar attributes can be similarly named in inherited classes. These capabilities of Java fulfill the OOP characteristic of polymorphism.

Overloading

Two methods with the same name but different signatures are said to be overloaded. Overloading is valid. Consider, for example, the overloaded methods `maintCustomer` shown in Figure 4.17. Note that the signatures of the

```
public class Customer
{

      int CustID;
      double AcctBal;

      void maintCustomer(int id){
            CustID = id;
      }
      void maintCustomer(int id, double ab){
            CustID = id;
            AcctBal = ab;
      }
}
```


FIGURE 4.17
Method maintCustomer is overloaded

two methods are different. (Remember, a method's signature is the number and type of parameters.)

The specific method invoked is determined by matching the passed arguments with the signatures of the overloaded methods. Consider the application shown in Figure 4.18.

```
public class UsingCustomer
{
      public static void main(String[] args)
      {
            Customer cust = new Customer();

            cust.maintCustomer(1001);
            cust.maintCustomer(2002, 22.22);
      }
}
```


FIGURE 4.18
Invoking overloaded methods

The method

```
void maintCustomer(int id){
    CustID = id;
}
```

is invoked by the statement

```
cust.maintCustomer(1001);
```

as the argument consists of a single `int` that matches with the signature of the invoked method. Similarly, the method

```
void maintCustomer(int id, double ab){
```

```
        CustID = id;
        AcctBal = ab;
    }
```

is invoked by the statement

```
    cust.maintCustomer(2002, 22.22);
```

Overloading methods can also be inherited. Consider the class `ForeignCustomer`, declared in Figure 4.19, in which declarations of the method `maintCustomer` have been inherited form the extend class `Customer` as well as declared in the class `ForeignCustomer`. Note that in each declaration of the method `maintCustomer` has a different signature.

```
public class ForeignCustomer extends Customer
{
      double ExchangeRate;

      void maintCustomer(int id, double ab, double er){
            CustID = id;
            AcctBal = ab;
            ExchangeRate = er;
      }
}
```

overloads maintCustomer method inherited from Customer

FIGURE 4.19
`ForeignCustomer` inherits overloaded methods and declares an overloaded method.

The invoking of inherited as well as declared overloaded methods is illustrated in the application `UsingForeignCustomer` shown in Figure 4.20.

```
public class UsingForeignCustomer
{
      public static void main(String[] args)
      {
           ForeignCustomer forCust = new ForeignCustomer();

           forCust.maintCustomer(1001);
           forCust.maintCustomer(2002, 22.22);
           forCust.maintCustomer(3003,33.33,3.0);
      }
}
```

FIGURE 4.20
Invoking inherited and declared overloaded methods of the class `ForeignCustomer`.

The statement

```
    forCust.maintCustomer(1001);
```

invokes the `maintCustomer` method declared in the `Customer` class

```
void maintCustomer(int id){
    CustID = id;
}
```

as the invoking statement contains a single `int` data type parameter matching the method's signature. Similarly, the statement

```
forCust.maintCustomer(3003,33.33,3.0);
```

invokes the `maintCustomer` method declared in the `ForeignCustomer` class.

```
void maintCustomer(int id, double ab, double er){
    CustID = id;
    AcctBal = ab;
    ExchangeRate = er;
}
```

Overriding

A method declared with the same name and signature as an inherited method is said to **override** the inherited method. Consider the version of `ForeignCustomer` shown in Figure 4.21. The method `maintCustomer` declared in `ForeignCustomer`,

```
void maintCustomer(int id, double ab){
    CustID = id;
    AcctBal = ab*ExchangeRate;
}
```

overloads the `maintCustomer` declared in `Customer` with the same signature

```
void maintCustomer(int id, double ab){
    CustID = id;
    AcctBal = ab;
}
```

Note that in the overriding `ForeignCustomer` declaration the account balance parameter passed in the method call is adjusted to account for exchange rate.

```
public class ForeignCustomer extends Customer
{
      double ExchangeRate;

      void maintCustomer(int id, double ab){
           CustID = id;
           AcctBal = ab*ExchangeRate;
      }
      void maintCustomer(int id, double ab, double er){
           CustID = id + 1000;
           AcctBal = ab*er;
           ExchangeRate = er;
      }
}
```

FIGURE 4.21
Overriding an inherited method.

The application in Figure 4.22, a version of `UsingForeignCustomer`, invokes the overriding `maintCustomer` method. The statement

```
forCust.maintCustomer(3003,33.33,3.0);
```

invokes the `maintCustomer` method declared in `ForeignCustomer` with a signature consisting of an `int`, `double`, and `double`.

```
public class UsingForeignCustomer
{
      public static void main(String[] args)
      {
           ForeignCustomer forCust = new ForeignCustomer();

           forCust.maintCustomer(3003,33.33,3.0);
           forCust.maintCustomer(2002, 22.22);
      }
}
```

FIGURE 4.22
Invoking an overriding method `maintCustomer`.

The statement

```
forCust.maintCustomer(2002, 22.22);
```

invokes the method `maintCustomer` declared in the class `ForeignCustomer` that overrides the `maintCustomer` method declared in the class `Customer` (see Figure 4.17). Note that the version of the method `maintCustomer` declared in `Customer` that is not overridden is inherited by the class `ForeignCustomer`. Note also that the declaration of a method that is not abstract of an overridden abstract method implements the inherited abstract method. A static inherited method cannot be overridden by an instance or non-static method.

Hiding

A field is said to hide an inherited field with the same name and different data type. Similarly a method declared static is said to hide inherited methods with the same name and signature. Consider the abbreviated version of the `Customer` class shown in Figure 4.23.

```
public class Customer
{
      int Discount;
      static int CreditLimit = 2000;
      static void maintCustomer(int cl){
           CreditLimit = cl;
      }
}
```

FIGURE 4.23
Class `Customer` declared with `int Discount` variable and `static` variable `CreditLimit` and method `maintCustomer`.

```
public class ForeignCustomer extends Customer
{
      double Discount;                    hides int Discount

      static void maintCustomer(int cl){
           CreditLimit = cl - 1000;            hides maintCustomer
      }                                        declared in class
      void setDiscount(double d){              Customer
           Discount = d;
      }
}
```

FIGURE 4.24
Class `ForeignCustomer` with hiding fields and methods.

Consider also the version of the class `ForeignCustomer` that extends the class `Customer` shown in Figure 4.24. The declaration of the `int` instance variable `Discount` declared in `Customer` is hidden by the declaration of the `double Discount` variable accessed in the class `ForeignCustomer`.

The method `maintCustomer` declared in the class `ForeignCustomer` hides the static method of the same name and signature declared in the class `Customer`. The implication of this hiding is illustrated in the application `UsingForeignCustomer` shown in Figure 4.25.

```
public class UsingForeignCustomer
{
      public static void main(String[] args)
      {
           ForeignCustomer forCust = new ForeignCustomer();

           forCust.setDiscount(21.21);
           forCust.maintCustomer(1001);

           Customer OtherForCust = new ForeignCustomer();
           OtherForCust.maintCustomer(4000);
      }
}
```

FIGURE 4.25
Accessing hiding variables and methods.

The method `setDiscount` invoked by the statement

```
forCust.setDiscount(21.21);
```

accesses the type `double` variable `Discount` declared in the class `ForeignCustomer` that hides the `int` type `Discount` variable inherited from the class `Customer`. Similarly, the `maintCustomer` method declared in the `ForeignCustomer` class, that hides the `maintCustomer` method declared in `Customer`, is invoked by the statement

```
forCust.maintCustomer(1001);
```

Consider now the creation of a new instance of `ForeignCustomer` with the statement

```
Customer OtherForCust = new ForeignCustomer();
```

In this case the reference to the `ForeignCustomer` object is assigned to a reference to a `Customer` object `OtherForCust`. When the `maintCustomer` method is invoked by the statement

```
OtherForCust.maintCustomer(4000);
```

the method declaration declared in the `Customer` class is executed as the type of the object determines the invocation of hidden methods. This compares to overridden methods, instance or non-static methods, in which invocation is determined by the class of the object.

Hiding an instance method with a static method is invalid.

`super` AND `this` KEYWORDS

The `this` keyword, used with instance methods and constructors, accesses a member of the current class. The `super` keyword, similarly used, refers to the accessed member of a super class.

Calling another Constructor

The class `Customer`, shown in Figure 4.26, has an overloaded constructor. The default constructor sets the variable `CustID` to the value 10001. The second constructor has a single parameter of type `double` representing the initial balance. The first line in the constructor uses the keyword `this` to invoke the default constructor of the current object. When control is returned, following execution of `this()`, the parameter value is assigned to the variable `AcctBal`.

Consider the application creating an instance of the `Customer` class as shown in Figure 4.27. An instance of the `Customer` class is created. The constructor with the signature consisting of a single `double`-type variable is called.

The first statement invokes to the default constructor. Control is passed to the default constructor in which the `CustID` variable is set to the value 10001. Note that the constructor called is determined by the signature in the `this` statement, which in this case is empty. When execution of the called constructor is completed, control returns to the next statement in the calling constructor, which in this case is

```
AcctBal = ab;
```

```
public class Customer
{
        int CustID;
        double AcctBal;

        Customer( )
        {
                CustID = 10001;
        }
        Customer(double ab)
        {
                this();
                AcctBal = ab;
        }
}
```


FIGURE 4.26
Constructor calling another constructor using `this` keyword.

```
public class UsingCustomer
{
        public static void main(String[] args)
        {
            Customer cust = new Customer(-99.01);
        }
}
```

FIGURE 4.27
Creating an object in which the constructor invokes a constructor.

This statement is executed, and control is returned to the main method. Note that the `this` statement must be the first statement in the constructor.

Referencing Superclass Constructors

The `super` keyword, with an optional list of parameters, can be used to invoke a superclass constructor. Consider again another version of the `Customer` class containing two constructors as shown in Figure 4.28. Note that each of the constructors has a different signature. In one case, the customer's identification code is passed as a parameter and assigned to the `CustID` variable. The account balance is the default value for doubles, zero real, when construction is completed. In the case of the other constructor, the account balance is passed as a parameter and assigned to the variable `AcctBal`. In this case, the customer identification number is given the default value integral zero.

Consider the class `SpecialCustomer`, a subclass of `Customer`, shown in Figure 4.29. The `SpecialCustomer` constructor calls the `Customer` class constructor `Customer(int id)` with the statement

```
super(id);
```

```
public class Customer
{
        int CustID;
        double AcctBal;

        Customer(int id)
        {
              CustID = id;
        }
        Customer(double ab)
        {
              AcctBal = ab;
        }
}
```

FIGURE 4.28
Class with overloaded constructors.

```
public class SpecialCustomer extends Customer
{
        SpecialCustomer(int id, double ab)
        {
                super(id);      ——— calling the superclass constructor
                AcctBal = ab;        Customer(int id)
        }
}
```

FIGURE 4.29
Subclass of `Customer`

Ambiguity regarding which of the overloaded `Customer` class constructors to invoke is resolved by matching the `super` keyword parameter list with the signature of the `Customer` class constructors. Because the passed parameter is `int` data type, the constructor `Customer(int id)` is called.

Consider the version of `UsingSpecialCustomer` shown in Figure 4.30. An instance of `SpecialCustomer` is created. The constructor `SpecialCustomer` is called during the creation. The first statement in the `SpecialCustomer` constructor accesses the `Customer` constructor `Customer(int id)` with the parameter value 711 using the `super` keyword.

```
public class UsingSpecialCustomer
{
        public static void main(String[] args)
        {
              SpecialCustomer cust = new SpecialCustomer(711, 99.00);
        }
}
```

FIGURE 4.30
Creating a class instance whose constructor calls a superclass constructor.

When the `Customer` constructor has completed execution, the `CustID` value is 711 and the `AcctBal` has the double default value 0.0. The second statement in the `SpecialCustomer` constructor is executed assigning the value 99.0 to the variable `AcctBal`.

As with the `this` keyword, the `super` keyword and optional parameter list must be the first statement in a constructor. The `super` keyword can also be used to reference hidden variables in a superclass.

INTERFACES

An **interface** is comparable to a template of a class that cannot be implemented because it is application dependent. Interfaces are useful in declaring fields and methods that will be used in many classes. For example, consider the declaration of constants describing attributes associated with processes common to many classes, or similarly, the declaration of common methods that need to be application-specific. Consider the case of a database interface with attributes describing the data transfer modes read, write, and read-write. One of these attributes is used to describe every data transaction. Thus, although the mode may change, the attributes describing the mode are constant and common to all classes involving data transfer. Similarly, methods required to transfer data include open, read, write, and close. These methods are also common to all data transfers but are not necessarily constant in that they may change depending on the specific requirements of the database or the application. The Java interface provides the means for declaring fields that are constant and methods that need to be implemented.

Interface Declaration

Declaration of the interface `DatabaseInterface` is shown in Figure 4.31. Declaration of an interface is similar to the declaration of a class. The declaration begins with an access modifier, `public` or `abstract`. An interface is implicitly abstract and the abstract modifier is no longer used. The interface keyword follows the access modifier. The interface keyword is followed by the interface identifier, a name unique to the package in which the interface is declared. In this case the identifier is `DatabaseInterface`. Scope of the declared interface is the entire package in which it is declared. An `extends` clause can be used to extend an interface, which is said to be the superinterface of the declared interface.

Both fields and methods are declared in the interface's body. The scope of an interface member, field or method, is the entire interface body in which it is declared. The members of an interface are implicitly `public`. Compare this with fields in a class, which are generally declared to be private and thus are accessible only within the class in which they are declared. If the interface is also declared public, the members are accessible outside the package in

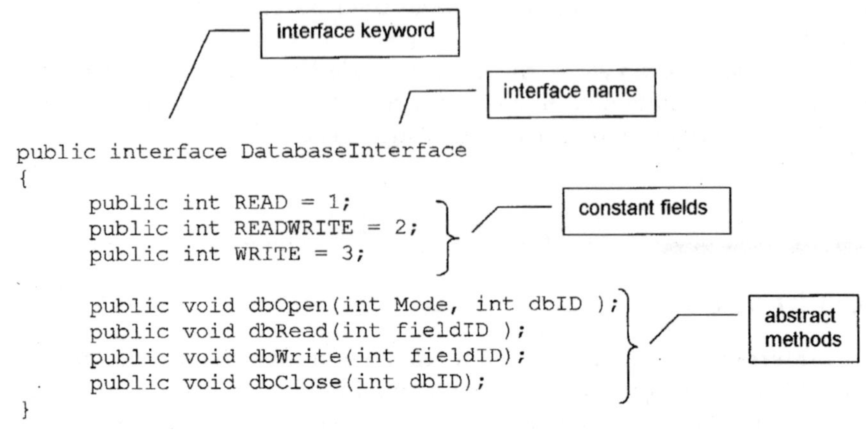


```
public interface DatabaseInterface
{
      public int READ = 1;
      public int READWRITE = 2;
      public int WRITE = 3;

      public void dbOpen(int Mode, int dbID );
      public void dbRead(int fieldID );
      public void dbWrite(int fieldID);
      public void dbClose(int dbID);
}
```

FIGURE 4.31
Declaration of the interface DatabaseInterface

which the interface is declared. Members of the interface also include members inherited from superinterfaces.

Every field declared in an interface is implicitly `public`, `static`, and `final` and must be initialized. Thus the statement

```
int READ = 1;
```

is equivalent to the statement

```
public static final int READ = 1;
```

Consequently, fields declared in an interface are constant. The field modifiers `synchronized`, `transient`, and `volatile` are invalid.

Methods declared in an interface are implicitly `public` and `abstract`. Thus the statement

```
void dbOpen(int Mode, int dbID );
```

is equivalent to the statement

```
public abstract void dbOpen(int Mode, int dbID );
```

The method modifiers `final`, `static`, `native`, and `synchronized` are invalid. Note that the interface method declaration does not include a body and the method is said to be incomplete.

Interfaces, like classes, inherit the fields and methods of a superclass. An interface can inherit more than one field with the same name. In this case, an attempt to reference the field within the interface body with a simple name is invalid, as the reference is ambiguous. A method is said to override any methods with the same name and signature in its superinterfaces. Interface methods can be overridden. If more than one method in a declaration or its superinterfaces have the same name, the method is said to be overloaded. Interface methods can be overloaded.

Implementing an Interface

An interface's fields and methods are incorporated into a class with the **`implements`** keyword in the class declaration. An example illustrating the implementation of the `DatabaseInterface` class in the `Customer` class is shown in Figure 4.32.

The `implements` keyword follows the class identifier. The interfaces implemented follow the keyword `implements`. More than one interface can be implemented. The methods declared in the interface are implemented in the class declaration implementing the interface. Although the implementations bodies are empty in this case, typically they include statements performing the relevant action. As all of the methods in the interface are abstract, all of the methods must be implemented or else the class will be abstract. Remember, an instance of an abstract class cannot be instantiated.

Using a Class that Implements an Interface

An example of an application that uses a class that implements an interface is shown in Figure 4.33.

The method `dbOpen` declared in the interface `DatabaseInterface` and implemented in the class `Customer` is invoked for the `Customer` class instance `cust` with the statement

```
cust.dbOpen(cust.READ, 711);
```

Similarly, the field `READ` declared in the interface `DatabaseInterface` and implemented in the class `Customer` is referenced for the instance `cust` with the expression

```
cust.READ
```

> Fields declared in an interface are constant. Methods declared in an interface are abstract and must be implemented in the class implementing the interface.

implements keyword

interface identifier

method implementaions

```
// Customer.java

public class Customer
                implements DatabaseInterface
{
      final int CreditRating = 10;
      double CreditLimit = 1000*CreditRating;
      protected int CustID;
      protected double AcctBal;
      public SalesPerson SlsPer;

      public void dbOpen(int Mode, int dbID ){ };
      public void dbRead(int fieldID ){ };
      public void dbWrite(int fieldID){ };
      public void dbClose(int dbID){ };

      Customer(int id, double ab, double cl, int r, int sp)
      {
            CustID = id;
            AcctBal = ab;
            CreditLimit = cl;
            SlsPer = new SalesPerson(r, sp);
      }

      Customer()
      {
      }

      public void setCust(int id, double ab)
      {
            CustID = id;
            AcctBal = ab;
      }

      public void setCustID(int id)
      {
            CustID = id;
            return ;
      }
      public void setAcctBal(double b)

      {
            AcctBal = b;
      }
      public void setCreditLimit(double cl)
      {
            CreditLimit = cl;
      }
      int getCustID( )
      {
            return CustID;
      }
      double getAcctBal( )
      {
            return AcctBal;
      }
      double getCreditLimit( )
      {
            return CreditLimit;
      }
      SalesPerson getSalesPerson( )
      {
```

FIGURE 4.32
Implementing the `DatabaseInterface`.

```
// UsingCustomer.java
 class UsingCustomer
{
      public static void main(String[] args)
      {
            Customer cust = new Customer(77,77.77,7777.77,5,99);

            cust.dbOpen(cust.READ, 711);

      }
}
```

interface field

interface method

FIGURE 4.33
Using a class that implements an interface.

ABSTRACT CLASSES

A method is declared abstract with the **abstract** modifier. An abstract method declaration specifies the method's return type, name, and signature but does not include an implementation. An example of an abstract method declaration is illustrated in Figure 4.34.

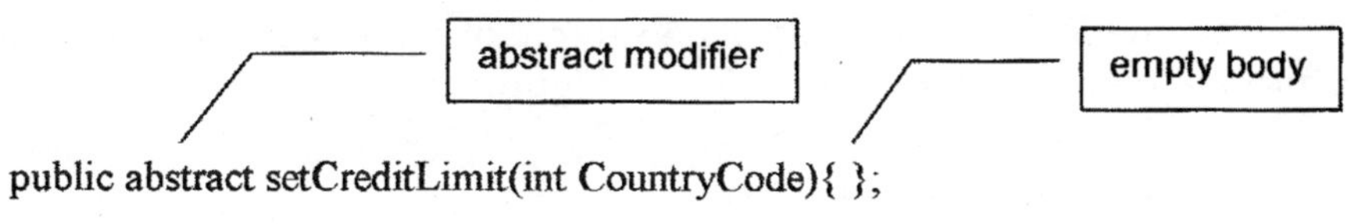

FIGURE 4.34
Declaring the abstract method `setCreditLimit`.

An abstract method is often used to mark a required method that can only be completed in a specific extension of the class. For example, determination of the credit limit for domestic customers is probably different than the determination for foreign customers.

Note that an abstract method cannot be private as private methods are not visible to subclasses. In addition, methods declared `static` and `final` cannot be declared `abstract`.

An abstract method can only be declared or inherited into a class declared abstract. A class is declared an abstract class with the `abstract` class modifier.

```
// Customer.java                    Abstract modifier

public abstract class Customer
{
      double CreditLimit;
      protected int CustID;                    abstract method
      protected double AcctBal;

      public abstract void setCreditLimit(int cc);

      public void setCreditLimit(double cl)
      {
            CreditLimit = cl;
      }
}
```

FIGURE 4.35
Declaration of an abstract class containing an abstract method.

The version of the `Customer` class shown in Figure 4.35 illustrates the declaration of an abstract method in an abstract class.

An abstract class is viewed as incomplete. An instance of an abstract class cannot be created. Hence, the statement

```
Customer cust = new Customer();
```

is invalid, as `Customer` is an abstract class.

> An instance of an abstract class cannot be instantiated.

A class should be declared abstract when it is intended that subclasses will implement the abstract methods. An example of an implementation of an abstract method in a subclass is shown in Figure 4.36.

```
// DomesticCustomer.java
 class DomesticCustomer extends Customer
{
      boolean SalesTax;
      void setSalesTax(boolean st){
            SalesTax = st;
      }
      boolean getSalesTax(){
            return SalesTax;
      }

      public void setCreditLimit(int cc)         abstract
      {                                          method
            CreditLimit = 5000*cc;               implementation
      }
 }
```

FIGURE 4.36
Implementing an abstract method in a subclass.

Because DomesticCustomer is complete—that is, contains no abstract methods or inherited abstract methods that have not been implemented—instances of DomesticCustomer can be created. For example, consider the application shown in Figure 4.37.

```
// UsingCustomer.java
class UsingCustomer
{
      public static void main(String[] args)
      {
            Customer docCust = new DomesticCustomer( );

            docCust.setCreditLimit(25);
      }
}
```

FIGURE 4.37
Application creating an instance of a class implementing an inherited abstract method.

The class DomesticCustomer is complete, contains no declared or inherited abstract methods, and thus a new instance can be created with the statement

Customer docCust = new DomesticCustomer( );

The implemented setCreditLimit method is invoked with the statement

```
docCust.setCreditLimit(25);
```

The implemented setCreditLimit method, with int data type signature, is invoked for the docCust instance of DomesticCustomer. The value of the parameter, 25, is multiplied by 5000 in the body of the method, and the product value of 125000 is assigned to CreditLimit.

An abstract class is like an interface in that they both contain abstract methods and cannot be used to create instances of an object. They differ, however, in that a class can be derived from only one abstract class with the extend keyword, whereas a class can implement many interfaces with the implements keyword. In this sense, an interface enables a sense of multiple inheritance. Moreover, a class can include fields that are "variable," whereas the fields in an interface are declared static and final and thus are constant. In addition, a class can contain methods that are not abstract as well as abstract, while an interface can contain only abstract methods.

HANDS-ON EXERCISE 4: HUMAN RESOURCE APPLICATION

Employees are characterized, in a simplified and basic manner, by an identification code, hourly rate, and employment status. This exercise presents the development and use of an `Employee` class. The exercise concludes with a challenge to develop and use a `Salesperson` class that inherits the fields and methods of the `Employee` class.

Step 1: Create a New Project

1. Start Visual J++.
2. Select the New Project... command in the File menu. The New Project dialog box is displayed.
3. Enter **`Human Resource`** in the Name text field.
4. Use the Browse button to enable navigation to the Location `C:\Hands-On\Human Resource`.
5. Click the Open button to create the new project.

Step 2: Create `Employee.java` File

1. Press the CTRL+D keys. The Add Item dialog box is displayed.
2. Select the Class folder.
3. Select the Class icon.
4. Enter **`Employee.java`** in the Name text field.
5. Click the Open button to create the new class.
6. The Code Editor is opened with the file `Employee.java`.

Step 3: Code the Employee Class

Enter code to create a frame window, checkboxes, checkbox group and handlers for the window events and item selection events.

```
public class Employee
{
        int EmployeeID;
        double HourlyRate;
        char EmploymentStatus;

        Employee(int id, double hr, char es){
                EmployeeID = id;
                HourlyRate = hr;
                EmploymentStatus = es;
        }

        int getEmployeeID( ){
                return EmployeeID;
        }

        double getHourlyRate( ){
```

```
                return HourlyRate;
        }

        char getEmploymentStatus( ){
                return EmploymentStatus;
        }

}
```

Step 4: Build the Basic Application

1. Press the CTRL+SHIFT+B keys.
2. Debug the code as required.

Step 5: Create NewHire.java File

1. Press the CTRL+D keys. The Add Item dialog box is displayed.
2. Select the Class folder.
3. Select the ClassMain icon.
4. Enter **`NewHire.java`** in the Name text field.
5. Click the Open button to create the new class.
6. The Code Editor is opened with the file `NewHire.java`.

Step 6: Code the NewHire Class

1. Begin to enter the `NewHire` code

```
public class NewHire
{
      public static void main(String[] args)
      {
            /** New employee data input
             */
            int newEmployeeNo = 21;
            double newHourlyRate = 35.70;
            /** Employee object created and new
             *  hire processing performed
             */
            Employee emp = new
      }
}
```

Visual J++ Statement Completion automatically performs reference look-up and lists relevant classes, fields, and methods, including signatures in a pop-up list box (Figure HO-4.1).

2. Scroll down the list to highlight `Employee`.

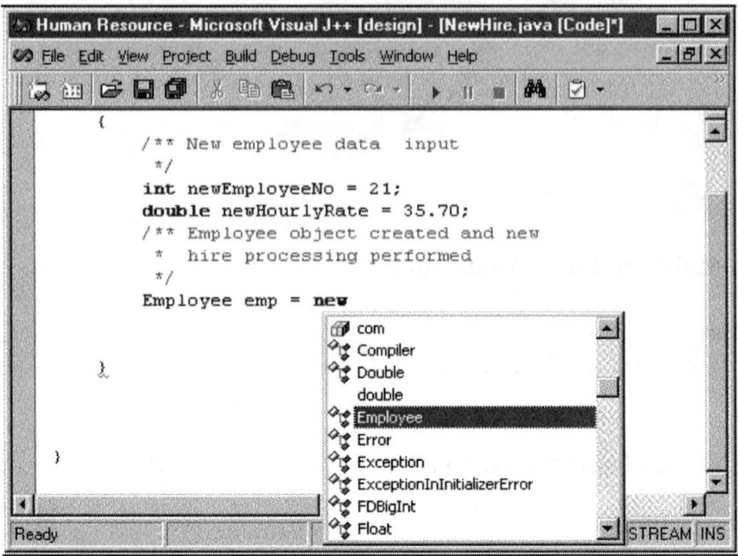

HO-4.1
Statement Completion lists relevant classes, fields, and methods.

3. Press the SPACE BAR, ENTER key, or TAB key to select `Employee`.
4. The Statement Completion pop-up list closes and the class `Employee` is entered in the code.
5. When the left parenthesis is typed, Statement Completion displays the constructor's parameter list, Figure HO-4.2.
6. Continue to enter the code shown in total below:

```
public class NewHire
{
     public static void main(String[] args)
     {
          /** New employee data  input
           */
          int newEmployeeNo = 21;
          double newHourlyRate = 35.70;
          /** Employee object created and new
           *  hire processing performed
           */
          Employee emp = new Employee(
                    newEmployeeNo, newHourlyRate,'P');
          /** Data regarding the newly hired
           *  employeed are output
           */
          int outEmployeeID =
                    emp.getEmployeeID();
          double outHourlyRate =
                    emp.getHourlyRate();
          char outEmployeeStatus =
                    emp.getEmploymentStatus();
     }
}
```

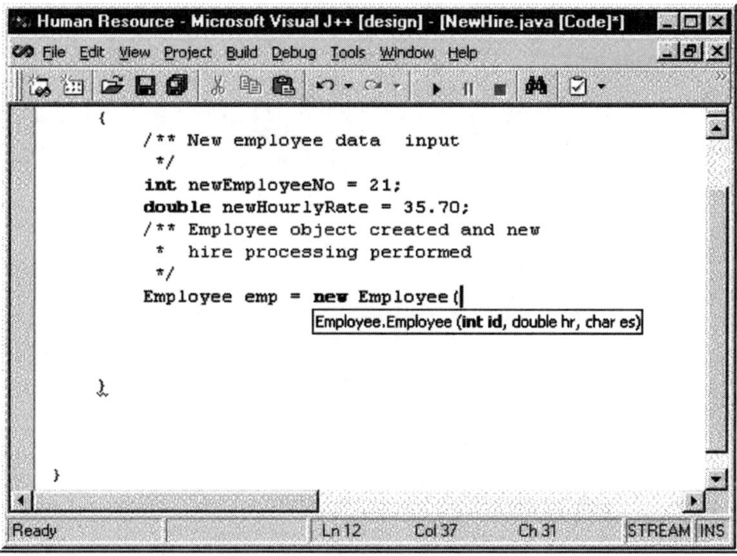

FIGURE HO-4.2
`Employee` constructor's parameter list displayed by Statement Completion.

Step 7: Build the Basic Application

1. Press the CTRL+SHIFT+B keys.
2. Debug the code as required.

Step 8: Set the Project Properties

1. Select the Human Resource Properties... option in the Project menu. The properties dialog box is displayed. Note that the Launch as Console Application is not checked (Figure HO-4.3).
2. Click the OK button to accept the default properties.

Step 9: Execute the Application

1. Position the cursor on the line declaring the variable `newEmployeeID` in the `NewHire` class' `main` method.
2. Press the CTRL+F10 keys. The application starts runs to the cursor. A yellow arrow is displayed to the left of the code line at which execution stops or breaks.
3. Select the Debug Windows option in the View menu. The Debug Windows submenu is displayed.

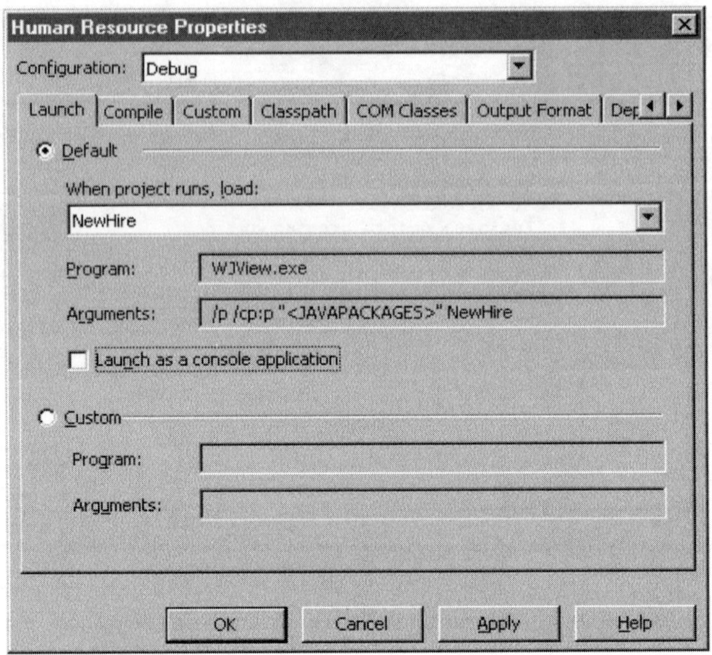

HO-4.3
Human Resource Properties dialog box.

4. Select the Locals option. The submenu and menu close. The Locals window is displayed, Figure HO14.4.

5. Repeatedly press the F11 key to step through the application code line by line.

   Note how the constructor and methods are called and control is passed.

6. Note the Locals window at the completion of the application (Figure HO-4.5).

Step 10: Save All the Project Files

Press the CTRL+SHIFT+5 keys.

Step 11: Close the Project

Select the Close All option in the File menu.

Step 12: Exit Visual J++ the Project

Select the Exit option in the File menu.

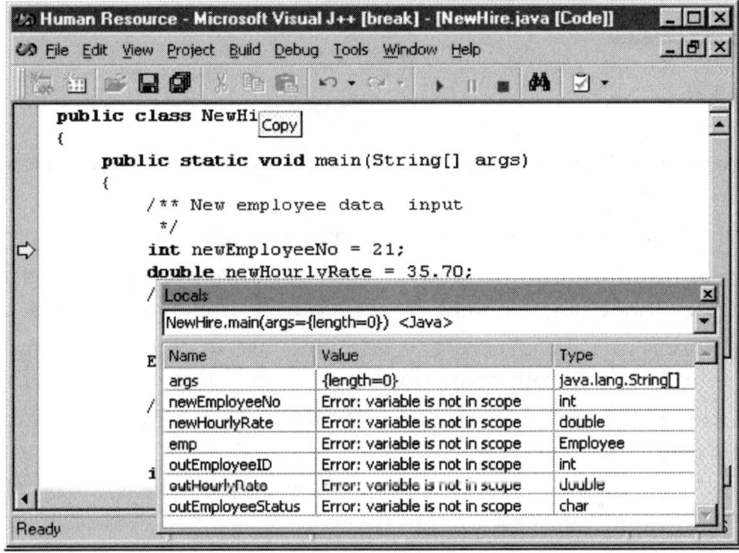

HO-4.4
Locals window open with status of local variables and references at current execution indicated by yellow arrow in Code Editor window.

Locals

NewHire.main(args={length=0}) <Java>

Name	Value	Type
args	{length=0}	java.lang.String[]
newEmployeeNo	21	int
newHourlyRate	35.700000000000	double
⊟ emp	{...}	Employee
java.lang.Obje	{...}	public java.lang.Object
EmployeeID	21	int
HourlyRate	35.700000000000	double
EmploymentSt	80 'P'	char
outEmployeeID	21	int
outHourlyRate	35.700000000000	double
outEmployeeStatu	80 'P'	char

HO-4.5
Locals window at completion of the `NewHire` application.

Challenge: Modify the application to include and use the class `Salesperson`. The class `Salesperson` inherits the class `Employee` and adds a variable representing the salesperson's commission rate.

SUMMARY

1. Java class data type enables the development of object-oriented-programming-based applications.
2. Private access modifiers limit field member access to public member.
3. `Extend` and `implement` keywords provide for the creation of new classes inheriting the fields and members of existing classes.
4. Methods may be overridden enabling use of the same name for methods with similar functionality.
5. `Super` and `this` keywords provide access to current and superclass constructors.
6. The relationship between superclasses and subclasses was developed graphically with hierarchy charts.
7. Interfaces and abstract class were compared.
8. Development of OOP-based applications was demonstrated.

KEY WORDS AND CONCEPTS

`abstract`
abstract class
abstract method
accessible
base class
default access
derived class
encapsulation
`extends`
`implements`
inheritance
`interface`
hide
polymorphism
protected
object
object-oriented programming (OOP)
object-oriented modeling (OOM)
package access
`public`
public access
subclass
`super`
superclass
`this`

REVIEW QUESTIONS AND EXERCISES

1. What is encapsulation?
2. What is polymorphism?
3. What is inheritance?
4. Are private members accessible externally?
5. Is access to a member with default access possible outside the package in which the member's type is declared?
6. Is access to a superclass class in an `extend` clause automatic?
7. Develop a class hierarchy that demonstrates that a class can be both a subclass and a superclass.
8. Is the following use of the keyword `this` valid?

```
public class UsingCustomer
{
    public static void main(String[] args)
    {
        Customer cust = new Customer(-99.01);
        this.AcctBal('d',100.01);
    }
}
```

9. Is the following use of the keyword `super` valid? If not, why?

```
public class SpecialCustomer extends Customer
{
    SpecialCustomer(int id, double ab)
    {
        int buf = 11;
        super();
    }
}
```

10. Is the following code valid? If not, why?

```
public class Customer
{
    int CustID;
    double AcctBal;
    void modCustomer(int id){
        CustID = id;
    }
    int modCustomer(int id){
        CustId = id;
```

```
            return id;
        }
    }
```

11. Consider the class `Customer` shown below:

```
public class Customer
{
    int CustID;
    double AcctBal;
    void maintCustomer(int id){
        CustID = id;
    }
    static void maintCustomer(int id, double ab){
        CustID = id;
        AcctBal = ab;
    }
}
```

    Is the declaration of the overriding `maintCustomer` method in the class `ForeignCustomer` valid?

```
public class ForeignCustomer extends Customer
{
    double ExchangeRate;

    void maintCustomer(int id, double ab){
        CustID = id;
        AcctBal = ab*ExchangeRate;
    }
    void maintCustomer(int id, double ab, double er){
        CustID = id + 1000;
        AcctBal = ab*er;
        ExchangeRate = er;
    }
}
```

12. What is an interface?
13. Compare abstract classes with interfaces.

PRACTICE WITH VISUAL J++

Develop a class that models a basic banking account. Include data members to represent the account number and account balance as well as methods to manipulate the data members. Extend the account class to model checking accounts and saving accounts. Demonstrate use of the class.

5

Expressions and Operators

OBJECTIVES

After reading this chapter, you will be able to:

1. Evaluate expressions comprised of operators and operands.
2. Use Java's operators for performing arithmetic, relational, and logical operations.
3. Understand operator precedence and its implication in expression evaluation.
4. Understand Java's conversion and promotion of data types in expression evaluation and assignment.
5. Demonstrate the ability to create expressions required in application implementation.

OVERVIEW

Identifiers reference data in memory. Manipulation of these data is expressed in statements in which the data identifiers, called operands or arguments, are related to each other by operators. For example, the statement

```
z = x + y;
```

instructs the computer to add the data represented by the identifiers x and y and assign the sum to the memory location associated with the identifier z. In the statement, x, y, and z are operands or arguments and the symbols + and = are operators.

Operators can operate on one, two, or three operands. Operators that operate on two arguments are called binary operators. An operator that operates on one argument is called a unary operator. An operator with three arguments is called a **ternary operator.**

Java contains operators for performing arithmetic, relational, and logical operations. The operators and their use in expressions are described below.

EVALUATING EXPRESSIONS

An expression is a sequence of operators and operands. Thus the statement `z = x + y;` contains the expression `x + y`, which is evaluated to a value that is the sum of the values of the variable represented by the identifiers `x` and `y`. This value is assigned to the variable `z`. Note that the result of the addition of `x` and `y`, the = operator and the variable `z` also form an expression. Java has well-defined rules for evaluating expressions. Some operations are always performed before other operations. This precedence of some operators over others is implied. For example, multiplication is always performed before addition. Consider the expression: `a + b * c`. The product of `b* c` is evaluated first. The result is added to `a`.

Parentheses are used to specify the order of evaluation; for example, in the expression `(a + b) *c`. The expression inside the parenthesis is evaluated first. Care must be taken to ensure that the intended evaluation of an expression is identical to the way that Java evaluates the expression, as the results can differ widely for different evaluation orders. Examples of the problems that can arise when the Java evaluation order is not the same as the program statement might imply are presented below.

An operand can be an expression. An operators operands are evaluated left to right; that is, the left operand of an operator is evaluated before the right operator is evaluated.

OPERATORS

Java contains operators for performing arithmetic, relational, and logical operations.

Simple Assignment

The assignment operator (=) changes the value of the operand to the left of the operator to the value of the operand to the right. In the example below, the variable `A` is assigned the value 711.

```
A = 711
```

If the type of the right operand is not convertible to the type of the left-hand side operator, a compile-time error results.

Assignment is right-associative, that is, groups from right to left. For example,

```
A = B = C
```

means

```
A = (B = C)
```

The assignment operator (=) sets the value of the variable on the left equal to the value of the expression on the right. Thus the statement

```
result = 5 + 3 * 2;
```

assigns the value 11 to the variable result. Assignment has the lowest precedence of the operators and thus is usually performed after all other operations.

> Note that the assignment operator in Java is not the equality operator in algebra. In Java, the expression on the right side of the assignment operator is completely evaluated and then the result is assigned to the variable on the left side.

Arithmetic Operators

The arithmetic operators are listed in Table 5.1.

TABLE 5.1

Arithmetic operators

Operator	Description	Association
*	Multiply	Left to right
/	Divide	
%	Remainder	
+	Addition	
-	Subtraction	

The arithmetic operators are similar to those in other languages and can be applied to expressions of any arithmetic type, for example, `char`, `int`, `long`, `double`, `float`. The two arguments need not be of the same type.

The **multiplicative** operators `*`, `/`, and `%` have the same precedence and are left-associative. The **additive** operators `+` and `-` have the same precedence and are left-associative. Operands of the arithmetic operators must be primitive numeric, with one exception: the + operator is also used to denote string concatenation.

> The multiplicative operators have precedence over the additive operators.

The **multiplication operator** (*) performs multiplication, producing the product of the operands. The **division operator** (/) performs division, producing the quotient of the operands. The left-hand operator is the dividend and the right-hand operator is the divisor. Note that the result of integer division is rounded toward zero. Note also that a run-time exception is thrown when the divisor is zero. The **remainder** operator (%) yields the remainder of a divide operation. The remainder operator is sometimes called the **modulus** operator. For example, the statement

```
remainder = 19 % 4;
```

assigns the value 3 to remainder as 19 divided by 4 is 4 with three left over or remaining. Similarly, the following statements result in the value 3 being assigned to remainder:

```
a = 19;
b = 4;
remainder = a % b.
```

The additive operators + and – have the same precedence and are left-associative. The **addition operator** (+) yields the sum of the operands when they are numeric. If one of the operands is a string, the operation produces a string concatenation. The **subtraction operator** (-) yields the difference between two operands. Examples illustrating use of the arithmetic operators are shown in Table 5.2.

TABLE 5.2

Using arithmetic operators

Expression	Explanation
`A = 1 + 2`	1 and 2 are summed.The result 3 is assigned to `A`.
`A = A - 0.5`	Assume `A` has the value 3. First, 0.5 is subtracted from `A`. Then the result 2.5 is assigned to `A`.
`C = 25 * A`	The number 25 is multiplied by `A`. The result is assigned to `C`.
`D = C / 4.3`	`C` is divided by 4.3. The result is assigned to `D`.
`Z = 27.8 % 3.5-`	The remainder of 27.8 divided by 3.5 is assigned to `Z`.

If one of the operands of the addition operator is a string, the operator performs string concatenation. For example,

```
Int A;
A = 10'
A + "is Ten"
```

results in the string expression

```
"10 is Ten"
```

The operand that is not a string is converted to a string at run-time. The two string operands are concatenated. The characters in the left-hand operand string precede the characters in the right-hand string in the resultant string.

Numerical Comparison Operators

The **numerical comparison operators** are relational operators and are used to compare the values of primitive numeric types. The relational operators are listed in Table 5.3.

TABLE 5.3

Relational operators

Operator	Description	Associatively
<	Less than	Left to right
<=	Less than or equal to	
>	Greater than	
>=	Greater than or equal to	

The numeric relational operators are used to compare primitive numeric types. That is, if the comparison is true the expression is assigned the value `true`. Similarly, if the expression is false, the expression is assigned the value `false`. Table 5.4 summarizes the meaning of the numeric comparison operators.

TABLE 5.4

Meaning of the numeric comparison operators

Expression	Meaning
`L < R`	`true` if L is less than R, `false` otherwise.
`L <= R`	`true` if L is less than or equal to R, `false` otherwise.
`L > R`	`true` if L is greater than R, `false` otherwise.
`L >= R`	`true` if L is greater than or equal to R, `false` otherwise.

Examples illustrating use of the numeric comparison operators are shown in Table 5.5.

TABLE 5.5

Use of numeric comparison operators

Expression	Meaning
`5 > 4`	`true`, 5 is greater than 4
`3 < (1 + 2)`	`false`, 3 is not less than 3
`3 <= (1 + 2)`	`true`

Assume that the variables `one`, `two`, and `three` are numeric primitive types. Note that the expression

```
one < two < three
```

results in an error, as the result of the comparison `one < two` results in a Boolean value; `true` if the numeric value of `one` is indeed less than the numeric value of an `two`. The result is the expression shown below:

```
true < three
```

Note that the < operator is not defined for Boolean types.

> The type of a relational expression is always Boolean.

In Java, there is a strict order in which the operations are performed. This order is defined by operator's precedence. In expressions involving both numeric comparison operators and arithmetic operators the arithmetic operations are performed first. Thus the expression

```
11 > 9 + 4
```

is `false`. A full discussion of operator precedence is presented in a later section of this chapter.

Equality Operators

The **equality operators** are used to compare two operands. The equality operators are of equal precedence and are left-associative. The equality operators are the **equal to** (==) and the **not equal to** (!=) operators.

TABLE 5.6

Equity operators

Operator	Meaning
==	`true` if left-hand operator is equal to the right-hand operator `false` otherwise.
!=	`true` if left-hand operator is not equal to the right-hand operator `false` otherwise.

Comparison of numeric operands is performed by the numerical equality operators == and != in which case both operands must be numeric. Comparison of Boolean operands is performed by the Boolean equality operators, also == and !=, and in which case both operands must be Boolean.

The equality operators are similar to the relational operators but have lower precedence. Consider the statement

```
3 < 5 == 4 < 6
```

The statement first evaluates the expressions containing the relational operators, which both evaluate to `true`. Thus, the entire expression evaluates to `true`.

Note that the expression

```
1 == 2 == 3
```

results in a compile-time error, because `1 == 2` is Boolean. Note that the comparison of this Boolean and the numeric 3 is invalid—both operands in equality operations must be Boolean or both must be numeric.

> Note that the assignment operator = and the equality operator == are often confused and used erroneously.

Bitwise and Logical Operators

The bitwise and logical operators include the **AND** operator (`&`), **exclusive OR** operator (`^`), and the **inclusive OR** operator (|). These operators are left-associative. The `&` operator has the highest precedence and | operator the lowest. If the operands are both primitive integral type, the operators are said to be integer bitwise operators, and the result is the same type as the operands. Similarly, if the operands are both Boolean type, the operators are said to be **Boolean logical operators** and the result is Boolean.

The **integer bitwise AND operator** (`&`) sets each result bit to 1 only if both of the corresponding bits in its operands are 1; otherwise, the result bit is set to 0. The **integer bitwise OR operator** (|) sets each of the result bits to 1 if either of the corresponding operand bits is 1; otherwise, the result bit is set to 0. The **integer bitwise exclusive OR operator** (^) sets each result bit to 1 if either of the corresponding operand bits is 1 and the other is 0; otherwise, the result bit is set to 0. The logical operation performed on the corresponding bits are shown in Table 5.7.

TABLE 5.7

Integer bitwise logical operations

x	y	&	|	^
0	0	0	0	0
0	1	0	1	1
1	0	0	1	1
1	1	1	1	0

The operators manipulate bits in integral type data. For example, let `attribute` be an integer type variable with value 1 and let `blinking` also be an integer type variable with decimal value 8. The memory locations are thus as follows.

```
                 blinking                attribute
----|----------------|-----|----------------|----
 ...|0000000000001000| ... |0000000000000111|...
----|----------------|-----|----------------|----
```

The expression

```
attribute = attribute | blinking
```

results in the following:

```
                blinking                     attribute
----|----------------|-----|----------------|----
... |0000000000001000| ... |0000000000001111|...
----|----------------|-----|----------------|----
```

Boolean logical operators, **AND** (&), **OR** (|), and **exclusive OR** (^), require that both operands be `Boolean`. Assume `x` and `y` are expressions. The results of using the logical operators on `x` and `y` are shown in Table 5.8.

TABLE 5.8

Boolean logical operations

x	y	&	|	^
true	true	true	false	false
true	false	false	false	true
false	true	false	false	true
false	false	false	true	false

The logical operators are evaluated from left to right. Therefore, the expression `x` is evaluated before expression `y` in the statement

```
x & y;
```

An example of a logical operation is the following:

```
Boolean       x, y, z;
x = 10;
y = 9;
(x != 11) & (y < 10)
```

The resulting value is true as `x` is not equal to `11`, resulting in a true value for the first expression, and `y` is less than 10, which results in a true value for the second expression. The result of a logical AND is `true` when both of the operands are `true`.

Note that the precedence of the bitwise and logical operators must be observed when being evaluated. Consider the following example:

```
0x00f0 ^ 0xff0f | 0xff00 & 0xf0f0
```

This expression is evaluated as follows: First the expression containing the & operator, `0xff00 & 0xf0f0`, is evaluated, producing the result `0xf000`

```
0x0ff0 ^ 0x0f00 | 0xf000
```

Then the expression containing the ^ operator, `0x0ff0 ^ 0x0f00`, is evaluated producing the result `0x00f0`

```
0x00f0 | 0xf000
```

Finally, the expression containing the | operator is evaluated, producing the result `0xf0f0`.

Conditional-AND Operator

The **conditional-AND operator** (**&&**) evaluates the right-hand operand only if the left-hand operand is `true` (Table 5.9). The operator is left-associative. Note that both operands must be type Boolean and the result is type Boolean.

TABLE 5.9

Conditional-AND operator

Left-hand operand	Result
`true`	Value of the right-hand operand
`false`	`false`

Consider the following:

```
(5 > 10) && (7 < 11)
```

The left-hand operand evaluates to `false`, as 5 is not greater than 10. Consequently, the conditional-AND expression evaluates to `false` although the right-hand side is `true`.

Consider also the following:

```
(5 < 10) && (7 > 11)
```

In this case, the left-hand operand evaluates to `true`, as 5 is Less than 10. Consequently, the conditional-

AND expression evaluates to the value of the right-hand operand, which is `false`: `7` is not greater than `11`. On the other hand:

```
(5 < 10) && (7 < 11)
```

In this case, the left-hand operand evaluates to `true`, as 5 is less than 10. The conditional-AND expression evaluates to the value of the right-hand operand, which is `true`.

Conditional-OR Operator

The **conditional-OR operator** (||) evaluates the right-hand operand only if the left-hand operand is `false` (Table 5.10). The operator is left-associative. Note that both operands must be type Boolean and the result is type Boolean.

TABLE 5.10

Conditional-OR operator

Left-hand operand	Result
`false`	Value of the right-hand operand
`true`	`false`

Consider the following:

```
(5 > 10) || (7 < 11)
```

The left-hand operand evaluates to `false`, as 5 is not greater than 10. Consequently, the conditional-OR expression is the value of the right-hand operand, which is `true`.

Consider also the following:

```
(5 < 10) || (7 < 11)
```

In this case, the left-hand operand evaluates to `true`, and thus the value of the conditional-OR expression is `false`.

Conditional Operator

The **conditional** operator (?:) is made up of three operands, the first of which is an expression, the second a value assigned to the result if the expression is true, and a third operand whose value is assigned to the result if the expression is false. The operand not chosen to be evaluated and assigned to the result is not evaluated.

For example, when the statement

```
result = (w < x) ? y : z;
```

is evaluated, the result is set to `y` if `w` is less than `x`. On the other hand, if `w` is greater than `x` the result is set to `z`.

The first operand expression of the conditional operator must be type `Boolean`. The second and third operands can both be numeric or the second and third operands can both be `Boolean`. The type of the conditional expression is generally the type of the expression evaluated.

The conditional operator (`?` :) is similar to the `if-else` construct:

```
if (w < x)
        result = y;
else
        result = z;
```

Postfix Operators

The postfix operators are the **postfix increment operator** (++) and the **postfix decrement operator** (--). A numeric identifier followed by the ++ operator is a postfix increment expression. Similarly, a numeric identifier followed by the -- operator is a postfix decrement expression. The result of both expressions is the type of the numeric identifier. After the numeric identifier is evaluated, its value is incremented or decremented by 1.

Thus, consider the following code fragment:

```
a = 10;
c = a++ + 1;
```

In the second statement the value of a is added to the numeric literal value 1 and the sum 11 is assigned to c. The value of a is then incremented by the postfix operator ++ from 10 to 11.

Unary Operators

The unary operators include the **prefix increment** (++), **prefix decrement** (--), **unary plus** (+), **unary minus** (-), **bitwise complement** (~), **logical complement** (!) and **cast** operators. The meaning of the unary operators is summarized in the Table 5.11

TABLE 5.11
Unary operators

Operator	Name	Unary expression value
++	prefix increment	Value of the operand plus 1.
--	prefix decrement	Value of the operand less 1.
+	unary plus	Value of the operand.
-	unary minus	Negation of the value of the operand.
~	bitwise complement	Bitwise complement of the operand.
!	logical complement	true if the operand is false and false if the operand is true.
(type)	cast	Type specified by the cast operator.

The unary operations group from right to left. Thus ++ -- a means ++ (--a); that is, the variable a is first decremented and then incremented. Permissible operand types for the unary operators are summarized in Table 5.12.

The **prefix increment operator** (++) increases the value of an expression by 1. The **prefix decrement operator** (--) decreases the value of an expression by 1. Incrementation or decrementation takes place before the expression's value

TABLE 5.12

Permissible unary operator operand type

Operator	Operand type
`++`	numeric
`--`	numeric
`+`	primitive numeric
`-`	primitive numeric
`~`	primitive integral
`!`	Boolean
`(type)`	numeric

is used. (Contrast this with the postfix increment and decrement operators, in which Incrementation or decrementation of the expression takes place after the expression's value is used.)

Note that the **parsing,** that is breaking up the statement into it constituent parts, generally precedes from left to right, but for some operators association is from right to left. Thus, unless care is taken to observe the rules of precedence and association, expressions can be evaluated with unexpected results. For example, how is the following expression evaluated?

```
a---b
```

Is it assumed to mean

```
(a--) - b.
```

or

```
a- (--b).
```

The use of parentheses avoids confusion and misstatement.

The **unary plus** and **unary minus operators** perform as in algebra. Note that the operand must be a primitive numeric type. Note that the result is a value and not a variable.

The **bitwise complement operator** (~), also known as the **tilde** and as the one's **complement operator,** changes each 1 bit in an expression to 0 and each 0 bit in an expression to 1.

If the variable `go` has the value 1, it is represented in memory by the following bit pattern:

```
                go
------|-------------------|--------
 . . .|  0000000000000001 |  . . .
------|-------------------|--------
```

The bitwise complement operation

```
~go;
```

results in a value with the following bit pattern:

```
------|-----------------|--------
. . . | 1111111111111110 | . . .
------|-----------------|--------
```

Note that each 0 bit was changed to 1 and every 1 bit was changed to 0. *Also note that the bitwise complement of 1 is not –1 or 0.*

Cast Operator

The **cast operator** converts data from one data type to another (Figure 5.1). In this figure, `dataType` is the data type of the cast operator result, and `dataN-ame` is the operand, the name of the datum to be converted.

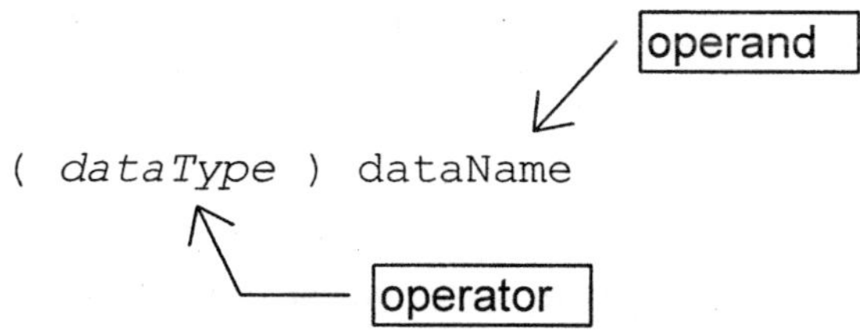

FIGURE 5.1
Elements of the cast operation.

An example is shown below:

```
int i;
double d;
d = (float) i
```

The value of `i` is converted into a double data type and the value assigned to `d`.

Shift Operators

The **shift operators,** also known as the **bitwise shift operators,** move the bits in the left operand to the left or right the number of bits specified in the right operand. The **left** shift operator is shown as <<, the **right** shift operator is shown as >>, and the **unsigned** right shift operator is shown as >>>.

Both operands must be primitive integral type. The shift range, specified by the right operand, is 0 to 31 if the left-hand operand is `int` and 0 to 63 if the left-hand operand is `long`.

When the bits are shifted left, the rightmost bits are padded with 0. An example illustrating the left bitwise shift operation is shown below. Given the value 8, the contents of the memory location of the operand number is as shown:

```
           number
----|----------------------------------|----
... | 00000000000000000000000000001000 | ...
----|----------------------------------|----
```

The value of the expression

```
number << 2
```

reflects the bits being shifted two positions to the right:

```
----|----------------------------------|----
... | 00000000000000000000000000100000 | ...
----|----------------------------------|----
```

Note that after the shift the memory location holds the number 32, which is `8 * (2 * 2)`. Shifting is a fast and efficient way of multiplying numbers by two or powers of two. Note that left shifting provides a quick method of dividing by two or powers of two in what is described as truncating integer division.

Compound Assignment Operators

Java has compound assignment operators, shown in Table 5.13.

TABLE 5.13

Compound operators—their use and meaning

Operator	Typical Use	Meaning
`*=`	`a *= b`	`a = a * b`
`/=`	`a */ b`	`a = a / b`
`%=`	`a %= b`	`a = a % b`
`+=`	`a += b`	`a = a + b`
`-=`	`a -= b`	`a = a - b`
`<<=`	`a <<= b`	`a = a << b`
`>>=`	`a >>= b`	`a = a >> b`
`>>>=`	`a >>>= b`	`**`
`&=`	`a &= b`	`a = a & b`
`^=`	`a ^= b`	`a = a ^ b`
`|=`	`a |= b`	`a = a | b`

The assignment operators are right-associative. The left-hand operand must be a variable. Both operands must be primitive type except for `+=`, which allows the right-hand operand to be a type `string`.

The expression is evaluated in four steps. First the left-hand operand is evaluated to obtain a variable. Second, the value of the left-hand operand is saved to a temporary memory location and the right-hand operand is evaluated. Third, the saved value of the left-hand operand and the value of the right-hand operand are evaluated according to the first operator of the compound operator. Fourth, and finally, the result is assigned to the left-hand operand, which is a variable.

Consider the expression

```
amt = amt * commission;
```

This statement can also be expressed as

```
amt *= commission;
```

using the *= compound assignment operator.

> The equal sign always follows the other sign in a compound assignment operator.

ORDER OF EVALUATION

The Java language has rules for evaluating expressions containing more than one operator. Each of the operators has been assigned a precedence level that determines when it is evaluated relative to the other operators in the expression. The rules for evaluating expressions are as follows:

1. Operators with higher precedence are evaluated before operators with lower precedence.
2. Operators with equal precedence are evaluated according to their associativity.
3. Expressions within parenthesis are evaluated first. When nested, the innermost parenthetical expression is evaluated before the outer expression. Java also has a subscript operator that is used in array object access and a membership operator that is used in field access and method invocation.

The operators are listed in Table 5.14. Operators with a higher precedence are listed above operators with a lower precedence. Operators grouped between the lines have the same precedence with the exception of the Bitwise and Logical and the Conditional operators which have the precedence shown.

TABLE 5.14

Operator precedence and association

Operator	Description	Associativity
( )	Parenthesis	Right to left
[ ]	Subscript	
.	Membership	
++	Postfix Increment	Right to left
--	Postfix Decrement	
++	Prefix Increment	Right to left
--	Prefix Decrement	
+	Unary Plus	
-	Unary Minus	
~	Bitwise Complement	
!	Logical Complement (Not)	
(type)	Cast	
*	Multiplication	Left to right
/	Division	
%	Remainder or Modulus	
+	Additive and String Concatenation	Left to right
-	Subtraction	
<<	Left Shift	Left to right
>>	Right Shift	
>>>	Unsigned Right Shift	
<	Less Than	Left to right
<=	Less Than Or Equal To	
>	Greater Than	
>=	Greater Than Or Equal To	
==	Equal	Left to right
!=	Not Equal (Inequality)	
&	Bitwise and Logical AND	Left to right
^	Bitwise and Logical Exclusive OR	Left to right
|	Bitwise and Logical OR	Left to right
&&	Conditional-AND	Left to right
||	Conditional-OR	Left to right
?:	Conditional	Right to left
=	Assignment	Right to left
*=, /=, %=, +=,-=,<<=, >>=,>>>=, &=,^=,|=	Compound Assignment	

CONVERSION AND PROMOTION

Java can automatically convert and promote from one data type to another in evaluating expressions and assigning values to a variable.

Conversion

Conversion to larger types, said to be **widening primitive conversions,** do not lose information regarding the overall magnitude of a numeric value. Widening primitive conversions are shown in Table 5.15.

TABLE 5.15

Widening conversions do not loose information

From	To
`byte`	`short, int, long, float or double`
`short`	`int, long, float or double`
`char`	`int, long, float or double`
`int`	`long, float or double`
`long`	`float or double`
`float`	`double`

Note that some precision may be lost in the conversion of an `int` or `long` value to a `float` as well as the conversion of a `long` value to a double.

On the other hand, **narrowing primitive conversions,** shown in Figure 5.16, may result in the loss of information about the magnitude of a numeric value as well as loss of precision.

TABLE 5.16

Narrowing conversions may loose information

From	To
`byte`	`char`
`short`	`byte or char`
`char`	`byte or short`
`int`	`byte, short or char`
`long`	`byte, short, char or int`
`float`	`byte, short, char, int or long`
`double`	`byte, short, char, int, long or float`

Promotion

Java automatically performs widening conversion, called **promotion,** to operands of an arithmetic operator. The promotion to a common type enables the operation to be performed. For example, assume the following declarations:

```
Int x = 4;
Double y = 3.11;
```

In evaluating the expression

```
x + y
```

Java automatically converts the value of the variable `x` to a double to enable the `+` operation on common types. Note that no information is lost in the promotion of the variable `x` from `int` to `double`. The value of the expression is 7.11.

Assignment Conversion

In the assignment of an expression to a variable, the type of the expression must be converted to the type of the variable. Assignment conversion occurs when the value of an expression is assigned to a variable. Java allows assignment conversion involving widening primitive conversion. In general, narrowing primitive conversion is invalid. Valid and invalid assignment conversions are shown in Figure 5.2.

```
public class TestAssignmentConversions
{
      public static void main(String[] args)
      {
            int x = 3;
            double y = 4.11;
            int intSum;
            double doubleSum;

            doubleSum = x + 5;

            intSum = x + y;

      }
}
```

valid assignment conversion

invalid assignment conversion

FIGURE 5.2
Valid and invalid assignment conversions.

Note that the value of `x` is converted to a `double` type and the resulting sum of `x` and `y` is a `double` type. The assignment of a type `double` value to an `int` type variable requires a narrowing primitive conversion and thus is an invalid assignment conversion.

Assignment involving widening primitive conversion is valid.

Tip for Former C/C++ Programmers:

Java/J++ does not have the following operators:
exponential
comma
sizeof

HANDS-ON EXERCISE 5: CONVERTING FAHRENHEIT TO CENTIGRADE

Formulas and equations are fundamental to business, engineering and science. The equation converting a temperature expressed in Fahrenheit to centigrade is typical.

C = (100.0/(212.0-32.0))*(F - 32.0);

This exercise develops a program to perform the conversion.

Visual J++ has essential key combinations that bring about the same action as selected menu options. Use of these keys is thought by many to be a "short cut" that saves program development time. Selected short-cut keys and the menu options they represent are summarized in Table HO-5.1.

TABLE HO-5.1

Selected short-cut keys

Keys	Action	Menu	Option
CTRL+N	Open New Project dialog box.	File	Open
CTRL+D	Open Add Item dialog box.	Project	Add Class
CTRL+SHIFT+B	Build the project.	Build	Build
F5	Starts application execution.	Debug	Start
SHIFT+F5	Stop application execution.	Debug	End
CTRL+SHIFT+S	Save all project files.	File	Save All

Step 1: Create a New Project

1. Start Visual J++.
2. Select the New Project... command in the File menu. The New Project dialog box is displayed (Figure HO-5.1).
3. Enter **`Temperature Conversion`** in the Name text field.
4. Use the Browse button to enable navigation to the Location `C:\Hands-On\Temperature Conversion`.
5. Click the Open button to create the new project.

Step 2: Create `TemperatureConversion.java` File

1. Press the CTRL+D keys. The Add Item dialog box is displayed.
2. Select the Class folder (Figure HO-5.2).

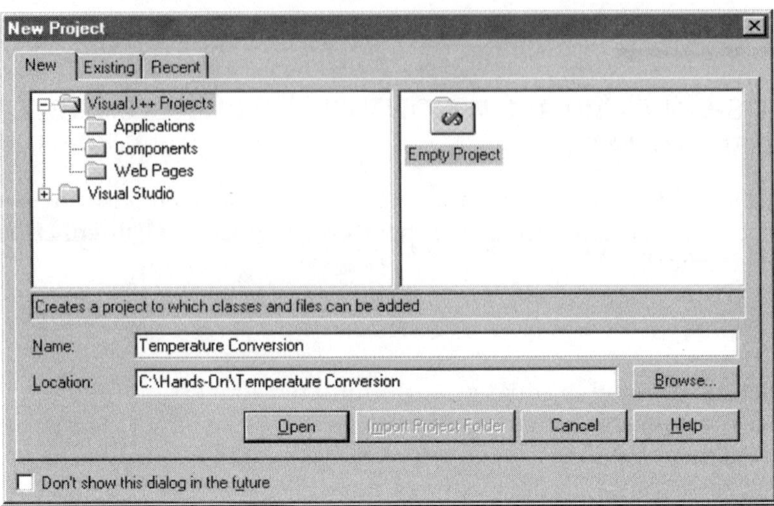

HO-5.1
New Project dialog box.

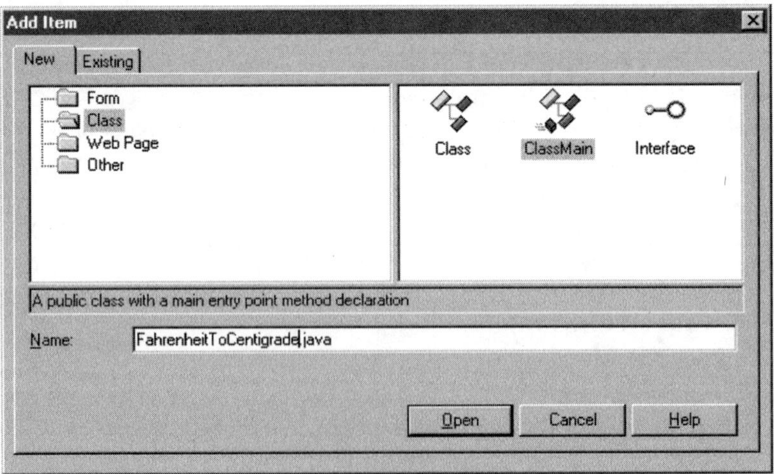

HO-5.2
Add Item dialog box opened with Class folder selected.

3. Select the ClassMain icon.
4. Enter FahrenheitToCentigrade.java in the Name text field.
5. Click the Open button to create the new class.
6. The Code Editor is opened with the file `FahrenheitToCentigrade.java` (Figure HO-5.3).

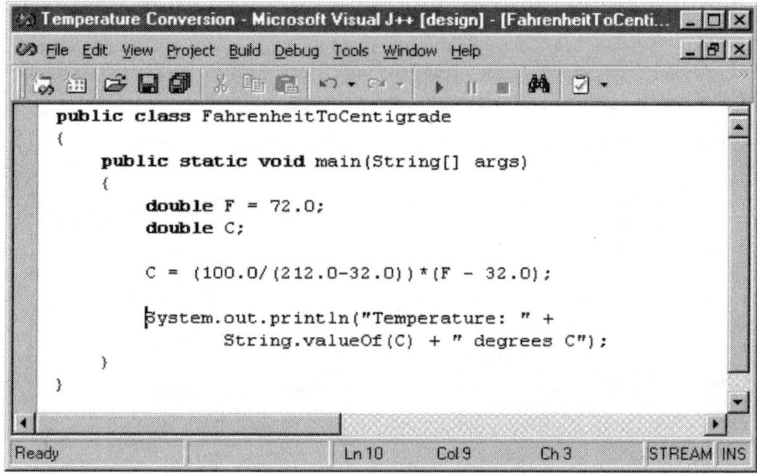

HO-5.3
Code editor opened with file `FahrenheitToCentigrade.java`.

Step 3: Code the Application

1. Enter the following statements to declare and initialize the variable `F` representing the temperature in Fahrenheit.

   ```
   double F = 72.0;
   ```

2. Declare the variable `C` representing the temperature in centigrade with the statement

   ```
   double C;
   ```

3. To convert the temperature in Fahrenheit to centigrade, enter the statement

   ```
   C = (100.0/(212.0-32.0))*(F - 32.0);
   ```

4. To display the result, enter the statement

   ```
   System.out.println("Temperature: " +
                   String.valueOf(C) + " degrees C");
   ```

Step 4: Build the Project

1. Press the keys CTRL+SHIFT+B to build the project.
2. Debug the code as required.

Step 5: Set the Project Properties

1. Select the Temperature Conversion Properties in the Project menu. The Properties dialog box is displayed.
2. Check the Launch as Console Application checkbox (Figure HO-5.4).

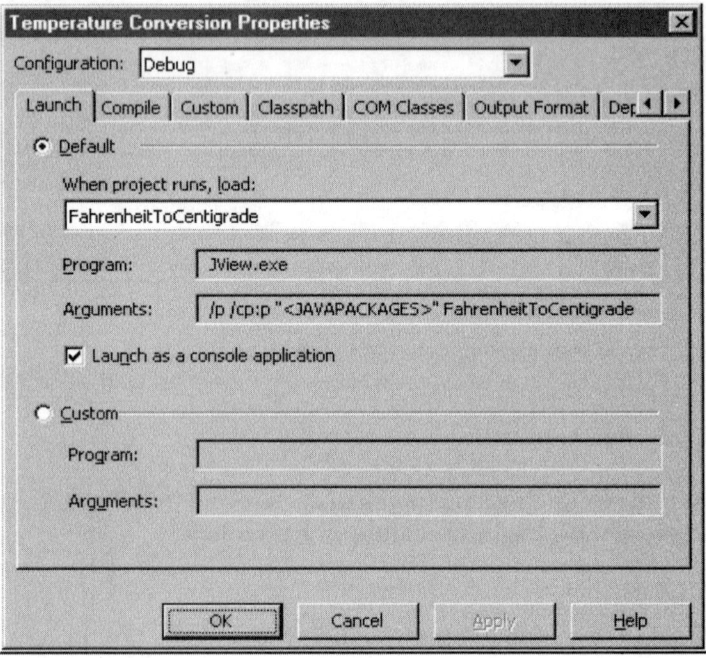

FIGURE HO-5.4
Temperature Conversion Properties dialog box checked to Launch as Console Application.

3. Click the Apply button.
4. Click the OK button.

Step 6: Execute the Application

Press the F5 key. The application executes and displays the temperature in centigrade (Figure HO-5.5).

Step 7: Save All the Project Files

Press the CTRL+SHIFT+F5 keys.

Step 8: Close the Project

Select the Close All option in the File menu.

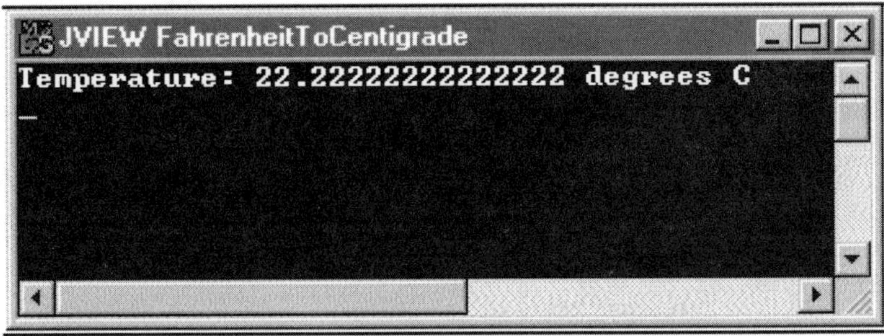

FIGURE HO-5.5
Application output.

Step 9: Exit Visual J++ the Project

Select the Exit option in the File menu.

Step 10: Develop a Centigrade to Fahrenheit Application

Create and test an application that will convert temperatures in centigrade to the equivalent temperature in Fahrenheit.

SUMMARY

1. Expressions relate operands with operators.
2. Operators describe arithmetic, relational, and logical operations performed on associated operands.
3. Expressions are evaluated according to operator precedence.
4. Promotion and conversion are automatically performed in expression evaluation and assignment.
5. Evaluation and assignment of expressions were demonstrated.

KEY WORDS AND CONCEPTS

- <
- <=
- >
- >=
- ==
- !=
- *
- /
- %
- +
- -
- =
- *=
- /=
- %=
- +=
- -=
- <<=
- >>=
- >>>=
- &=
- ^=
- |=
- --
- ++
- ~
- |
- &
- ^
- &&
- ||
- !
- assignment conversion
- binary operator
- conversion
- narrowing primitive conversion
- operand
- operator
- promotion
- ternary operator
- unary operator
- widening primitive conversion

REVIEW QUESTIONS AND EXERCISES

1. Given the following:

```
A = 1
B = 2
C = 3
```

what is the value of `A` after the statement is executed?

```
A = B = C = 711
```

2. Given the following:

```
char c
int i
Boolean b
long l
```

which of the following result in compile time errors?

```
c = i
b = i
i = c
b = c
l = b
c = "ABCDE"
l = "long"
b = "TRUE"
```

3. What message is produced by the following code? Try it!

```
Class Test {
    Public static void main(String[] args) {
            double dividen, divisor, quotient;
            dividen = 1.1;
            divisor = 2.2;
            quotient = dividen/(divisor - divisor);
    }
}
```

4. What is the result? Try it!

```
Class Test {
    Public static void main(String[] args) {
            double remainder;
            remainder = 72.27 % 4.1;
    }
}
```

5. What message is output by the following code? Try it!

```
Class Test {
    Public static void main(String[] args) {
            System.out.println("Pi is " + 3.14);
    }
}
```

6. What message is output by the following code? Try it!

```
Class Test {
    Public static void main(String[] args) {
            double Pi;
            Pi = 3.14;
            System.out.println("Pi is " + Pi);
    }
}
```

7. Which of the following expressions are valid?

```
10 < 20
11 >= "Eleven"
true < false
7 - 11 <= 11 + 7
5 =< 10
"z" > "a"
```

8. Which of the following expressions are valid?

```
10 == 20
11 != "Eleven"
true == 1
7 - 11 == 11 + 7
```

9. Which of the following are valid?

```
(10 == 5) & true
0x0005 | 5
5 | "five"
(5 - 2) ^ 3
```

10. Evaluate the following:

	Result
`0xff0f | 0xf0f0 & 0xf000 ^ 0x0ff0`	________
`true | false & true ^ false`	________

11. Which of the following expressions are valid (assume a, b and c are numeric type):

```
10 && 20
true && false
(5 < 10) && (c = 21)
(a >= b) && (a == c)
```

12. Evaluate the following:

	Result
`(0xff0f < 0xf0f0) && (0xf000 < 0x0ff0)`	________
`(0xf0f0 < 0xffof) && (0xf000 < 0x0ff0)`	________

13. Which of the following expressions are valid (assume a, b, and c are numeric type)?

```
10 || 20
true || false
(5 < 10) || (c = 21)
(a >= b) || (a == c)
```

14. Evaluate the following:

	Result
`(0xff0f < 0xf0f0) || (0xf000 < 0x0ff0)`	________
`(0xf0f0 < 0xffof) || (0xf000 < 0x0ff0)`	________

15. Which of the following expressions are valid (assume a, b and c are numeric type)?

```
a ? b : a
(a < b) ? false : 10
true ? a < b : (c < 21)
(a <= b) ? (a - 21 * c) : (a - 21 * c)
```

16. Evaluate the following where a = 10, b = 5.5, and c = –7:

Result

```
C < b ? a : a - 9                    __________
C < b ? C < b : b < c                __________
```

17. Which of the following expressions are valid (assume `a`, `b`, and `c` are numeric type)?

```
a++
(a < b)--
a++ --
10--
```

18. Evaluate the following where a = 10, b = 5.5, and c = –7?

Result

```
a--                                  __________
a++ --                               __________
a + b++ - c                          __________
```

19. Which of the following expressions are valid (assume `a` is numeric, b is Boolean, c is numeric integral, and d is a double)?

```
+a
~b
!b
~c
-d
```

20. Evaluate the following where a = 10, b = 5.5, c = -7 and x is `true`:

Result

```
-a                                   __________
~7                                   __________
!x                                   __________
(int) b                              __________
```

21. Show how to quickly and efficiently multiply the `int` variable `x` by 4.
22. Why is the term *truncating integer division* associated with the right shift operator?
23. Evaluate the following where `amt` = 10, `factor` = 4 and `location` = 20:

Result

```
amt << factor                        __________
location>>4                          __________
```

24. Given the following

```
int 1 = 7;
char c = 'c';
double d = 21.21;
boolena b = true;
long 1 = 10000.01;
```

identify the valid assignment conversions

	Valid	Invalid
`b = 1;`	______	______
`1 = i * c;`	______	______
`c = b;`	______	______
`d = 1 * d + i;`	______	______
`i = 1;`	______	______
`1 = i;`	______	______

25. Why can information be lost in a narrowing conversion?

PRACTICE WITH VISUAL J++

Develop and test Java application evaluating the following algebraic expressions:

a. Perimeter of a rectangle: `p = 21 + 2w`
b. Area of a circle: `a = 3.14r2`
c. Area of a triangle: `a = bh/2`
e. Straight line: `y = ax + b`
f. Commission: Com = `SaleAmt * ComRate`
g. Profit: `Profit = GrossRevenue - OperatingCost`

Part 2

Programming Techniques

6

Program Flow

OBJECTIVES

After reading this chapter, you will be able to:

1. Understand the program control of flow concept.
2. Program decision-making and iterative procedures.
3. Demonstrate use of program flow control techniques in developing applications.

OVERVIEW

The order in which the steps are taken is often spoken of as the program's logical flow. Collectively, the steps and the order in which they are executed are often referred to as a procedure. Generally then, a **procedure** is simply a detailed statement of exactly what is to be done. When asking how to solve a problem, the question asked is, What is the procedure? Technically, computers compute; that is, computers perform a computation. In this case, the question asked is, What is the procedure for computing the answer? Procedures implemented on a computer are referred to as **programs.**

In simple procedures, the steps required to realize an answer involve performing one step right after another in sequence until all are done. More complicated procedures involve performing different steps depending on the answer to a question. For example, is the customer new or is the customer an old one? Many procedures involve repetitively performing the same procedure again and again for each member of a specified group. An example of this is the computation of pay checks in which a similar process must be executed for each member of the firm—if chaos is not to ensue on payday! Prescribing which step is to be taken next is embodied in the concept of **program**

control. The stepping from one element of a program to another is referred to as **program flow.** The statements enabling control of program flow are referred to as **control structures.**

This chapter presents Java statements that enable program flow.

BLOCKS AND STATEMENTS

The steps in a program are expressed as statements. **Statements** specify how the programs objects are to be manipulated. Statements are processed sequentially, except when the statement specifies iteration or that control should be transferred to another statement.

Empty Statement

The **empty statement** does absolutely nothing, but is a valid statement. The empty statement is shown below:

Note that the empty statement does not have an expression to be calculated and is often referred to as a null statement.

Labeled Statement

Statements can be labeled. A **label** is an identifier followed by a colon. The labeled statement is the statement immediately following the colon. The code fragment shown in Figure 6.1 illustrates a label and labeled statement.

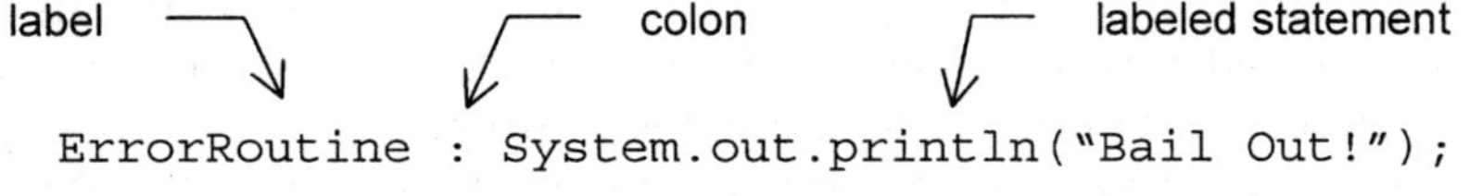

FIGURE 6.1
Illustrating a labeled statement.

In this case the label is `ErrorRoutine`. The labeled statement is `System.out.println("Bail Out!");`.

A **compound statement** consists of one or more statements enclosed by curly braces (`{ }`). The code fragment in Figure 6.2 illustrates compound statements. A compound statement can be empty. A **`block`** is a sequence of statements enclosed in braces (`{ }`). A block is executed by executing each of the statements in order, first to last, top to bottom, left to right.

```
    Block1:
    {
      cout << "Block 1 Statement 1" << "\n";
      cout << "Block 1 Statement 2" << "\n";
    }

    Block2:
    {

    }
}
```

empty compound statement

empty compound statement

FIGURE 6.2
Illustrating compound statements.

```
public class SequentialExecution

{
    public static void main(String[] args)
    {
        int x = 1, y = 2, z = 0;

        z = x + y;

        System.out.println("Z = " + z);
    }
}
```

executed first

executed next

executed last

FIGURE 6.3A
Statement are executed sequentially within a block.

Consider the application shown in Figure 6.3a. The program is the method `main`. The program consists of four statements. The first statement defines and initializes three variables: `x`, `y`, and `z`. This statement is executed first. The following statement, calculation of `z` as the sum of `x` and `y`, is executed next. The statement following the calculation is then executed and displays the text "`Z:` " followed by the value of z (Figure 6.3b).

Finally, the program reaches the closing brace of the `main` method and the programs ends. Every time the program is executed, the statements are executed in the same order. The way in which statements are executed is referred to as the program's **flow.** The flow shown above is one in which the statements are executed sequentially. Variations in this sequential execution are provided by statements providing loop and decision capability. That is, the flow can be altered by loop statements such that the same statements can be executed over and over again until a specified condition is met. Similarly, the flow can be altered by decision statements, in which case the next statement executed depends on the value of a specific condition. As such the pro-

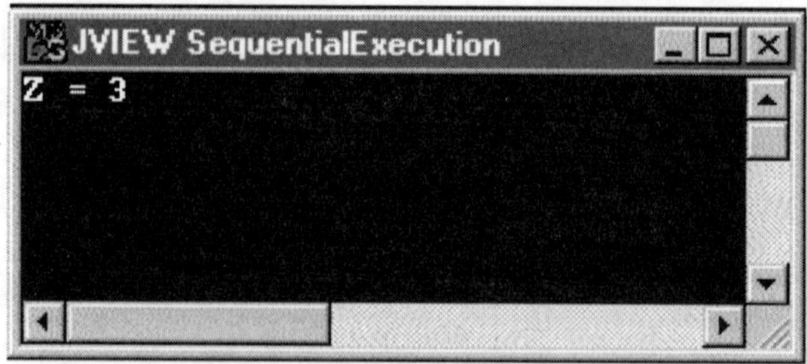

FIGURE 6.3B
Output of the application illustrated in 6.3a.

```
public class BlocksAndScope
{
     public static void main(String[] args)
     {
     //    BlockA
           {
                int x = 3;

           //   BlockB
                {
                     int y = 3;
                     y = y + x;
                }
           //   BlockC
                {
                     int x = 4
                     y = y + x;
                }
           //   BlockD
                {
                     ;
                {
           }
     }
}
```

scope limited to BlockB

already defined in method

undefined in BlockC

empty statement

FIGURE 6.4
Scope of a method variable is limited to the block in which it is defined.

gram's flow can be controlled to enable different courses of action and repetitive processing based on values in the program itself. As such, the actual order of statement execution can be controlled, and is referred to as the **control of flow.**

Compound statements are commonly called blocks. The scope of a variable is limited to the block in which it is defined. Consider the program shown in Figure 6.4. The variable `x` is declared in BlockA. Thus it is visible, or within scope, in BlockB, which is within BlockA. However, the variable `y` declared in BlockB is not visible in BlockC.

A statement can also be null. The **null statement** is a statement without an expression to be evaluated.

A good programming practice is to limit a program module to one page of code.

JUMP STATEMENTS

Jump statements result in the immediate transfer of control to another statement in the program or return control from a method. Use of `continue` and `break` statements will be illustrated in the sections of this chapter dealing with loops. Use of the `return` statement was illustrated in Chapter 3, Class Data Type, Methods, Members.

`return` Statement

The **`return`** statement transfers control from the current method to the calling method. The syntax for the `return` statement is

```
return   ( iReturnCode );
```

`return` is a required keyword beginning the statement. The return code is an expression that evaluates to a value. The value then is returned to the calling method or in the case of the main method to the operating system.

Methods with a return type other than void must return a value. A method with return type void cannot return a value and thus cannot contain a return statement. A method can have any number of return statements.

`break` Statement

The **`break`** statement causes control to transfer to the statement following the end of the innermost loop in which the break occurs. The break statement ends only one loop!

`continue` Statement

The **`continue`** statement transfers control to the end of the loop. When loops are nested, control is passed to the continuation statement of the current loop. The continue statement is often used to bypass sections of code and continue with the next iteration of the loop.

Java does not have a `goto` statement.

DECISION STATEMENTS

Decision making is an essential element of problem-solving procedures. Java provides control structures that implement decisions.

Decision in Procedures

Most procedures involve many decisions. For example, consider the point in the payroll process when the pay is calculated. If the employee has not worked more than 40 hours, then pay is calculated using the following formula:

```
dPay = dRegularHours * dRate
```

On the other hand, if the employee has worked more than 40 hours, then the pay is calculated using the following formula:

```
dPay = dRegularHours * dRate + 1.5 * dOvertime-
Hours * dRate
```

This process is illustrated in the flowchart shown in Figure 6.5.

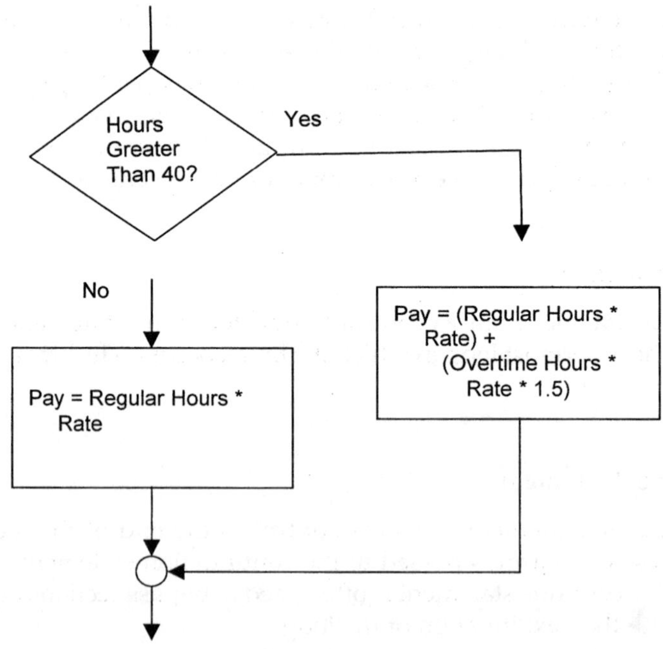

FIGURE 6.5
Pay calculation flowchart.

At the critical point, the question is asked, Are the hours greater than 40? If the answer is no, the decision is made to calculate pay as simply regulars hours times the pay rate. If on the other hand the answer is yes, then the decision is made to calculate pay as the regular hours times the rate plus the overtime hours at one and one-half times the regular rate.

Regardless of which method is used, the process continues using the calculated pay. Note that in the flow chart, the decision point is depicted as a diamond. The process of calculating the pay is represented in boxes. After the pay is calculated the payroll process continues on the same path.

This type of decision is implemented in Java using the `if` statement.

This is not the only decision encountered in the payroll process. After an employee's pay is calculated the question must be asked, Should pay be calculated for another employee? Imagine what would happen on payday if pay is processed for only one employee in a company that has many employees! In the old days, a clerk started with a stack of time cards: one for each employee. The time cards contained the employee's number, name, and hours worked. Typically these cards were stacked in an in-tray. The clerk removed a card from the in-tray and calculated the employee's pay. The card and paycheck were placed in an out-tray. When the in-tray is empty, time cards for all employees have been processed and the clerk is finished processing payroll. In computer terms, the in-tray and out-trays are files. An empty file is said to be at end-of-file. The question whether paychecks for all employees have been processed is asked with the expression: End Of File?

The flowchart in Figure 6.6 depicts the payroll process. First the data for the first employee is input. The pay is processed. The questions is asked: Have paychecks been processed for all employees? If the answer is yes, the payroll process ends. If the answer is no, data for the next employee is input and the pay process is continued.

Representing Questions and Answers in Java Programs

Questions in programs are expressed as `Boolean` expressions. The `Boolean` value `true` represents a yes answer and the `Boolean` value `false` represents a no answer. For example, if the variable `dRegHrs` represents an employees regular hours, then the expression

```
dRegHrs > 40
```

has the `Boolean` value `true` when the employee has worked more than 40 hours and `false` when the employee has worked 40 hours or less.

Questions can also be expressed by evaluating the value of a `Boolean` variable. For example, a `Boolean` variable `bEOF` is often used to characterize a file's end-of-file condition. `bEOF` is true when the file is at end and false otherwise. Thus, in a typical payroll process as long as there are time cards and paychecks to be processed, `bEOF` is false. When time cards and paychecks for all employee's have been processed, `bEOF` is true. The question, have paychecks for all employees been processed is asked with the expression

```
bEOF
```

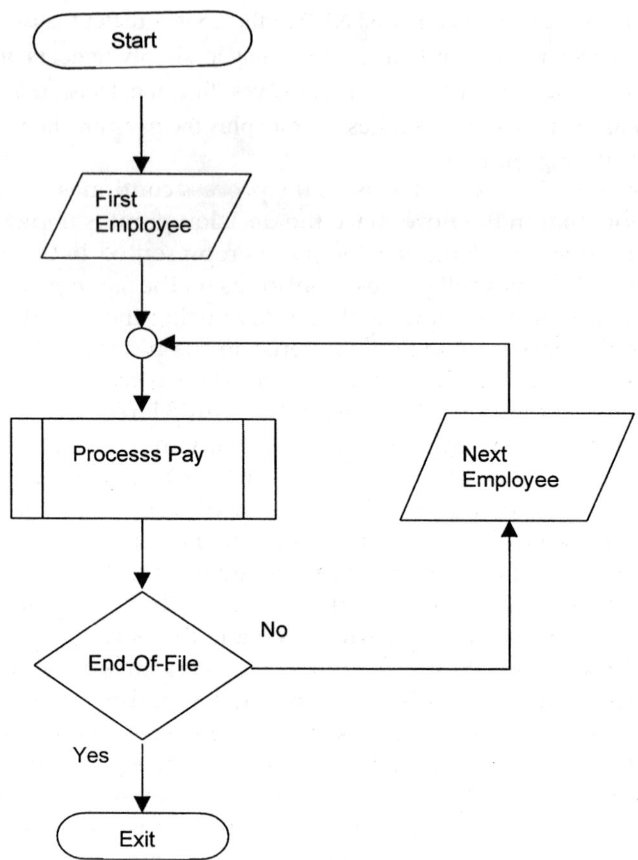

FIGURE 6.6
Flowchart illustrating process for computing gross pay.

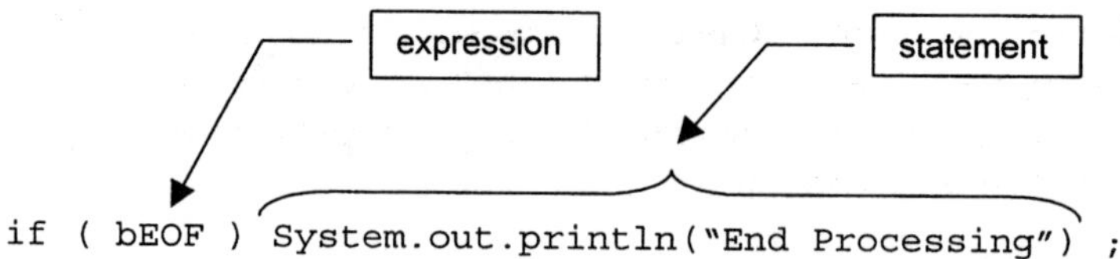

FIGURE 6.7
The if statement.

`if` Statement

The **`if` statement** is illustrated below in Figure 6.7. When the expression evaluates to the value `true` the statement is executed. After the statement is executed control passes to the following statement. If the expression evaluates to `false` the statement is not executed and control passes immediately to the following statement. The statement is often referred to as the **`if` statement body.** The body can be empty, a single statement or a group of statements. If the body is empty or consists of a single statement, the `if` statement must end with a semicolon. For example,

```
if ( bEOF );
```

is an example of an empty `if` statement. If the body consists of several statements they must be enclosed in braces as a block of code.

```
if ( bEOF ) {
    System.out.println("Statement 1");
    System.out.println("Statement 2");
}
```

A semicolon is not placed after the closing brace. However, statements within the braces must be terminated with a semicolon. The flow chart in Figure 6.8 illustrates the flow of control in an `if` statement.

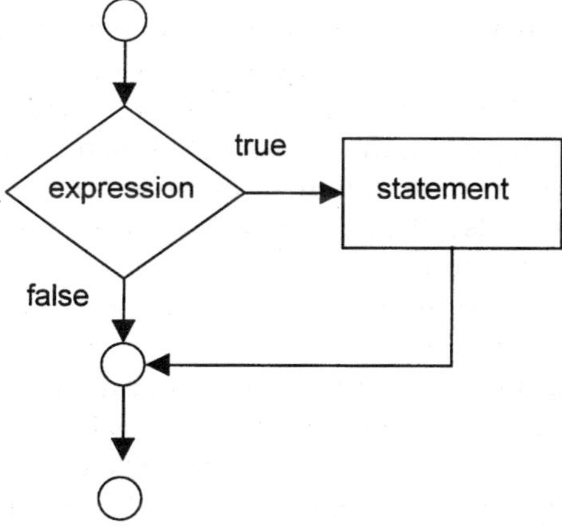

FIGURE 6.8
Flowchart for Java `if` statement.

The if statement is often referred to as the "if-then" statement. For example: If the expression is true, then perform the statement! Note that the "then," unlike in some languages, is not included in the Java statement.

In using the `if` statement be sure the expression is type `Boolean` and remember that control is transferred to the following statement both when the `if` statement is performed and when the `if` statement is not performed.

> Tip for Basic and Visual Basic Programmers:
> `then` is not an element of the Java `if` statement.

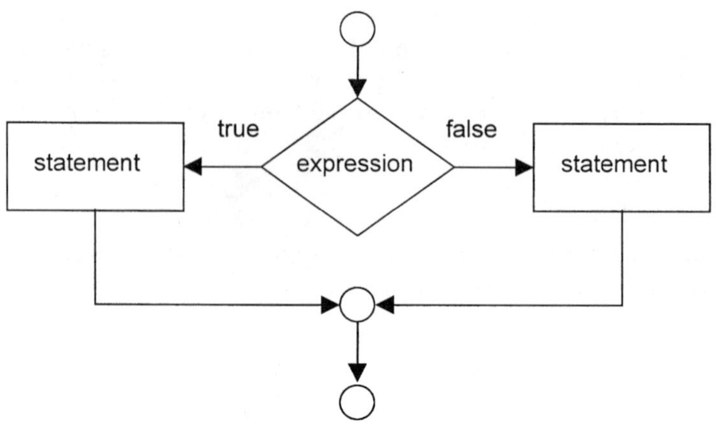

FIGURE 6.9
`if-else` statement flowchart.

`if-else` Statement

The **`if-else`** statement is used with the `if` statement and provides for programming either–or situations. That is, perform one set of actions if the expression is true but if the expression is false, perform a different set of actions. A flowchart depicting the `if-else` statement is shown in Figure 6.9.

If the expression is true, the statement in the left box is performed. If however, the expression is false, the statement in the right box is performed. Note that the both paths lead to the same point. The `if-else` structure is said to have a single entrance and single exit. The `if-else` statement is often referred to as the "if-then-else" statement.

Consider the program statements required to calculate pay based on hours worked and pay rate. If the hours worked is less than or equal to forty, the pay is simply the pay rate times the hours worked. However, if the hours worked exceeds forty, pay is calculated as pay rate times the hours plus a premium for hours exceeding, often one-half the regular rate. This programming circumstance is implemented with the `if-else` statement as illustrated with the following code fragment:

Note that in the example, shown in Figure 6.10, statements following the `if` and `else` keywords are both simple and thus need not be enclosed in braces. If however, several statements are to be performed in either case the multiple statements must be enclosed in braces. Remember, the statements following `if` and `else` keywords can both be a null statement terminated with a semicolon, a simple statement terminated with a semicolon, or multiple statements each of which is terminated with a semicolon and all enclosed in braces.

> Multiple statements following `if` or `else` must be coded as a block.

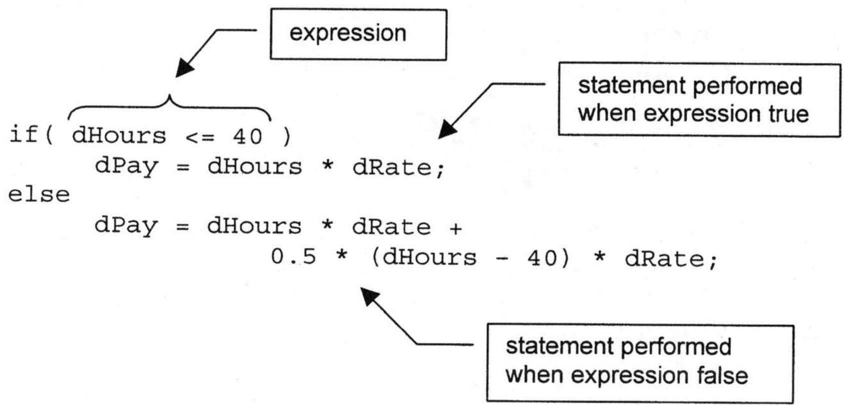

FIGURE 6.10
Calculating pay.

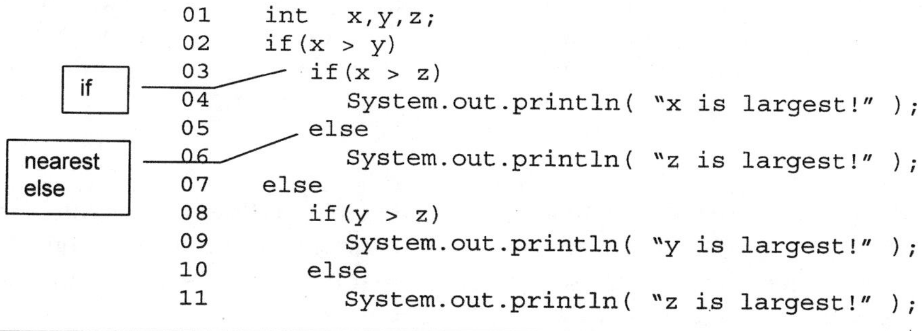

FIGURE 6.11
Matching `else` with nearest `if`.

Nesting `if` Statements

`if` statements and `if-else` statements can be nested. In nested `if` statements each `else` statement is associated with the closest preceding `if` statement that has not already been associated with an `else`. Indentation is used to improve readability. Indentation is ignored by the compiler but is helpful in matching `if`s and `else`s.

Consider the code fragment shown in Figure 6.11. In this example three values are compared. If `x > y > z` then expression `x > y` is `true` and the statement following the `if` is performed. In this case, the statement is another `if`

```
if(x > z)
```

which also evaluates to `true`. The statement following this `if` is then performed and

```
x is largest!
```

is displayed. Control passes to the statement following the statement on line 11.

Consider now the case in which `z > x > y`. In this case the condition `x .> y` is `true` and control passes to the `if` statement at line 3. In this case however, the condition `x > z` is `false` so the statement following the else is performed.

```
z is largest!
```

is displayed. Again, control passes to the statement following statement 11.

Consider now the case y > x > z. In this case the condition x .> y is `false` and control passes to the `else` at line 7. Remember that an `else` is associated with the preceding `if`. Hence the `else` at line 5 is associated with the `if` at line 3. The statement following the `else` at line 7 is performed. This statement contains an `if` expression,

```
if(y > z)
```

which evaluates to `true`. Consequently,

```
z is largest!
```

is displayed. Again, control passes to the statement following statement 11.

Finally, when `z > y > x`, the condition `x > y` is `false` and control passes to `else` at line 7. In this case however, the condition `y > z` is `false` so the statement

```
z is largest!
```

is displayed. Again, control passes to the statement following the statement.

`If` and `else` statements can be nested to test for multiple conditions. For example, the code shown in Figure 6.12 can be used to execute different actions depending on the value of a transaction code.

```
if (cTransCd=='a')
   System.out.println ("Add new customer to file");
else if (cTransCD == 'c')
   System.out.println("Update customer's existing record");
else if(cTransCd =='d'
   System.out.println("Delete customer's record");
```

FIGURE 6.12
Example of nested `if` statements

In this case if the value of the transaction code is 'a',

```
Add new customer to file
```

is displayed. If the transaction code is not `'a'` the first `if` expression evaluates to `false` and control passes to the `else` clause. The statement following an

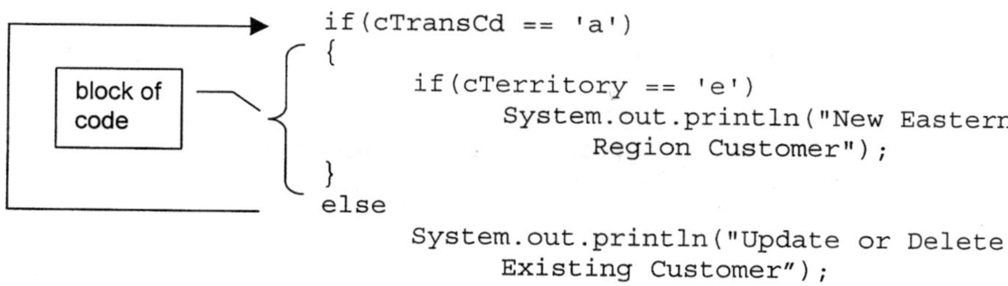

FIGURE 6.13
Using braces to tie an `else` to an `if`.

`else` can be another `if`, as in this case. The `if` expression is evaluated. If the transaction code is `'c'`,

```
Update customer's existing record
```

is displayed. If the transaction code is not `'c'` then the expression evaluates to `false` and control passes to the `else` clause. This `else` is also followed by an `if` statement. The expression is evaluated. If the transaction code is `'d'`

```
Delete customer's record
```

is displayed. If the transaction code is not `'d'` then nothing is displayed.

This style of indenting saves space and is used by many programmers.

Braces can be used to tie an `else` to an `if` that is separated from the `else` by another `if` as shown in Figure 6.13. This code will display

```
New Eastern Region Customer
```

when the variable `cTransCd` has the value `'a'` for add and the variable `cTerritory` has the value `'e'` for eastern. When `cTransCd` code has the value `'a'` and `cTerritory` has a value other than `'e'` nothing is displayed. Whenever the transaction code is not `'a'`, as when the transaction code variable has the value `'c'` for change or `'d'` for delete, the statement

```
Update or Delete Existing Customer
```

is displayed regardless of the value of the territory.

> Use of indentation and parenthesis in the coding of complicated `if-else` logic is a good programming practice and a welcome tool in debugging!

`switch` Statement

The **`switch`** statement provides for programming decisions with multiple outcomes. A flowchart illustrating the switch logical structure is shown in Figure 6.14. The switch is entered with an expression value. In the flowchart

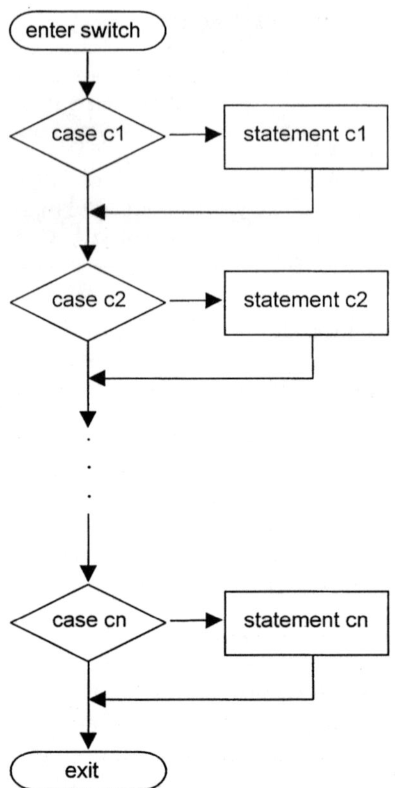

FIGURE 6.14
The `switch` statement enables selection from a number of alternatives.

the first case value is c1. If the expression equals the case value statements s1 thru statement sn are executed and the switch is exited. If the expression does not equal the case value the next case comparison is performed. Again if the expression equals the case value, in this instance c2, statements s2 through sn are executed and the switch is exited. Again if the expression does not equal c2, statement s2 is not executed, and the expression is compared with the next case value. The process is repeated until control reaches the exit.

Consider the classical problem of what to do with customer data that has just been input. Typically this data is referred to as a transaction. The character of the data is represented by the value of the data's transaction code. If the data has been input with a transaction code of `1`, the data is assumed to represent a new customer, in which case the data is added to the database. This process of adding is also referred to with the terms "save" and "insert." On the other hand, if the transaction code is `2` the data is assumed to reflect new information regarding an old customer, in which case the data in the customer's database record is revised. This action of revision is also referred to with the terms "change," "modify,"

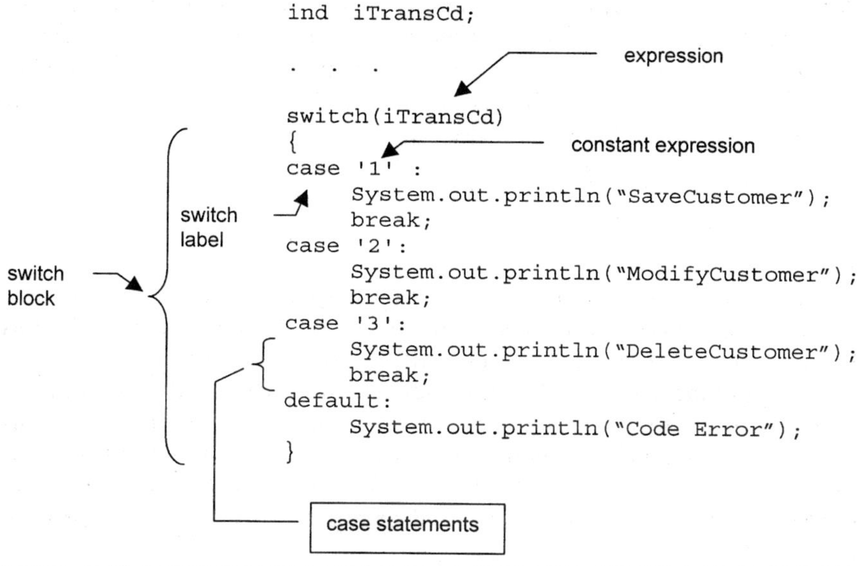

FIGURE 6.15
The `switch` statement modeling the add–update–delete decision process.

or "update." Finally, if the transaction code value is `3`, the customer data should be removed from the database. In this case the record regarding the customer specified in the transaction is removed or deleted from the database. Note the decision process has multiple outcomes--three to be exact. Java's `switch` statement efficiently models this decision process.

In the example shown in Figure 6.15, the switch statement starts with the keyword `switch` followed by an expression enclosed in parentheses. The expression type must be `char`, `byte`, `short`, or `int`. In this case, the expression is the integral variable `cTransCd`, which is an `int` data type. The expression is followed by the switch block enclosed in parentheses. The switch block contains alternatives. Each alternative begins with the keyword `case` followed by a constant which is followed by a colon. The alternatives are evaluated sequentially. When the expression, in this case `cTransCd`, is equal to the constant value following the case keyword, in this case `'1'`, the statements following the case, constant, and colon are executed. Note that the last statement is a break statement that transfer control to the statement immediately following the switch statement. Thus if the value of `cTransCd` is `'1'`,

```
SaveCustomer
```

is displayed. Consider the case now where the value of `cTransCd` is `'2'`—in this case

```
ModifyCustomer
```

is displayed.

Note that a `break` statement transfers control to the bottom of the switch. In the absence of `break` statements, the case statements in all following cases will be executed one after another until the bottom of the switch is reached. Thus if the break statements were not included and the `cTransCd` value was `'2'` the display would be

```
ModifyCustomer
    DeleteCustomer
CodeError
```

A switch can have as many case alternatives as needed. Each alternative must be labeled and contain a constant value. The switch statement can have a default value. The `default` value will be executed if the test expression result does not match any of the case label constant values. If no default case is written and the test expression result does not match any of the case label constants, control passes immediately to the statement following the switch. Generally the default case is put at the bottom of the switch and the other case alternatives are arranged in numerical or alphabetical order for program readability.

> The switch statement provides for programming decision procedures with multiple outcomes.

LOOPING STATEMENTS

Iterative procedures, commonly called **loops,** can be programmed using the `for`, `while`, and `do while` statements. This section presents these three statements and shows how they are used for constructing iterative procedures.

`for` Statement

The **for** statement is used when a block of code must be repeated a specific number of times. Consider the program in Figure 6.16a, which illustrates the `for` statement when used to print the numbers from 1 to 4.

The **initializing expression** is used to initialize the loop parameters. Most loops contain a **loop counter,** which is used to determine how many iterations of the loop are performed. In the example above, the variable `i` is the loop counter and is initialized to the value one. The **test expression** is used to terminate the loop. The test expression is evaluated at the beginning of every loop iteration. The statements comprising the loop's action are performed as long as the test expression evaluates to the value true. For example, in the case above the loop is repeated as long as the counter `i` is less than five. At

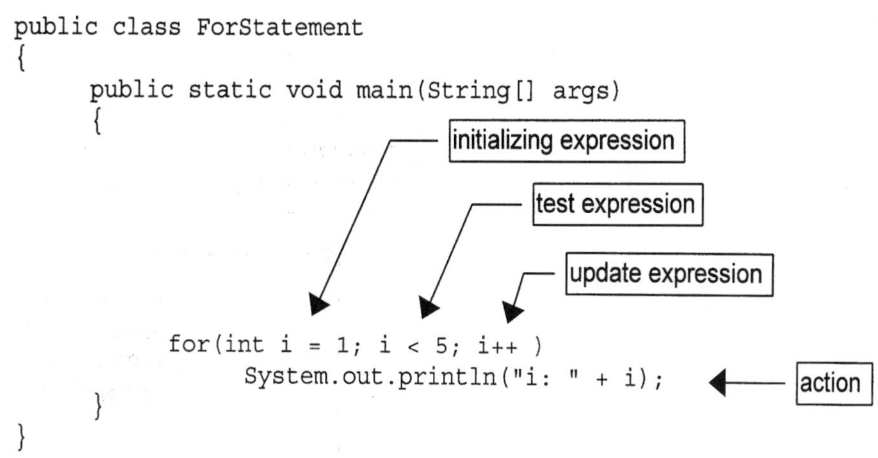

FIGURE 6.16A
Counting from 1 to 4 with a `for` loop.

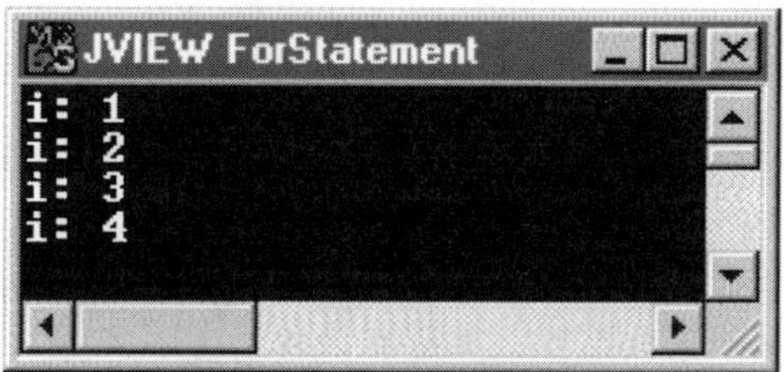

FIGURE 6.16B
Output of a program counting from 1 to 4 with a `for` loop.

the conclusion of each loop iteration the **update expression** is evaluated. This expression is used to modify the loop's parameters. In the example above the counter is incremented. The output is displayed in Figure 6.16b.

A flowchart illustrating the functionality of the `for` statement is shown in Figure 6.17. The initializing and update expressions can be multiple expressions. For many procedures it is necessary to count backward from some predetermined value while also counting forward to another predetermined value. The program shown in Figure 6.18a illustrates this procedure and the use of multiple initializing and update expressions. The output is shown in Figure 6.18b.

The initializing expression, test expression, and update expression can be omitted, although the two semicolons must always be present. If the initializing expression is absent the initialization is not done as part of the `for` loop. If the test expression is absent the test is assumed to evaluate as true. If the update expression is absent, modification of the loop parameters is not performed as part of the `for` statement. However, this does not preclude initial-

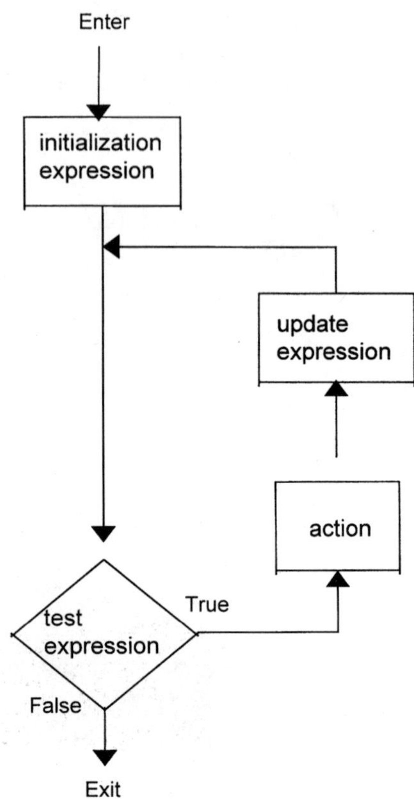

FIGURE 6.17
`for` statement flowchart.

izing the parameters prior to executing the `for`, testing the loops parameters as part of the loop's action and exiting the loop with a `break` or `exit` statement or modifying the loops parameters as part of the loop's action. For example,

```
for( ; ; )
    System.out.println(" Forever!! ");
```

prints `Forever` once initiated.

When the loop's action consists of only one statements the statement can be terminated with a semicolon. When the loop's action or body consists of multiple statements the statements must be enclosed in brackets and each statement must be terminated with a semicolon.

The `break` statement can be used to cause an immediate exit from a `for` loop. For example, consider the program shown in Figure 6.19a. The output displayed is shown in Figure 6.19b. When a `break` statement is encountered control transfers to the statement immediately following the `for` statement.

The `continue` statement can cause suspension of the loop's action in the current iteration and the immediate update of the loop's parameters followed by the initiation of the nest loop iteration. Consider the code shown in

```
public class ForStatement
{
      public static void main(String[] args)
      {
            int i = 1;
            int y;

            for( i = 1, y = i + 3; i < 5; i++, y-- )
                  System.out.println("i: " + i + "  y: " + y);
      }
}
```

multiple initialization expressions

multiple update expressions

FIGURE 6.18A
`for` loop counting forward and backward and utilizing multiple expressions.

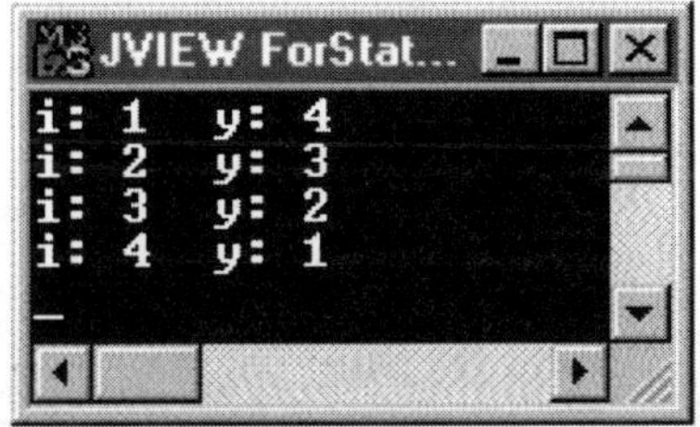

FIGURE 6.18B
Output of program counting forward and backward and utilizing multiple expressions.

Figure 6.20a. In this case, the `continue` statement is executed when the counter `i` has the value 2. The loop action is suspended, the counter is updated, and the next iteration begun. The output is shown in Figure 6.20b.

> Note the difference between = and == .

Note that the output shown in Figure 6.20b does not show the counter value 2, as the loop action, which in this case included the display, was suspended for the counter value 2.

`while` Statement

The **`while`** statement provides for continuously executing a body of statements as long as a test expression is true. The `while` statement syntax is shown in Figure 6.21.

The statement begins with the keyword `while`. First, the expression, which must be enclosed in parentheses, is evaluated. If the expression is `true`, the statements comprising the body of the `while` statement are executed. This process is repeated as long as the expression is `true`. When the expression is `false`, control skips to the statement immediately following the `while` state-

```
public class ForStatement
{
      public static void main(String[] args)
      {
            int i = 1;
            for( i = 1; i < 5; i++)
            {
                  System.out.println("i:  " + i);
                  if (i == 3)
                  {
                     System.out.println("Break executing!");
                     break;
                  }
            }
            System.out.println("for loop exited!");
      }
}
```


FIGURE 6.19A
Using the break statement to exit a `for` loop.

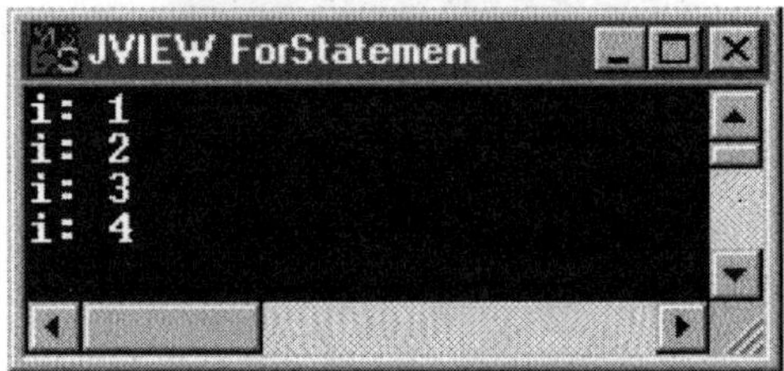

FIGURE 6.19B
Output of program using the `break` statement to exit a `for` loop.

```
public class ForStatement
{
      public static void main(String[] args)
      {
            int i = 1;
            for( i = 1; i < 5; i++)
            {
                  if (i == 2)continue;
                  System.out.println("i:  " + i);

            }
      }
}
```


FIGURE 6.20A
Using the `continue` statement to suspend the current loop iteration.

ment. This process is illustrated in the flowchart shown in Figure 6.22. Note that if the body consists of only one statement, the braces can be eliminated.

An example of a program using the `while` statement is shown in Figure 6.23a. In this example, the variable `i` is assigned the value 1. Thus the expres-

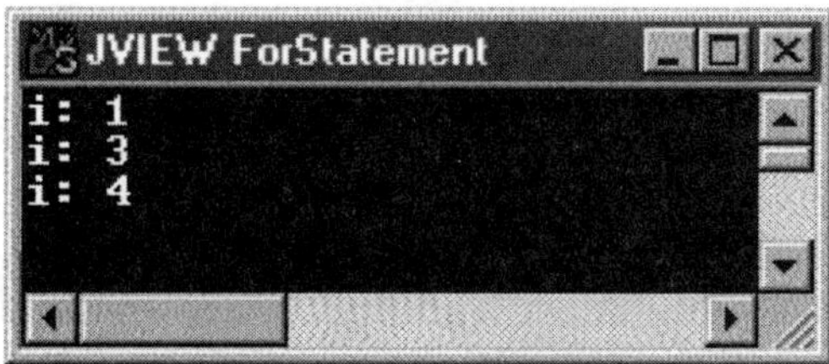

FIGURE 6.20B
Output of program using the `continue` statement to suspend the current loop iteration.

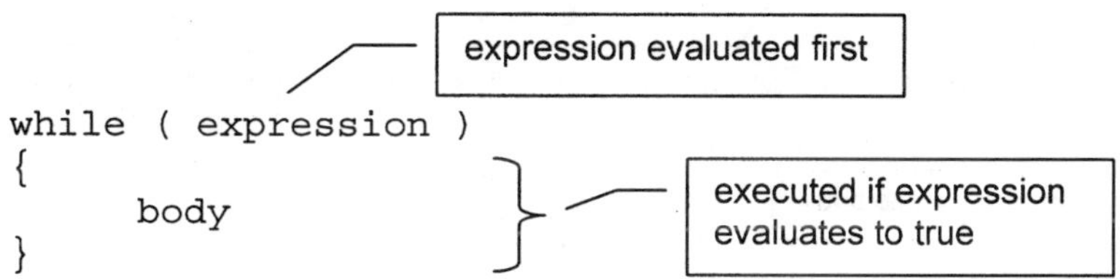

FIGURE 6.21
`while` statement syntax.

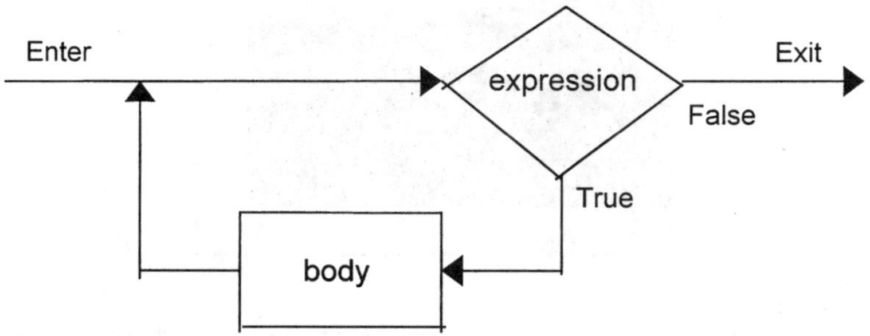

FIGURE 6.22
`while` statement flowchart.

sion in the `while` statement evaluates as `true`. The statements comprising the body are executed. First the value of `i` is added to `sum`. Next, the value of the variables `i` and `sum` are displayed. Finally, the value of `i` is increased by 1. The `while` is again evaluated as true; the value of `i` is 2, which is less than 6. Again, the statements comprising the body are executed. The value of `I` and `sum` are displayed and then `i` is increased by one to 3. Again, the expression is evaluated as true resulting in the values of `i` and `sum` being displayed and `i` being increased to 4. Once again, the expression is evaluated as true resulting in the values of `i` and `sum` being displayed and `i` being increased to 5. Still again, the expression is evaluated as true resulting in the values of `i` and `sum` being displayed and `i` being increased to 6. Once more, the expression is evaluated. In this case the expression evaluates to false as the value of `i` is not less than 6. Control skips to the statement following the `while` body and

```
Sum: 15
```

```
public class WhileStatement
{
    public static void main(String[] args)
    {
        int sum = 0;
        int i = 1;
        while( i < 6)
        {
            sum = sum + i;
            System.out.println("i: " + i + "    sum: " + sum);
            i = i + 1;
        }
        System.out.println("Sum: " + sum);
    }
}
```

expression evaluated

i incremented each iteration

while body

FIGURE 6.23A
Program using a `while` statement to compute the sum of the first five numbers.

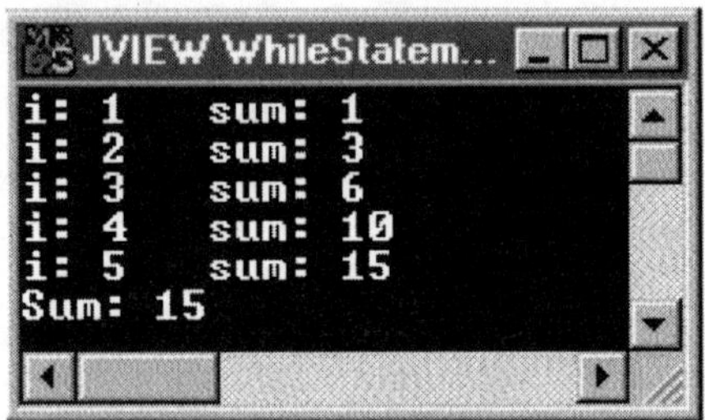

FIGURE 6.23B
Output of program using a `while` statement to compute the sum of the first five numbers.

is displayed (Figure 6.23b).

Note that the following `while` loop never executes:

```
while ( 7 >= 11)
{
}
```

The while body is executed if the expression evaluates to true.

do Statement

The **do** statement is similar to the `while` statement except that the expression is tested after the body of the loop has been executed. Thus the body of the loop is executed once even if the expression is false when the loop begins. If the expression is true the body is performed again followed by another

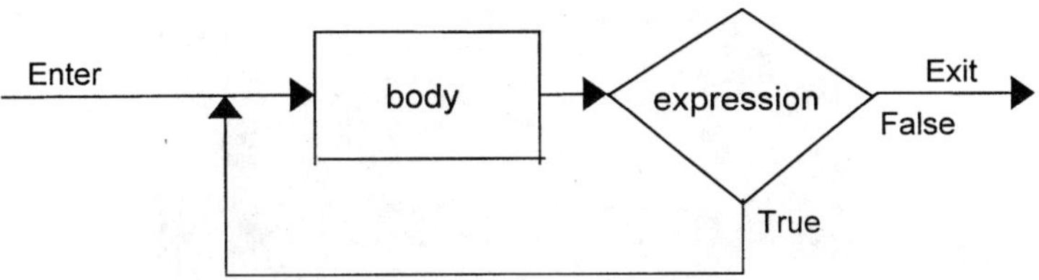

FIGURE 6.24
do statement flowchart.

evaluation of the expression (Figure 6.24). This looping continues until the expression is determined to be false, at which time control is transferred to the statement immediately following the do statement. This is in contrast to the body of the `while` loop, which is executed only if the expression is determined to be true.

The `continue` and `break` statements are used to control flow in the do statement just as in the `while` statement.

The expression

$$(i + 1)^n$$

where `i` represents interest rate and `n` number of periods, is used in financial analysis. The program, shown in Figure 6.25a, illustrates evaluation of the expression using a do statement loop.

```
public class DoStatement
{
      public static void main(String[] args)
      {
            double i = 0.075;
            int n = 1;
            double prod = 1.0;
            do {
                  prod = prod * (i + 1);
                  System.out.println(n + "     " + prod);
                  n = n + 1;

            }while (n < 21);
      }
}
```

do body

expression evaluated

FIGURE 6.25A
Program calculating a financial expression with a do loop.

The do loop is entered following initialization of the variables `i`, `n` and `prod`. Prior to executing the body of the do loop the expression is evaluated. Upon entry, the value of `n` is 1 and hence the expression evaluates to true and the body of the loop is executed. The value of `n` is incremented in the body of the loop each time it is performed. After 20 iterations the value of `n` is incremented to value 21. When the twenty-first iteration of the do loop is

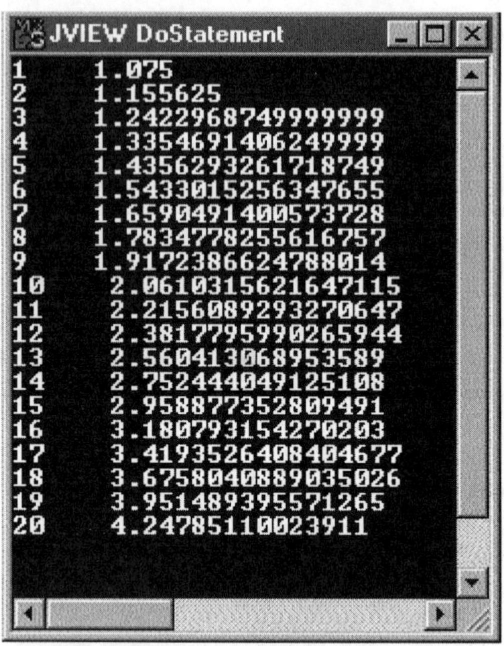

FIGURE 6.25B
Output of a program calculating a financial expression with a `do` loop.

attempted the expression evaluates to value `false` and control is transferred to the statement following the `do` statement. During each iteration the value of the expression is displayed. The output of the program evaluating a financial expression is shown in Figure 6.25b.

Because the `do` statement includes the keyword `while` it often is referred to as the `do...while` statement. Note that the while test expression is terminated with a semicolon.

HANDS-ON EXERCISE 6: FINANCIAL APPLICATION

Auto loans, mortgages, and similar financial agreements all use the principle of compounding interest in computing the value of money. The basic principal is expressed by the formula

$$S = P\,(1 + R)^{n}$$

where **P** is the principal amount, **S** is the future value of the principle amount, **n** is the number of periods, and **R** is the interest rate per period. This exercise develops an application that uses the equation to compute the future value of $1000 invested for thirty years at an annual interest rate of 6% compounded monthly.

Visual J++ has key combination that bring about the same action as selected Debug menu options. Selected Debug menu "short-cut" keys and the menu options they represent are summarized in Table HO-6.1.

TABLE HO6.1

Selected Debug menu short cut keys

Keys	Action	Option
CTRL+F10	Start application and run to cursor	Run to Cursor
F11	Perform the next operation	Step Into
F10	Step over the current method	Step Over
SHIFT+F11	Step out of the current method.	Step Out
SHIFT+F5	Stop application execution	End
F5	Start application execution	Start

Step 1: Create a New Project

1. tart Visual J++.
2. Select the New Project... command in the File menu. The New Project dialog box is displayed.
3. Enter **`Financial Application`** in the Name text field.
4. Use the Browse button to enable navigation to the Location `C:\Hands-On\Financial Application`.
5. Click the Open button to create the new project.

Step 2: Create `FutureValue.java` File

1. Press the CTRL+D keys. The Add Item dialog box is displayed.
2. Select the Class folder.
3. Select the ClassMain icon.
4. Enter **`FutureValue.java`** in the Name text field.
5. Click the Open button to create the new class.
6. The Code Editor is opened with the file `FutureValue.java` (Figure HO-6.1).

Step 3: Code the Application

1. Enter the following statements to declare and initialize the variable `P` representing the current value of the amount to be invested, `S` the future value of the money invested, and `R` the interest rate.

```
double P = 1000.00;
```

```
double S = P;
double R = 0.06/12;
```

2. Note that `P` is initialized to `1000`, the amount invested and `R` is initialized to the monthly value of 6% (0.06) annual interest. Note also that initial future value, `S`, is equal to the principal value, `P`.
3. Enter a `for` loop to compute the principal amount `30 * 12` times, which is equivalent to once a month for 30 years. In each iteration, the future value `S` is increased by the amount (`1 + R`).

```
for(int N = 1; N <= 30 * 12; N++)
    S = S * (1 + R);
```

4. Enter the statement

```
System.out.println("Future Value: " +
                   String.valueOf(S);
```

to display the result.

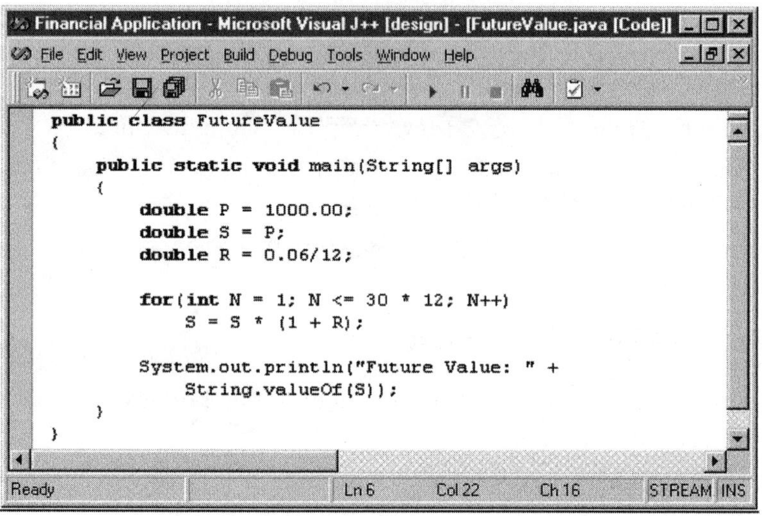

HO-6.1
Code Editor opened with file `FutureValue.java`.

Step 4: Build the Project

1. Press the keys CTRL+SHIFT+B to build the project.
2. Debug the code as required.

Step 5: Set the Project Properties

1. Select the Future Value Properties option in the Project menu. The properties dialog box is displayed.
2. Check the Launch as Console Application checkbox.
3. Click the Apply button.
4. Click the OK button.

Step 6: Step through Execution of the Application

1. Position the cursor on the code line containing the for statement;
2. Press the CTRL+F10 keys. The application starts and runs to the cursor. A yellow arrow is displayed in the left margin of the Code Editor, indicating the next statement to be executed.
3. Press F11 repeatedly to iterate through the `for` statement.
4. Position the cursor over a variable identifier. The current value of the variable is displayed (Figure HO-6.2).

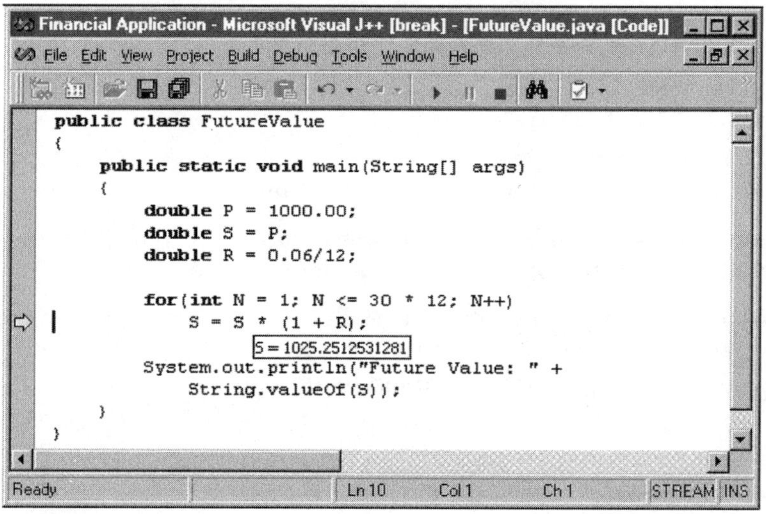
Financial Application - Microsoft Visual J++ [break] - [FutureValue.java [Code]]
File Edit View Project Build Debug Tools Window Help
public class FutureValue
{
public static void main(String[] args)
{
double P = 1000.00;
double S = P;
double R = 0.06/12;
for(int N = 1; N <= 30 * 12; N++)
S = S * (1 + R);
S = 1025.2512531281
System.out.println("Future Value: " +
String.valueOf(S));
}
}
Ready
Ln 10
Col 1
Ch 1
STREAM
INS

FIGURE HO6.2
Code Editor in break mode displaying the value of variable `S` during the sixth iteration of the `for` loop.

5. Press SHIFT+F5 to end the application. The future value of the principal amount compounded monthly at a 6% annual interest rate for 30 years is shown (Figure HO-6.3).

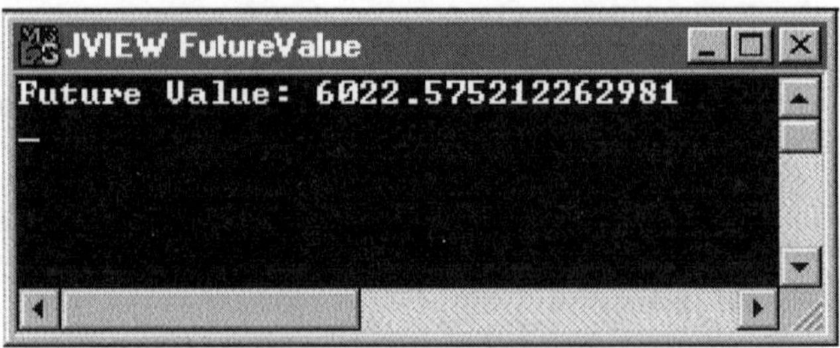

HO6.3
Application output.

Step 7: Save All the Project Files

Press the CTRL+SHIFT+F5 keys.

Step 8: Close the Project

Select the Close All option in the File menu.

Step 9: Exit Visual J++ the Project

Select the Exit option in the File menu.

Step 10: Develop a Current Value Application

Create and test an application that determines the current value of $10,000 20 years from now assuming a 4% annual interest rate.

Challenge: Research the formula and develop an application to compute value that is accumulated over thirty years by investing $100 monthly at an annual interest rate of 6% that is computed monthly.

SUMMARY

1. Java provides for control of the program flow; that is, the order in which statements are executed.
2. Statements are normally executed sequentially.
3. The `if`, `if-else`, and `switch` statements transfer control, enabling the execution of alternative statements, based on the evaluated value of a test expression.

4. The `for`, `while`, and `do` statements provide for performing statements multiple times based on the evaluated value of a text expression.
5. Jump statements, `continue` and `break`, enable the immediate transfer of control.

KEY WORDS AND CONCEPTS

Action
block
`break`
case
compound statement
`continue`
control structures
control of flow
default
`do`
`do...while`
empty statement
flow
`for`
`if`
`if-else`
initializing expression
jump
label
label statements
loop counter
loops
null statement
procedure
program
program control
program flow
`return`
statements
`switch`
test expression
update expression
`while`

REVIEW QUESTIONS AND EXERCISES

1. Is the following `if` statement true?

```
int x = 7;
if ( x == 21) ;
```

2. Assume `iNo` is integer type and has value 0. Is the following `if` statement valid?

```
if ( iNo );
```

3. What is displayed when `bTestValue` has value false?

```
if ( btestValue );
    System.out.println( "Why am I displayed?" );
```

4. Consider the following code:

```
if(cTransCd == 'a')
  if(cTerritory == 'e')
    System.out.println("New Eastern Region Customer");
else
    System.out.println("Update or Delete Existing
        Customer");
```

What is displayed for the following variable values:

CtransCd	Cterritory	
a	e	______
a	w	______
c	e	______
d	w	______

5. What is the output?

```
char x = 'y';
if( x =='n')
    System.out.println("x = n);
else
    System.out.println("x not = n");
```

6. What is the output?

```
char x = 'n';
if( x == 'n')
    System.out.println("x = n");
else;
    System.out.println("Always printed -- Why?");
```

7. Consider the following code:

```
if(cTransCd =='a')
    System.out.println("Add new customer to file");
else if(cTransCd =='c')
    System.out.println("Update existing customer's record");
else if(cTransCd =='d')
    System.out.println("Delete customer's record");
else
    System.out.println("Transaction code error");
```

a. What is displayed when `iTransCd` has the value `'c'`?

b. What is displayed when `iTransCd` has the value `'x'`?

8. What is displayed if `cTransCd` equals `'a'` and `cTerritory` does not equal `'s'`?

```
if(cTransCd=='a')
    if(cTerritory=='e')

        System.out.println("New Eastern Region
Customer");
else
System.out.println("Update or Delete Existing
Customer");
```

9. What method(s) are executed when `cTransCd` has the value `'d'`?

```
switch(cTransCd)
{
    case 'a':
        SaveCustomer( );
    case 'c':
        odifyCustomer( );
    case 'd':
        DeleteCustomer( );
    default:
        TransCodeError(.);
}
```

what procedure is executed if `cTransCd` has the value `'x'`? What statements must be added to limit execution to statements specified in the selected case?

10. What is displayed?

```
for (i = 1; i < 7; i++)
{   if ( i < 4)
        continue;
    else
    {
        System.out.println(" i = " + i );
        break;
    }
}
```

11. What is displayed?

```
for (i = 1; i < 7; i++)
{   if ( i < 4)
        continue;
    else
    {
        System.out.println(" i = " + i );
    }
}
```

12. Will the following code display the numbers 1 through 7?

```
for( int i = 1; ;i++)
{
    System.out.println( i );
    if( i > 7) break;
}
```

13. Is the following `for` valid?

```
int x,y;
for(x=0,y=0; x>10,y>15; x++,y++)
    System.out.println( x + " " + y );
```

14. What is the output?

```
int x = 0;
while ( x >= 10 )
{

    System.out.println(" x = " + x );
    x = x - 1;
}
```

15. What is the output?

```
int x = 0;
while ( x < 20 )
{

    System.out.println("x = " + x );
    x = x + 5;
}
```

16. What is wrong with the following code, which is intended to list the numbers from 1 to 10?

```
int x = 1;
while (x < 11)
    System.out.println("x = " + x );
    x = x + 1;
```

17. What is displayed?

```
int i = 0;
do
{
    i++;

if(i == 2) continue;

        System.out.println("i = " + i );

} while( I < 3);
```

18. What is displayed?

```
int i = 1;
do
{
    if(i == 3) break;
    System.out.println("i = " + i );
        i++;
} while( i < 10);
```

PRACTICE WITH VISUAL J++

1. Develop a program to select the smallest of three numbers. Test the program with various values. Does the program always select the proper value as smallest?
2. Develop a program to count backward by increments of 5 starting at 105 and ending at 80. Test the program.
3. Write a program that converts an amount of U.S. money to the comparable amount of French, Italian, Canadian, or English currency based on the user response to a multiple-choice selection. Hard code the U.S. amount and user choice—that is, declared and initialize variables representing the U.S. amount and the user's choice. Test the program for various amounts and choices.

7

Data Handling Techniques and Structures

OBJECTIVES

After reading this chapter, you will be able to:

1. Understand and use linked lists.
2. Understand and use arrays.
3. Demonstrate the use of linked lists and arrays in building application involving tabular and list data structures.

OVERVIEW

Linked lists and arrays enable the efficient handling of tabular and list data structures so often found in applications. Examples of lists and arrays include customers waiting for service, documents in a print queue, shipping costs by weight and zone and high, low and closing stock prices for a thirty-one day period; and medicinal dosage by weight.

Some lists and tables are **static** in that the number of entries is fixed. Static lists and tables are modeled with an array data structure. A list of postage amounts by zone or bus fares by destination are examples of static lists. Other lists are **dynamic** in that the number of entries is ever-changing. Examples of a dynamic list are the number of documents in a print queue or number of items in an order. Dynamic lists and tables are modeled with the linked list data structure.

This chapter presents the Java statements enabling the programmatic implementation of lists and tables.

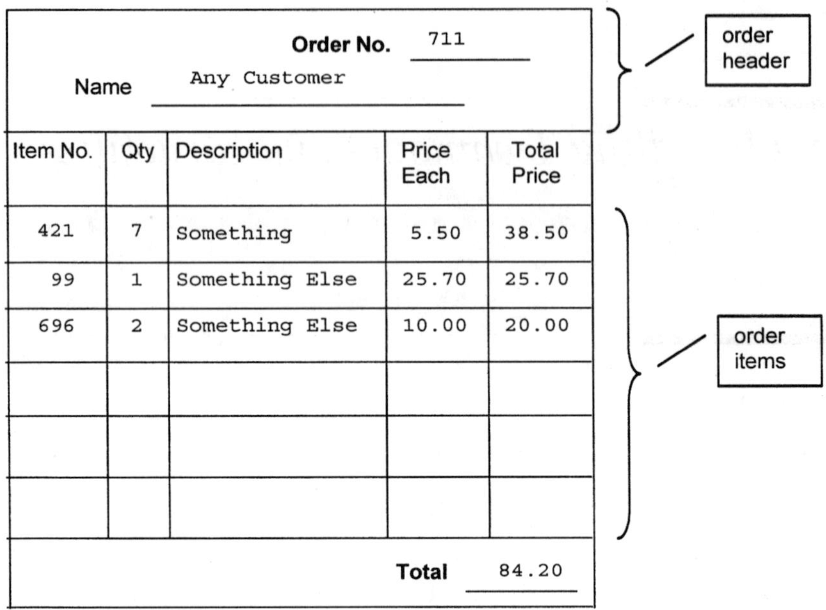

FIGURE 7.1
Simplified order.

LINKED LISTS

An order illustrates the type of data structure efficiently implemented with linked lists. A simplified order is shown in Figure 7.1. The order specifies three items. The item number, quantity, description, price each, and total price are specified for each. Although three items are shown in the order, the number of item can vary. Some customers will order one item, some will order several, and some might order some of every item stocked. Thus a structure is required that will enable the number of items to vary. Note that most order forms have space for recording a fixed number of items. The form shown has space for six items. A linked list can be used to implement a data structure containing a variable number of items. The lines in the order are referred to as the **order items.**

The order also contains other information such as invoice number, customer name, and total order amount, which are handled as a group and referred to as **order header** or **order header data.**

List Structure

A **list** can be formally described as a collection of ordered elements. In other words, a first element, followed by a second element, followed by a third element, followed by subsequent elements in order. A computer **linked list** consists of elements called **nodes,** each of which contains a reference to the next node. By convention, the last node contains a null reference to the next node. A basic list structure is shown in Figure 7.2.

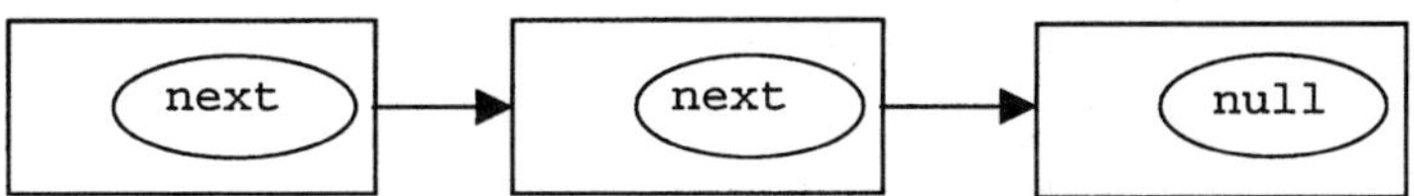

FIGURE 7.2
Nodes containing a reference to next node.

The first node is referred to as the **head** and the last node is referred to as the **tail.** Note that each list node contains a reference to the next list node. A list node is modeled by the class `Node` shown in Figure 7.3.

```
class Node {
     String nodeName;
     Node next;
     void setNext(Node n){

          next = n;
     }
     Node getNext(){
          return next;
     }
     void setNodeName(String ln){
          nodeName = ln;
     }
     String getNodeName(){
          return nodeName;
     }
}
```

pointer to next list node

FIGURE 7.3
Class modeling a linked list node.

The class contains one data member, `nodeName`. `nodeName` is a `String` data type. Class `String` is a built-in Java class that models strings and is examined in Chapter 9. Note that the class also contains a reference to the next node called `next`. Two member methods `setNext( )` and `getNext( )` are included for manipulating the `next` data member. The methods `setNodeName( )` and `getNodeName( )` enable manipulation of the data member `nodeName`. These members are representative of the members used to model a list node. In the case of an order item node, fields and methods for item number, quantity, description, price each, and total price are added.

```
class List {
     String listName;
     Node head;
     void setListName(String ln){
          listName = ln;
     }
     String getListName(){
          return listName;
     }
     void setHead(Node n){
          head = n;
     }
     Node getHead(){
          return head;
     }
}
```

pointer to first list node

FIGURE 7.4
Class modeling a container.

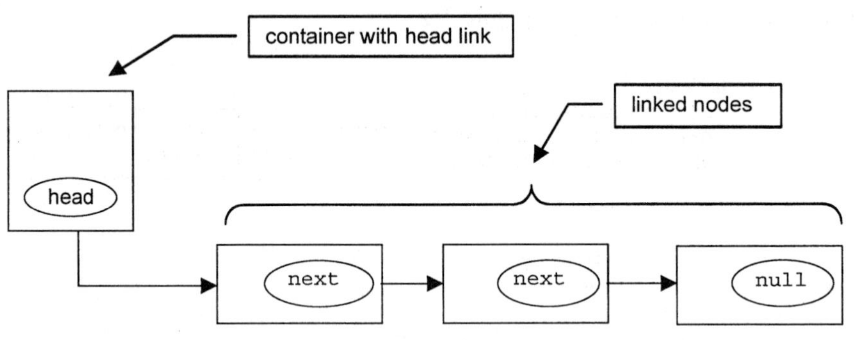

FIGURE 7.5
Basic linked list structure.

In order to use the list a reference to the first node or head is required. An object containing the reference to the list's head is referred to as a **container.** A container is modeled by the class `List`, shown in Figure 7.4. Note the `head` reference. The relationship of the head member to the linked list is shown in Figure 7.5.

Creating and Initializing a List

The statements

```
List listObj = new List();
```

```
listObj.setListName("List of Nodes");
listObj.setHead(null);
```

create an object of type `List` named `listObj`. The object's data member, `listName`, is set to "`List of Nodes`" by invoking the method `setListName`. The reference to the head of the list is set equal to `null`, indicating an empty list.

The statements

```
listObj.head = new Node();
listObj.head.setNodeName("Lou");
listObj.head.setNext(null);
```

create a `Node` object. The data member head in the `List` object `listObj` is assigned the reference value of the new node. Consequently, the node is positioned at the head of the list. The methods `setNodeName( )` and `setNext( )` are invoked to set the value of the new node's `nodeName` and `next` value. Note that the node's `next` value is set to `null`, indicating that the node is the last node in the list. The initialized list and node are shown in Figure 7.6.

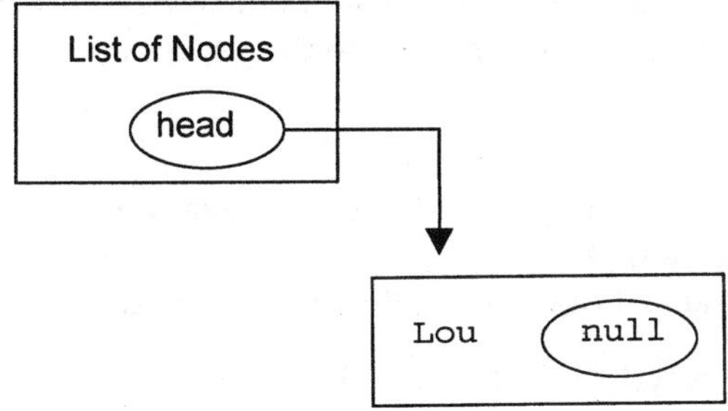

FIGURE 7.6
Initialized list with one node.

Stepping through the List

The following code can be used step through the list's nodes. First a `Node` reference, called `currentNode`, is declared and set equal to the head of the list.

```
Node currentNode = listObj.head;
```

Remember, an empty list and last node in a list are both indicated by a null reference. Thus if the `currentNode` reference is not `null`, the reference `currentNode`, at this time, points to the first node in the list—that is, the list head. The `currentNode` reference can be set to reference the next node in the list with the statement

```
currentNode = currentNode.getNext( );
```

This statement first retrieves the value of the reference to the next node by invoking the `getNext( )` method of the current node. The value returned, a reference to the next node, is assigned to the reference `currentNode`. Thus after the statement has been executed, `currentNode` references the next node in the list. The next node is now the current node. Note that executing this statement always results in the reference of the next node becoming the reference of the current node. This is true for all the nodes in the list.

Note, however, that if the reference to the next node is `null`, the current node is the last node in the list. Thus while the expression

```
currentNode!==null
```

is true, the current node is not the last node in the list. Note also that the statement is true for the reference to the head of the list as well. That is, if `currentNode` is a reference to the head of an empty list, the expression is false.

Consequently, the code

```
while (currentNode!==null) {
    currentNode = currentNode.getNext( );
}
```

will set `currentNode` to reference subsequent nodes in the list until the last node in the list is reached, in which case the expression

```
currentNode!==null
```

is false. (Remember that the `while` statement evaluates the expression before executing the loop body.)

A reference to the last node in the list is calculated using the following:

```
Node currentNode = listObj.head;
Node tail;
while (currentNode!==null) {
    tail = currentNode;
    currentNode = currentNode.getNext( );
}
```

When the looping ends the reference `tail` points to the last node in the list. This calculation is lengthy if the list is long, thus most container classes contain a reference to the list's tail as well as the list's head, thereby eliminating the need for this calculation.

The code

```
Node currentNode = listObj.head;
while (currentNode!==null) {
    System.out.println(currentNode.getNodeName( )
);
    currentNode = currentNode.getNext( );
}
```

displays the node name of each node in the list. This process can be generalized to access the fields in a node used to model a specific application.

Inserting a Node at the End of the List

Inserting a new node at the tail of the list can be viewed as a three-step process (Figure 7.7). The container and nodes in the list at the start of the insertion, Step 0, are displayed in Figure 7.7. Note that a reference to the tail of the list has been established, as described above, and that it points to the last node in the list.

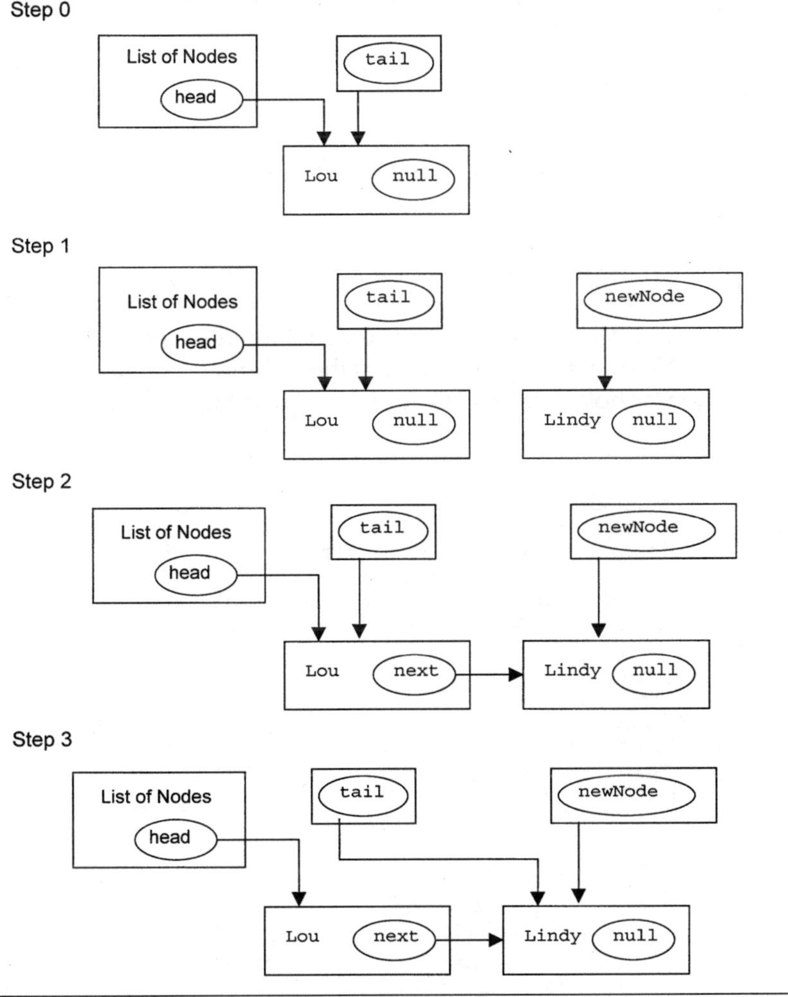

FIGURE 7.7
Inserting a node at the list tail.

The statement

```
Node newNode = new Node( );
```

declares a reference newNode, creates a new Node object, and assigns the value returned by the new instance statement to the reference newNode. This is shown as Step 1 in Figure 7.7.

The statements

```
newNode.setNodeName( "Lindy" );
newNode.setNext( null );
```

initialize the node name and next data members in the new node.

The next reference value in the last node of the list is changed to the new node with the statement

```
tail.setNext( newNode );
```

This is illustrated in Step 2 of Figure 7.7.

Finally, the reference to the tail is changed to the new node with the statement

```
tail = newNode;
```

This is represented in Step 3 of Figure 7.7.

A program illustrating the creation of a list container and the addition of two nodes, and displaying information contained in the nodes, is given in Figure 7.8a. Note that the program uses the techniques developed in the previous sections. The programs output is shown in Figure 7.8b.

```
class Node {
       String nodeName;
       Node next;
       void setNext(Node n){ next = n; }
       Node getNext(){ return next; }
       void setNodeName(String ln){ nodeName = ln; }
       String getNodeName(){ return nodeName; }
}
class List {
       String listName;
       Node head;
       void setListName(String ln){ listName = ln; }
       String getListName(){ return listName; }
       void setHead(Node n){ head = n; }
       Node getHead(){ return head; }
}
public class LinkedList {
       public static void main(String[] args) {
              // create  and initialize list
              List listObj = new List();
              listObj.setListName("List of Nodes");
              listObj.setHead(null);
              System.out.println(listObj.getListName() );
              // add first node
              listObj.head = new Node();
              listObj.head.setNodeName("Lou");
```

```
                listObj.head.setNext(null);
                // calculate reference to last node
                Node tail;
                Node currentNode;
                tail = currentNode = listObj.head;
                while (currentNode!=null){
                       tail = currentNode;
                       currentNode = currentNode.getNext();
                }
                // insert node at list tail
                Node newNode = new Node( );
                newNode.setNodeName( "Lindy" );
                newNode.setNext( null );
                tail.setNext( newNode );
                tail = newNode;
                // list nodes
                currentNode = listObj.head;
                while (currentNode!=null){
                       System.out.println(currentNode.getNodeName()
);
                       currentNode = currentNode.getNext();
                }
        }
}
```

FIGURE 7.8A
Java program to create and initialize a list, add two nodes, and display the node names.

```
List of Nodes
Lou
Lindy
```

FIGURE 7.8B
Output of Java program to create and initialize a list, add two nodes and display the node names.

Constructors Add to Programming Efficiency

Constructors that will improve programming efficiency can be added to the container and node classes. For example, the `Node` constructor

```
Node ( String nn ){
    nodeName = nn;
    next = null;
}
```

initializes the `nodeName` to the parameter `nn` and the `next` reference to `null`.

A similar constructor in the `List` class initializes the `listName` and sets the `head` and `tail` references to `null`.

```
List ( String ln ){
    listName = nn;
    head = null;
    tail = null;
}
```

The tail reference, which eliminates the need to step through the list to find the tail every time a node is inserted at the tail, is also initialized.

Insertion Methods Add to Programming Efficiency

A method to insert a node at the tail of the list is shown in Figure 7.9. The method consists of two blocks of code. One block is executed when the list is empty. If the list is empty, the expression

```
head == null
```

evaluates as true and the statement

```
head = tail = new Node( nn );
```

is executed. Note that the Node constructor is called by the new instance creation. The constructor initializes the node name to the passed parameter and the next reference to null.

```
void InsertAtTail( String nn ) {
      if ( head == null )
            head = tail = new Node( nn );
      else {
            Node newNode;
            newNode = new Node( nn );
            tail.setNext( newNode );
            tail = newNode;
      }
}
```


FIGURE 7.9
Method to insert a node at the list tail.

When the list has nodes, that is head is not equal to null, the other code block is performed. First, a new Node reference newNode is declared. Then another instance of the node class is created. The Node constructor is passed the name reference, nn, as a parameter. Following creation and initialization, a value returned by the new instance expression is assigned to the reference newNode. At this point, the new Node object exists but it is not part of the existing list. Remember, the tail reference points to the last node in the existing list. The setNext method is called for this instance with the statement

```
tail.setNext( newNode );
```

Note that the reference to the new node is passed as a parameter. When the method has finished executing, the next reference in the existing list points to the new node, which becomes the last node in the list. The tail reference is modified to point to the new node, which is now the last node in the list, with the statement

```
tail = newNode;
```

A program using the constructors and `InsertAtTail` method to construct a list is shown in Figure 7.10a. Note the `DisplayList( )` method that has been added to the `List` class. The program output is shown in Figure 7.10b.

```
class Node {
      String nodeName;
      Node next;
      Node ( String nn ) {
           nodeName = nn;
           next = null;
      }
      void setNext(Node n){ next = n; }
      Node getNext(){ return next; }
      void setNodeName(String ln){ nodeName = ln; }
      String getNodeName(){ return nodeName; }
}
class List {
      String listName;
      Node head;
      Node tail;
      List ( String ln ) {
           listName = ln;
           head = tail = null;
      }
      void setListName(String ln){ listName = ln; }
      String getListName(){ return listName; }
      void setHead(Node n){ head = n; }
      Node getHead(){ return head; }
      void InsertAtTail( String nn ) {
           if ( head == null )
                head = tail = new Node( nn );
           else {
                Node newNode;
                newNode = new Node( nn );
                tail.setNext( newNode );
                tail = newNode;
           }
      }
      void DisplayList( ) {
           Node  currentNode = head;
           while (currentNode!=null){
                System.out.println(currentNode.getNodeName() );
                currentNode = currentNode.getNext();
           }
      }
}
public class LinkedList
{
      public static void main(String[] args)
      {
           // create  and initialize list
           List listObj = new List("List of Nodes");
           System.out.println(listObj.getListName() );
           // add first node
           listObj.InsertAtTail("Lou");
           // insert nodes at list tail
           listObj.InsertAtTail("Lindy");
           listObj.InsertAtTail("Larry");
           // list nodes
```

```
            listObj.DisplayList();
        }
}
```

FIGURE 7.10A
Java program to create and display a list that uses constructors and `InsertAtTail` method.

```
List of Nodes
Lou
Lindy
Larry
```

FIGURE 7.10B
Output of Java program to create and display a list using constructors and `InsertAtTail` method.

ARRAYS

A table of train fares cost by destination and type of ticket is an example of a data structure that can be modeled by an array. Consider the simplified fare table shown in Figure 7.11.

ZONES	TICKET TYPE	ONE-WAY	ROUND TRIP EXC	WEEKLY
	TERMINAL	HOB/NWK	HOB/NWK	HOB/NWK
9	AVENEL	3.35	5.00	28.50
10	WOODBRIDGE	3.60	5.50	30.50
12	PERTH AMBOY	4.45	6.75	38.00
13	SOUTH AMBOY	5.00	7.50	42.50
15	MATAWAN	5.75	8.75	49.00
16	HAZLET	6.10	9.25	52.00
17	MIDDLETOWN	6.25	9.50	53.00
18	RED BANK	6.40	9.50	54.50
19	LITTLE SILVER	6.40	9.50	54.50
20	LONG BRANCH	6.65	10.00	56.50
21	ASBURY PARK	6.65	10.00	56.50
22	BELMAR	6.95	10.50	59.00
23	BAY HEAD JUNCTION	6.95	10.50	59.00

FIGURE 7.11
Fares table.

The table entries are called **elements.** For example, AVENEL is the value of an element. Likewise, 54.50 is also the value of an element. Note that some of the elements are type integral, some are type string, and others are type double. A table of the type shown is **static** in that the number of elements does not change. For example, rarely are new stations built and rarely are the ticket types changed. This type of data structure is effectively modeled by an array data type. Note that the static character of the data structure does not preclude change in the value of the elements. In fact, most fares change seasonally, reflecting operating cost changes as well as governmental tax and subsidy changes.

Array Structure

An **array** is an ordered arrangement of elements or variables referred to as the **components** of the array. The terms "elements" and "components" are used interchangeably. All elements of an array must be of the same type. An array is graphically represented in Figure 7.12. The array shown consist of seven `int` elements.

11	31	42	99	7	654	21

FIGURE 7.12
Array structure of seven type `int` elements.

The component variables of an array do not have names but are accessed by their position in the array. The number of elements in an array is said to be the **length of the array.** Thus, the array shown has length 7. The number of elements can be zero, in which case the array is said to be **empty** and of length zero.

Java arrays are objects. The type of a Java array is the same as the type of the elements. Note that the component type may itself be an array type. The elements of an array can be any type, primitive or reference. Thus there can be arrays of integers, arrays of Booleans, or arrays of references to objects.

A single-dimension array is often referred to as a **vector.**

Declaring an Array Variable

An array variable holds a reference to an array object—that is, an array data structure. An array declaration begins with a data type, followed by the reference variable identifier, and the **element operator** or **brackets** (**[ ]**) either prefixing or suffixing the identifier. The following statement declares an array variable of type `long`:

```
long [ ]  longArray;
```

An array variable of `String` objects is created with the statement

```
String [ ]  stringArray;
```

The brackets are often referred to as the **index operator.** Note that the brackets can suffix the reference, and thus the preceding array variables can be declared with the following statements:

```
long  longArray [ ];
String  stringArray [ ];
```

Note that the preceding declarations did not create array objects (that is, allocate space in memory). Note also that the declarations did not specify the array length.

Allocating Memory

Memory for an array is allocated with the `new` operator. For example, the statement

```
long longArray = new long[ 7 ];
```

allocates memory for seven components of type `long`. The value assigned by the new instance expression is assigned to reference `longArray`.

The length of an array cannot be changed once it is created. However, an array variable can be assigned to an array object of different length.

The memory allocation is also made with the following statements:

```
long longArray [  ];
longArray = new long[ 7 ];
```

The allocated array object is illustrated in Figure 7.13. Java array components are initialized when memory is allocated. The initial value depends on type (Table 7.1).

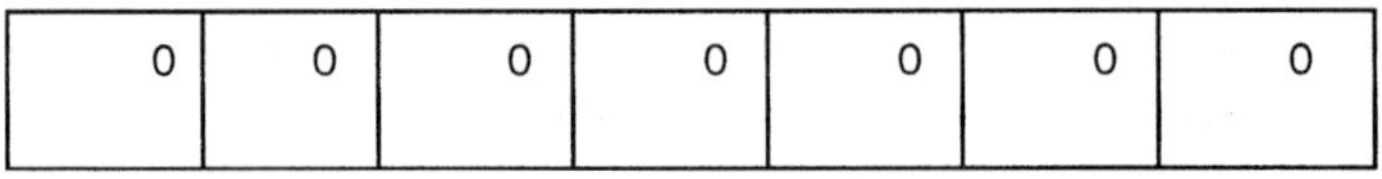

0	0	0	0	0	0	0

FIGURE 7.13
Type `long` array with seven elements with default initialization.

TABLE 7.1 DEFAULT INITIAL VALUE OF ARRAY COMPONENTS

Type	Value
`byte`	`0`
`boolean`	`false`
`char`	`0`
`long`	`0`
`double`	`0`
`Reference`	`null`

Array initialization may be specified in the declaration, as shown in Figure 7.14a. In this case the declaration creates a seven-element array with initial values as specified in the **array initializer,** which is a comma-separated list of expressions enclosed by braces.

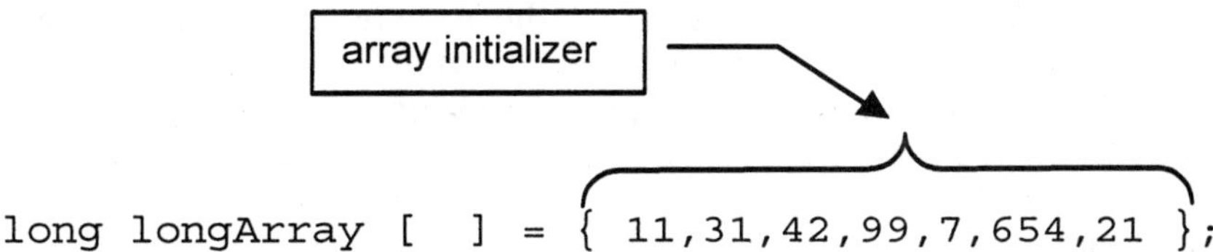

FIGURE 7.14A
Creating and initializing an array.

11	31	42	99	7	654	21

FIGURE 7.14B
The array created and initialized.

The length of the array is equal to the number of expressions. The reserved word `new` is implied. The array created is shown in Figure 7.14b.

Accessing Array Components

An element of the array is accessed by an array access expression consisting an array reference followed by an **indexing expression** enclosed in brackets (`[ ]`). The indexing expression is often simply called the array **index** and its value the **index value.** The first array element has the index value 0. Thus, the last element of a five-element array has index value 4. Accessing of the third element of the array `longArray` is illustrated in Figure 7.15.

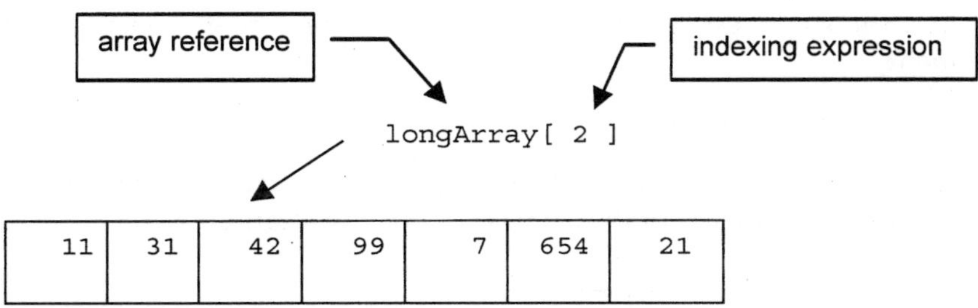

FIGURE 7.15
Accessing the third element of an array.

This numbering arrangement has caused many programming errors! Note that the third array element is accessed with the index value 2.

Arrays must be indexed by `int` values. Note that `byte`, `char`, and `short` are valid types as they are promoted to `int`.

Values of array components are assigned by simply assigning a value to a referenced element. The statement

```
longArray [ 2 ] = 96 ;
```

assigns the value 96 to the third element of the array referenced by `longArray`. An attempt to access an array with an index value less than zero or greater than or equal to the length of the array will result in a run time exception: `IndexOutOfBoundsException`. (Exceptions are examined in Chapter 8.)

Arrays of Objects

Arrays of objects can also be declared. Memory allocation of an array of objects only creates an array of references to the objects. Note the references are all assigned the `null` value. For example, the statements

```
String stringArray [  ];
stringArray = new String[ 3 ];
```

create an array of type `String` references and allocate memory for three components. The array is shown graphically in Figure 7.16. The references are initialized to `null`.

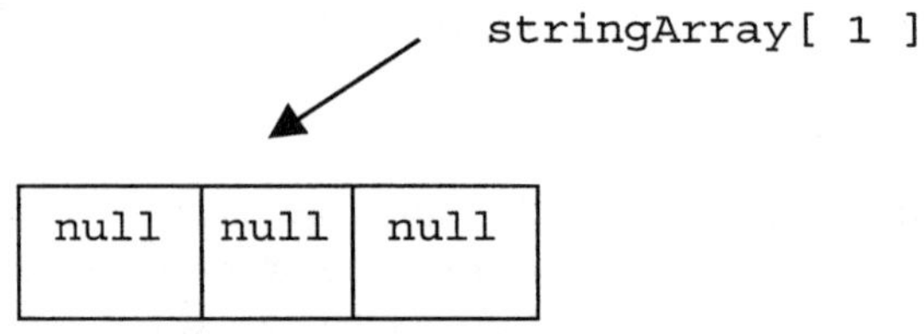

FIGURE 7.16
Array of string object references.

Objects referenced by the array references are created with the new operator and their address assigned to the array elements. For example, the statement

```
String stringArray [ 1 ] = { "Larry" };
```

creates a string object containing the string `"Larry"` and assigns the reference to the second component of the array `stringArray` (Figure 7.17).

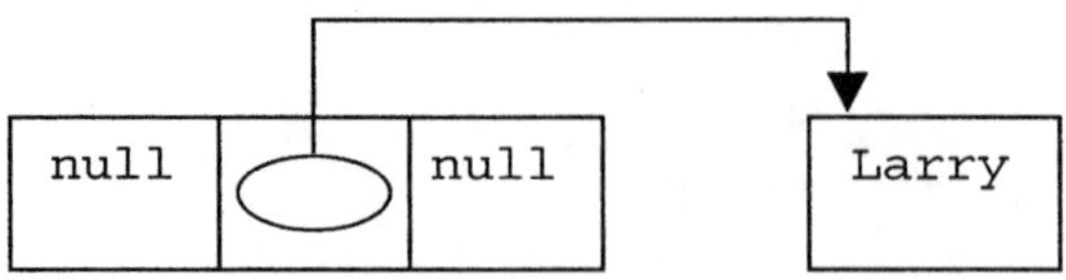

FIGURE 7.17
Component two of the array references a string.

> *An array is not a string!* The Java array is not a sequence of characters terminated with a `null` as found in C++. In Java, *a string is a String!*

Multi-Subscripted Arrays

Java array elements can themselves be arrays, enabling the modeling of multidimensional data structures. For example, consider the statements

```
int tableArray [  ][ ];
tableArray = {{11,12,13,13},{21,22,23,24},{31,32,33,34}};
```

An array `tableArray` is created with type array components. The components are type `int` arrays. By convention the first index associates with the highest level component. Thus the first index or subscript refers to the array components. The second index or subscript, refers to the next level component, in this case the `int` elements. The data structure modeled by `tableArray` is shown in Figure 7.18.

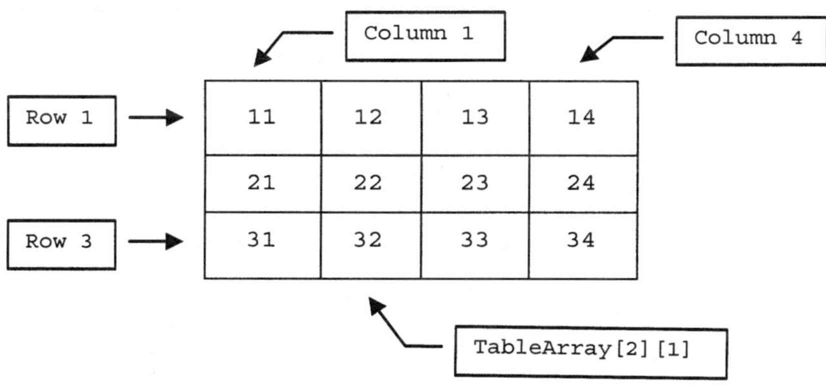

FIGURE 7.18
Graphical interpretation of the `tableArray` object.

By convention the first table row and column are identified as row 1 and column 1. However, the first element of an array is accessed with the subscript value 0. Thus the element accessed with the `tableArray[2][1]` corresponds to the tabular element in row 3 and column 2.

> Although Java does not support multidimensional arrays, multi-subscripting enables the modeling of multidimensional data structures.

Using Arrays

Finding the largest element in an array and exchanging it with the last element is a basic and often-used procedure. Consider the array

```
int searchArray[ ] = {11,42,75,19,53,35,01,13,13};
```

The largest element is 75 and the last element is 13. The `FindAndExchange` class, Figure 7.19a, loops through the elements of the array searching for the largest element. Whenever an element larger than the current element is found, the value and location are saved in temporary variable: `largestElement` and `largestElementLocation`. After all the elements in the array have been compared with the variable `largestElement`, the value of the last element in the array is moved to the location of the largest element and the largest element value is moved to the last element of the array.

```
public class FindAndExchange
{
      public static void main(String[] args)
      {

           int searchArray[ ] = {11,42,75,19,53,35,01,13,13};

           int largestElement,largestElementLocation;
           largestElement = searchArray[0];
           largestElementLocation = 0;

           System.out.println("Length: " + searchArray.length);

           for(int i =0;i<searchArray.length;i++) {
                if (searchArray[i] > largestElement){
                     largestElement = searchArray[i];
                     largestElementLocation = i;
                }
           }

           searchArray[largestElementLocation] =
                searchArray[ searchArray.length-1 ];
           searchArray[ searchArray.length-1 ] = largestElement;

           for(int i =0;i<searchArray.length;i++) {
                System.out.println(searchArray[i]);
           }
      }
}
```

FIGURE 7.19A
`FindAndExchange` class.

The array is declared, and memory allocated and initialized with the statement

```
int searchArray[ ] = {11,42,75,19,53,35,01,13,13};
```

The variables `largestElement` and `largestElementLocation` are defined.

```
int largestElement,largestElementLocation;
largestElement = searchArray[0];
largestElementLocation = 0;
```

At the start the first element of the array is the largest element, so far. Hence, the value of the first element is assigned to the variable `largestElement` and the array index value 0 is assigned to `largestElementLocation`.

The length of the array is output using the **length** member of the `array` class. All Java arrays are subclasses of the `array` class and thus inherit the methods of the array class.

```
System.out.println("Length: " + searchArray.length);
```

The number 9 is output.

The for statement

```
for(int i =0;i<searchArray.length;i++) {

}
```

loops through the subscript values used to access the elements of the array. The `for` loop starts with the value `i = 0`. The value of `i` is incremented by one until as long as `i<searchArray.length`. Remember that the last element of the array is accessed by a subscript one less than the length.

In the loop, the value of the array element is compared with the value of the `largestElement`.

```
if (searchArray[i] > largestElement){
    largestElement = searchArray[i];
    largestElementLocation = i;
}
```

When a larger value is encountered, the value is assigned to the variable `largestElement` and its location is assigned to `largestElementLocation`.

When the loop is completed, all of the elements have been examined and the largest value and location of the largest value are contained in `largest-Element` and `largestElementLocation`. At this point,

```
searchArray[largestElementLocation] =
    searchArray[ searchArray.length-1 ];
searchArray[ searchArray.length-1 ] = largestElement;
```

the value of the last element of the array is moved to the location of the largest element uncovered during the search. Then the largest value uncovered during the array is moved to the last element.

The array element values are displayed with the statements

```
for(int i =0;i<searchArray.length;i++) {

    System.out.println(searchArray[i]);
}
```

See Figure 7.19b for the output of the FindAndExchange class.

```
Length: 9
11
42
13
19
53
35
01
13
75
```

FIGURE 7.19B
Output of `FindAndExchange` class.

Consider now a similar find and exchange over the element accessed with subscript 0 through `length-2`. In this case the loop results in the examination of all the array element except the last, which is the largest. The largest found is thus the next to largest value in the array. This element is exchanged with the next to last array element. At this point, the last element in the array is the largest and the next to last element is the next to largest. This process can be extended to the elements of the array from the first to `length-3`. The results of sequential finds and exchanges are illustrated in Figure 7.20.

	Examination Range				
Start	0 to length-1	0 to length-2	0 to length-3	0 to length-4	0 to length-5
11	11	11	11	11	11
42	42	42	01	01	01
75	13	13	13	13	13
19	19	19	19	19	13
53	53	13	13	13	19
35	35	35	35	35	35
01	01	01	42	42	42
13	13	53	53	53	53
13	75	75	75	75	75

FIGURE 7.20
Results of subsequent find and exchanges after examination.

If the process continues, the array will be sorted in **ascending sort order,** that is smallest first to largest last.

SUMMARY

1. Linked lists can be used to model dynamic list applications.
2. Arrays can be used to model tabular applications.

KEY WORDS AND CONCEPTS

`[ ]`
array
array initializer
ascending sort order
brackets
components
container
dynamic (lists and tables)
element operator
elements
empty
head
index
index operator
index value
indexing expression
length (member of array class)
length of the array
linked list
nodes
order items
order header
order header data
`static`
static (lists and tables)
tail
vector

REVIEW QUESTIONS AND EXERCISES

1. Modify the program in Figure 7.9 to perform the following:
   a. Add a node at the head
   b. Delete a node at the tail
   c. Insert nodes in node name order
   d. Display the node names backward
   e. Delete a node with a user input node name (the program should handle the case in which a node with the name input does not exist)
2. What index value is used to access the first element in an array? __________
3. Which of the following are valid array index types?

```
byte
boolean
char
long
double
int
```

4. Given

```
String stringArray [  ];
```

```
stringArray = new String[ 3 ];
```

and

```
int i;
i = 3;
```

is the assignment valid? Why?

5. Review the progress of sorting shown in Figure 7.20. What is the value of `n` such that the array will be sorted after the examination of the elements `0` through `length-n`? Is it when `n` is equal to length? Is it when `n` is equal to `length-1`? Is it when `n` is equal to `length-2`? Explain your answer.
6. Revise the `FindAndExchange` program to sort the array elements in largest first and smallest last order, commonly called **descending sort order**.

PRACTICE WITH VISUAL J++

Implement the order as a linked list. Include the order header items in the list container. Each node should correspond to an item in the list and include the item number, quantity, price each, and total price. (Note that the item description is not included.)

Build a menu-driven user interface that enables the user to perform the following operations:

- Enter order item.
- Delete order item.
- Display order items.
- Compute the total of the items ordered.

8

Exceptions

OBJECTIVES

After reading this chapter, you will be able to:

1. Understand an alternative to the traditional method of handling exceptional conditions, which generally resulted in program termination.
2. Throw and catch exception objects in Java programs.
3. Understand the Java Throwable, Errors, Exception, and Runtime-Exception classes.
4. Use the Exception subclasses to enable recovery from exceptional conditions.
5. Develop classes inherited from class Throwable.

OVERVIEW

Traditionally, exceptional conditions have been handled by the system with program termination. Java provides an alternative to the traditional method of handling exceptional conditions. Instead of passing exceptions to the system, control is passed to an exception handler. In fact, when the exceptional condition is encountered, an exception is *thrown*. The object is *caught* by an exception handler. The handler is passed an object characterizing the exceptional condition. The object is type `Throwable` or a subclass of `Throwable`. Exception subclasses have been defined regarding exceptional condition most often encountered by programmers. These conditions include arithmetic exceptions such as divide by zero, input and output exceptions such as end of file condition, and data manipulation exceptions such as array indexes that are negative or exceeding the size of the array. Exception class can be derived to reflect custom exception object requirements. Implementation of the exception handling is implemented with the `try` statement including the `catch` and `finally` clauses.

TRADITIONAL EXCEPTION HANDLING

Problems can occur during program execution. For example, most programmers will test data input by a user to ensure that data used in an arithmetic operation will not lead to a divide by zero situation. Dividing by zero will lead to problems that will result in program abortion. However, in an effort not to waste time, programmers rarely test data from special files or databases that promise the data will not result in divide by zero problems. However, the exception is the rule, and eventually the errant data will surface and a divide by zero problem will abort. Problems of this type arise at the most inopportune times, generally when the data processing or computation must be done. Generally when the program aborts, the entire program must be run again. This can be done only when the errant data has been located and corrected.

Other problems arise when needed resources are exhausted. This situation often arises when a memory resource is depleted. The memory resource can be a file or database as well as the system's memory. Most computer users have encountered the abortion of a copy procedure when the computer's A-drive becomes full. In this case the system simply aborts the copy process and the user only looses the time spent in the failed copy process. When the system runs out of memory, however, the consequences can be more dire in that the system can stop operating. In these situations the user can loose all of the computations performed. Who has not lost a word processing document during an unplanned system shutdown?

Still other problems can arise due to the system's inability to find and load needed program components. This situation to leads to system shutdown.

In all these scenarios, the problem has resulted in an abnormal termination of the program. When the program encounters a problem the system is notified. It is impossible for the system to know the rules for solving problems associated with all programs; thus it simply stops the program abruptly, a condition known as **abnormal termination.** Often the system prints a cryptic message describing the possible problem and list the contents of selected memory locations.

In any case the result is abnormal termination of the program, always resulting in a loss of time and a need to rerun the program, and often a permanent loss of data.

AN ALTERNATE EXCEPTION-HANDLING METHOD

When a problem is encountered, an exception is thrown. When the exception is thrown, the program block throwing the exception expires. The exception is caught by a handler that assumes control. Control is not passed back to the block throwing the exception (Figure 8.1).

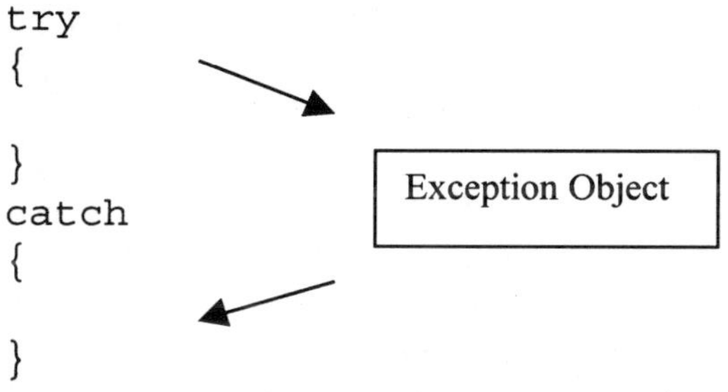

FIGURE 8.1
Throwing and catching an exception.

Note that program control is not transferred to the system, as in the traditional manner, but instead to a block of code within the application. This enables the execution of recovery procedures.

> When an exceptional condition is transferred, control is not passed to the system but to a block of code within the application enabling recovery.

HANDLING EXCEPTIONS IN JAVA

The Java exception handling enables separation of error-handling procedures from the application-related procedures. When exceptional conditions are encountered, an exception object is thrown from the point at which the exception occurred to the point at which the exception is handled. The exception is thrown in a *try block*. The exception is caught in a *catch block*. When an exception is thrown, control passes from the try block to the catch block in which the exception is caught. Note that an exception class object is passed as the parameter of the catch. This object contains information about the exception.

Throwable Objects

All classes modeling exceptional conditions that can be thrown, by the `throw` statement, are derived from the class **`Throwable`**. The class contains a trail referred to as *stack trace* that summarizing methods called leading to the exception condition. The class also contains a detailed message regarding the exception. These data are accessed with member methods (Table 8.1).

TABLE 8.1

Methods of the class `Throwable`

Name	Explanation
`getMessage( )`	Print the detail message, generally describing the exception.
`PrintStackTrace( )`	Print the back trace to the standard stream.

The keyword `throw` is used within a try block to indicate an exceptional condition that requires normal processing to be stopped and exception procedures to be undertaken. When a `throw` statement is encountered, an exception object characterizing the exceptional condition is created and program control is immediately transferred to the nearest `catch` statement that can handle the exception.

The statement

```
throw new Throwable("Thrown in method2\n");
```

throws an object of type `Throwable`. The string `"Thrown in method2"` is assigned to the thrown objects method by the class `Throwable` constructor.

The `throws` keyword is used in a method declaration to list exceptions that the method can throw. For example,

```
public static void method2() throws Throwable
{
    throw new Throwable("Thrown inmethod2\n");}
}
```

declares `method2( )` that throws an exception of type `Throwable`. Note, a method can be declared to throw an exception even if the exception is not explicitly thrown.

```
public static void method1() throws Throwable
{
    method2();
}
```

The object must be thrown within a `try` block.

```
    try  {
                method1();
    }
```

Even though the `throw` statement is not explicitly stated in the try block, `method2( )` which is called by `method1( )` can throw the exception. Note that both `method1( )` and `method2( )` have been declared with the `throws Throwable clause`.

The object is caught in the nearest `catch` statement. The `catch` parameter clause must specify the type of the object. For example

```
catch (Throwable  e)  {
      System.out.println("getMessage():  "
            +  e.getMessage()
                  + "printStacktrace():  ");
      e.printStackTrace();
   }
```

Typically, procedures to recover from the error are coded in the catch block. In this case the thrown object's specific message and backtrail are displayed.

A complete program illustrating the throwing and catching of a thrown object is presented in Figure 8.2a. The program's output is shown in Figure 8.2b. The message `"Thrown in method2" is passed` to the `Throwable` constructor to initialize the message in the object shown. This message is accessed by invoking the `getMessage( )` member with the object caught in the `catch` block (lines 10 through 15) and displayed (line 12). Note that the backtrail clearly indicates that the exception occurred in `method1( )` of the class Test. Note also that the backtrail also indicates that `method2( )` was called by `method1( )` in `main( )` of the class Test. The backtrail is accessed by invoking the `printStackTrace()` (line 14).

```
import java.lang.Throwable;
 public class Test
 {
       public static void main(String args[])
       {
             try  {
                         method1();
             }
             catch (Throwable  e)  {
                   System.out.println("getMessage():  "
                         +  e.getMessage()
                               + "printStackTrace():  ");
                   e.printStackTrace();
             }
       }
       public static void method1() throws Throwable
       {
             method2();
       }
       public static void method2() throws Throwable
       {
             throw new Throwable("Thrown in method2 \n");
       }
 }
```

FIGURE 8.2A
The throwing and catching of a thrown object.

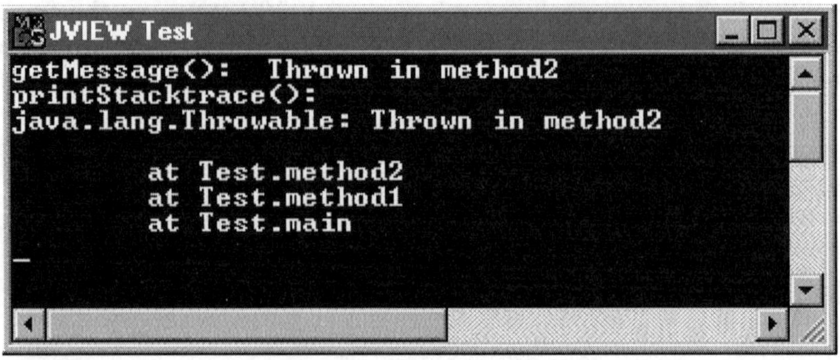

```
JVIEW Test
getMessage():  Thrown in method2
printStacktrace():
java.lang.Throwable: Thrown in method2

        at Test.method2
        at Test.method1
        at Test.main

_
```

FIGURE 8.2B
The throwing and catching of a thrown object—output.

Extending Class `Throwable`

Consider the point in an application at which an exceptional condition might be encountered. At that point an object should be thrown that can be caught and control transferred to procedures that may enable resolution of the exceptional condition. Consider the class `MyThrowable` that defines the object to be thrown. The class `MyThrowable` is defined with the statements

```
class MyThrowable extends Throwable {
    MyThrowable(){ super(); }
    MyThrowable(String s) { super(s); }
}
```

Note that `MyThrowable` extends the class `Throwable` and thereby inherits the members of `Throwable`. The class has two constructors enabling the specification of an object with no message or an object with a message, as a `String`, at instantiation.

An object `MyThrowable` is thrown with the statement

```
throw new MyThrowable("MyThrowable Exception Con-
dition");
```

The object must be thrown within a `try` block. The object is caught in the nearest `catch` statement. The `catch` parameter clause must specify the type of the object. For example,

```
    catch (Throwable t) {
        System.out.println( "Exception Thrown : "
+
                t.getClass() +
                "\nException Message: " +
                t.getMessage());
    }
```

Program control transfers to the `catch` block. Note that the member functions inherited from the class Throwable can be used to determine the class and message of the object thrown. An example illustrating the use of a defined `Throwable` class is illustrated in Figure 8.3.

```
class MyThrowable extends Throwable {
      MyThrowable(){ super(); }
      MyThrowable(String s) { super(s); }
}

public class TestThrowable {
      public static void main(String[] args) {
           try {
                throw new MyThrowable("MyThrowable Exception
                            Condition");
           }
           catch (Throwable t) {
                System.out.println( "Exception Thrown : " +
                            t.getClass() +
                            "\nException Message: " +
                            t.getMessage());
           }
      }
}
```

FIGURE 8.3
Defining and using a `Throwable` object.

Class `Exception`

The class `Exception` is derived from the class `Throwable`. Class **`Exception`** objects are thrown for exceptional conditions that can be programmatically handled by an application. Examples of exceptions that can be handled programmatically include divide by zero, improper array, and end of file conditions.

Exceptions that can be checked by the compiler are called **checked exceptions.** Checked exceptions commonly encountered in input and output operations of application programs are summarized in Table 8.2.

TABLE 8.2
Selected checked exception classes in the java.io.IOException package

Exception	Description
EOFExceptions	End of file encountered.
FileNotFoundException	Specified file not found.

Checked exception classes are also defined in the `java.lang` and the `java.net` packages.

Typically, new exception classes are defined as subclasses of `Exception` in order to take advantage of Java's compile-time exception handler checking.

Class `RuntimeException` is derived from the class `Exception`. Objects of the class `RuntimeException` and its subclasses are thrown for conditions that normally are handled by the application program, but because of their nature cannot be checked at compile time. Selected `RuntimeException` subclasses are summarized in Table 8.3.

TABLE 8.3
Selected `RuntimeException` classes

Class	Description
`ArithmeticException`	Arithmetic situations such as divide by zero.
`ArrayStoreException`	Stores an inappropriate type to an array.
`IllegalArgumentException`	Invalid method argument.
`IndexOutOfBoundsException`	Index out of range.

Handling Exceptions

The statements that might give rise to an exception must be enclosed within a `try` statement. The thrown exception is caught in the nearest `catch` block. An example illustrating the handling of an exception is shown in Figure 8.4a. The program produces the output shown in Figure 8.4b.

```
import java.lang.ArithmeticException;
public class Test
{
      public static void main(String args[])
      {
           try  {
                     int  no1 = 1, no2  =  0;
                     no1 / no2;
           }  catch (ArithmeticException  e)  {
                System.out.println("exception:  "  +
e.getMessage());
                e.printStackTrace();
           }
      }
}
```

FIGURE 8.4A
Handling an exception

Another example is provided in Figure 8.5. The program produces the following output:

```
class java.lang.ArithmeticException
null
```

Statements that give rise to exceptions must be enclosed in a `try` block and the thrown object caught in a `catch` block.

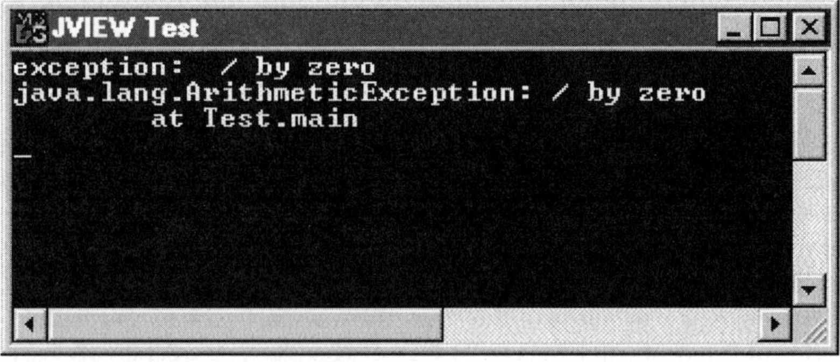

FIGURE 8.4B
Handling an exception—output.

```
import java.io.*;
class Test {
      public static void main(String[] args)
                                          throws IOException {
            try {
                  int q = 10;
                  int d = 0;
                  int r = q/d;
                  int    x = System.in.read();
            }
            catch (Exception e) {
                  System.out.println(e.getClass() + "\n" +
                        e.getMessage());
            }
      }
}
```

FIGURE 8.5
Handling an exception.

The class `Error` is also derived from the class `Exceptions`. Class **`Errors`** objects are thrown for exceptions that are generally not programmatically correctable; for example, loading errors, linkage errors, and virtual machine errors such as out of memory errors, which cannot be corrected in an application program and call for program termination.

`Try` Statement

Consider a `try` block in which data are input, arithmetically manipulated, and assigned to an array element. Presumably, the data are valid, but presumptions are like assumptions in that reality often falls short of the idealization. In this case, a variety of exceptions can be thrown: input[nd]output exceptions, array assignments exceptions, and arithmetic exceptions. The **`try-catch`** statement provides for selecting recovery procedures based on the type of exception thrown. In addition, a `finally` clause provides for specifying a block of code that will be executed after the try block or any catch block that might be executed.

First consider the try-catch without the finally clause. If the try block completes without an exception, the next sequential statement follows. If on the other hand an exception is thrown, control passes to the nearest catch block. The catch blocks are also referred to as **exception handlers.** Each catch has one parameter, referred to as an **exception parameter,** that consists of the type of the exception with which the block is associated and an identifier that is used to reference the exception object caught. The parameter type must be class Throwable or a subclass of the class Throwable. In the example below, the parameter of the first catch is type ArithmeticException. When an ArithmeticException is thrown in the try block it is caught by the catch block whose parameter specifies ArithmeticException. When an exception is caught, execution continues as stated in the catch block. The parameter identifier, called an exception parameter, is used to access the thrown object in the catch block. The scope of the exception parameter is the catch block in which it declared.

If on the other hand, the exception thrown is type MyException, it is processed in the catch block with the MyException parameter type (Figure 8.6).

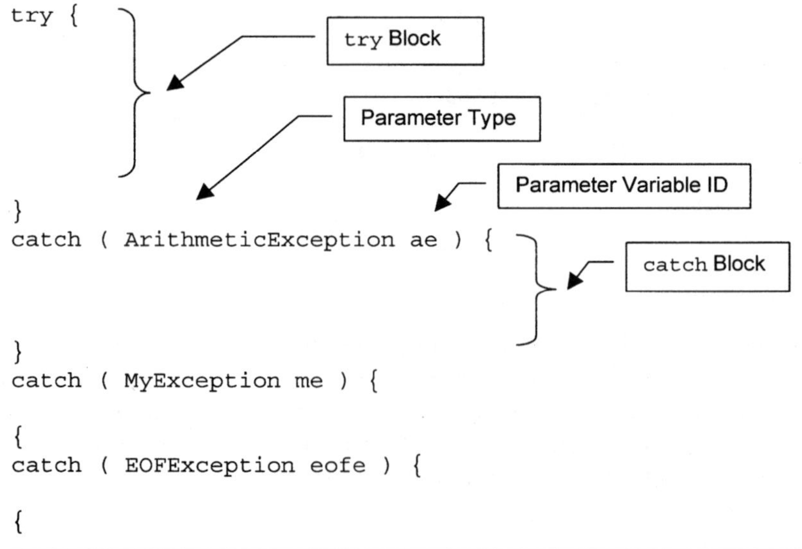

FIGURE 8.6
catch block with MyException parameter type.

Consider now the try-catch with a finally clause. If the try block completes normally, that is an exception is not thrown, then the finally block is executed. That is, after normal completion of the try block, program control transfers to the finally block. When the finally block is completed, control transfers to the next statement. If, however, an exception is thrown, control passes to the appropriate catch clause. After completion of the catch block,

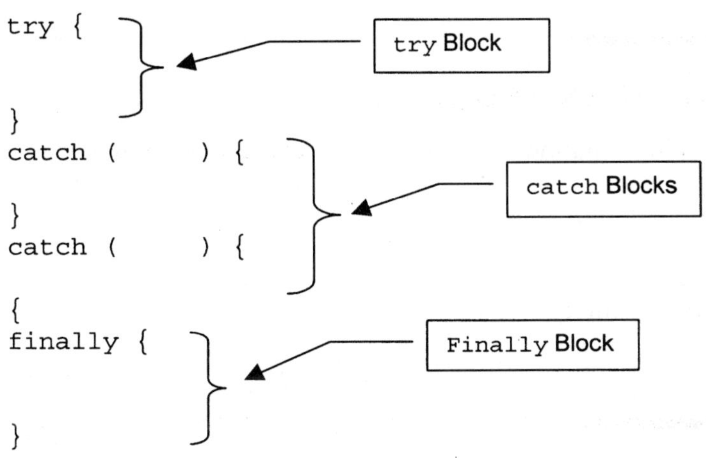

FIGURE 8.7
Illustrating the `finally` block.

program control is transferred to the `finally` block. When the `finally` block is completed, control transfers to the next statement (Figure 8.7).

> When errors cannot be handled in a simple manner, throw an exception!

SUMMARY

1. `Try` and `catch` blocks are the basis for exception handling.
2. Exceptions are thrown in a `try` block.
3. Class `Exception` objects represent conditions that can be handled by the application program.
4. Exceptions are caught by a `catch` block with an exception parameter associated with the exception.
5. Class `Throwable` represents exception conditions that can be thrown.
6. Checked exceptions are checked by the compiler.
7. Class `RuntimeExceptions` objects are thrown for conditions that cannot be checked by the compiler.
8. `try`, `catch`, and `finally` keywords enable exception handling.

KEY WORDS AND CONCEPTS

abnormal termination
`catch`
checked exception
`Errors`
`Exception`
exception handlers
exception parameter
`finally`
`throw`
`throws`
`Throwable`
`try`

REVIEW QUESTIONS AND EXERCISES

1. List three different types of problems that can lead to abnormal termination.
2. Which of the following statements can throw exceptions? Why?

```
cent = (far - 32) * 212 / 100;
area = 3.14 * r * r;
whatever = largeArray[ iDaysfromStart];
buf = System.in.read( );
arrayOfObjects[ 21 ] = (newType) CastableDatum;
```

3. What is output?

```
public class TestThrowable
{
    public static void main(String[] args)
    {
         try {
              throw new ArithmeticException("MyThrowable
        Exception Condition");
         }
         catch (Throwable t) {
              System.out.println( "Exception Thrown : "
                      + t.getClass() + "\nException Mes-
               sage: " +
                      t.getMessage());
         }
    }
}
```

4. What happens when this program is executed?

```
public class TestException
{

    public static void main(String[] args)
    {
         try {
                   int d = 1, r;
                   r = d / (d-d);
         }
         catch (IndexOutOfBoundsException e) {
                   System.out.println( e.ge
Class() +
                   "   " + e.getMessage());
         }
    }
}
```

5. List five runtime exceptions:
   1. ______
   2. ______
   3. ______
   4. ______
   5. ______
6. Are runtime exceptions checked? ________ What are checked exceptions?
7. What type of exceptions are characterized by the class `Errors`?

PRACTICE WITH VISUAL J++

Develop and test and application that throws an exception when all of the parameters of a called method are not valid. This type of error-handling condition is commonly used in the development of software building blocks intended for reuse in the development of complex applications.

4. What happens when the program is executed?

9

Packages

OBJECTIVES

After reading this chapter, you will be able to:

1. Understand the Java package concept and its use in organizing Java programs.
2. Understand the content of standard Java packages.
3. Use types declared in other packages.
4. Use the `Math`, `String`, `System`, and Random classes.
5. Demonstrate simulation of random processes.

OVERVIEW

Java classes and interfaces are organized in a hierarchical manner and grouped in packages. Access to classes and interfaces in other packages is affected with `import` statements. The many classes and interfaces declared in the Java API packages, particularly, `java.lang`, `java.util` and `java.io` classes, help to make Java a rich and robust language. The `Math`, `String`, `System`, and `Random` classes are widely used in the development of "real world" applications. Java provides for the modeling of random processes with the `Random` class.

JAVA API PACKAGES

Java programs are organized as packages. Members of the packages are classes and interface types. A type is accessible outside the package if it is declared public. Packages included in the Java API are summarized in Table 9.1.

TABLE 9.1

Java API packages

Package	Enables
`applet`	Applets, that is, programs embedded inside another application such as a Web page.
`awt`	Graphical user interfaces including frames and windows; controls such as button, text fields, scrollbars, and labels; lines and points; fonts, color, and layout.
`awt.datatransfer`	Cut, copy, and paste data transfer using a clipboard.
`awt.event`	Event-oriented programming environment including graphical user interface components, keyboard, and mouse.
`awt.image`	Image manipulation.
`beans`	Creation of reusable objects, called "beans" similar to ActiveX and Delphi controls that can be used in Java programs without internal modification.
`io`	Data input and output.
`lang`	Java program basics including wrappers, strings, exception handling, mathematical operations, and threads.
`lang.reflect`	Creating and accessing arrays, as well as providing information regarding constructors, fields, and methods of classes and interfaces.
`math`	Arbitrary-precision decimal and integer numbers for arithmetic, scale manipulation, comparison, format conversion, and hashing operations.
`net`	Network-related programming including modeling of datagram packets, sockets, and URLs.
`security`	Security methods including identifies, keys, digital signatures algorithms, and pseudo-random number generation.
`security.acl`	Access control lists including management of owners, permissions, and groups.
`security.interfaces`	Interfaces to digital signal algorithm public and private keys, key parameters, and key pair generating objects.
sql	Database access including driver management, access to data retrieved, and time stamps.
`text`	Text formatting including dates, messages, and numbers.
`util`	Implementing utility objects including date management including time zone consideration and Gregorian calendars; stack and vector data structures; and random number generation.
`util.zip`	File compression and decompression.

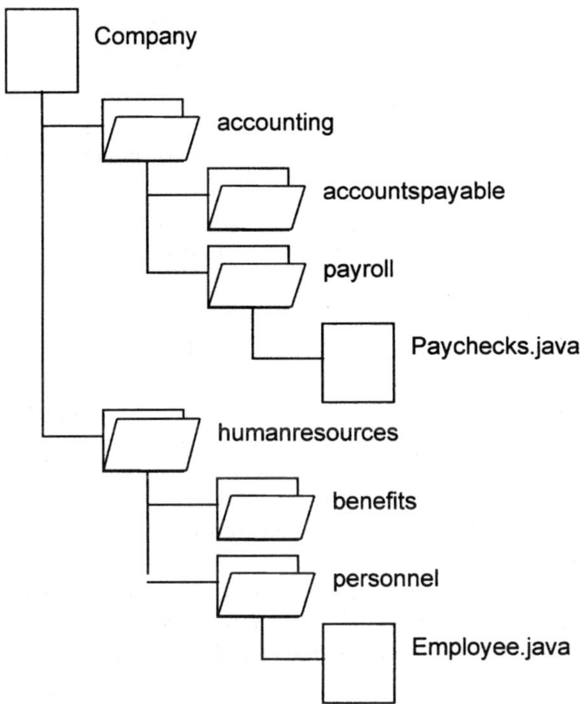

FIGURE 9.1
Java programs are organized in a hierarchical structure.

The Java packages `java.lang`, `java.util`, and `java.io` are always available to a Java program. Often, the Java packages `java.applet`, `java.awt`, and `java.net` are also automatically made available to a Java program.

When organized in a file, packages correspond to file folders. A simplified company's organization of Java programs is illustrated in Figure 9.1. The organization of the files is hierarchical. Both the `accountspayable` and `payroll` packages are contained in the package `accounting`. A package that is contained in another package is said to be a **subpackage.** A package that contains subpackages is said to be a containing package. Both the `accountspayable` and `payroll` packages are subpackages of the containing package `accounting`. The Java classes and interfaces are declared in packages. The class `Employee` is declared in the package `Personnel`, which is a subpackage of the containing package `humanresources`.

A package's **simple name** is a single identifier. The simple name of the package `accounting` is `accounting`. Similarly, the simple name of the package `payroll` is `payroll`. The **fully qualified name** of a package that is a subpackage of another package, called the containing package, is the fully qualified name of the containing package, followed by the "." Token, which in turn is followed by the simple name of the subpackage. Thus the fully qualified name of the package payroll is `accounting.payroll`. Note that the fully qual-

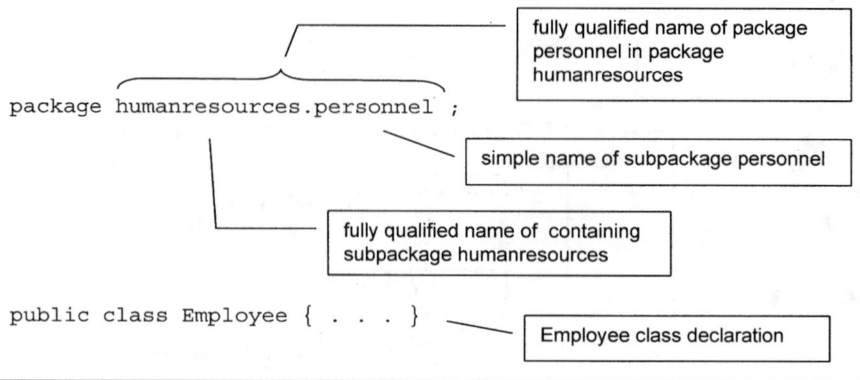

FIGURE 9.2
Class `Employee` declaration in package personnel.

ified name of a package that is not a subpackage is its simple name. Thus both the simple name and the fully qualified name of the package `accounting` is `accounting`.

The naming structure for packages is hierarchical. A **package declaration** specifies the name of the package. The keyword `package` begins a package declaration. The package keyword is followed by the fully qualified package name. For example, the package payroll is declared with the statement

```
package humanresources.payroll;
```

A package declaration and data type declaration are shown in Figure 9.2.

A type declared in another package can be referred to with its fully qualified name. Thus the type `Employee` declared in the package `personnel` can be referenced in the `accounting` package with its fully qualified name

```
Humanresources.personnel.Employee
```

On the other hand, the reference

```
Employee
```

is invalid.

An **import declaration** allows a type declared in another package to be referred to by its simple name. The `import` keyword begins the import declaration. The `import` keyword is followed by the **fully qualified name** of the type. The fully qualified type name is formed in the same manner as the fully qualified package name. An `import` declaration for the type `Employee` is shown in Figure 9.4.

The import declaration of a single type is said to be a **single type import declaration.**

A **type import on demand declaration** provides for the types declared in the package to be imported as needed and referred to by simple name. The

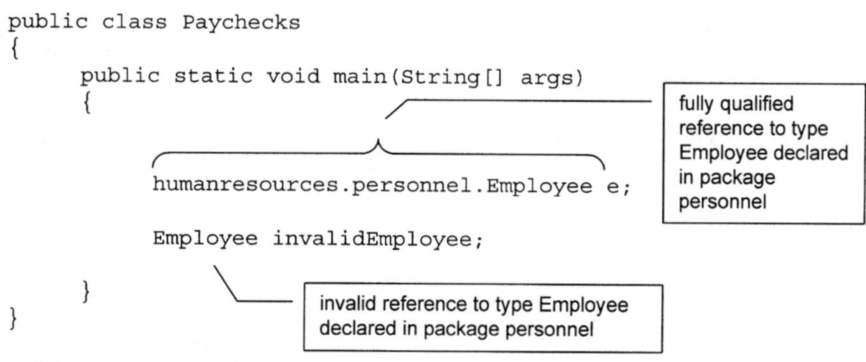

FIGURE 9.3
Referencing a type declared in another package.

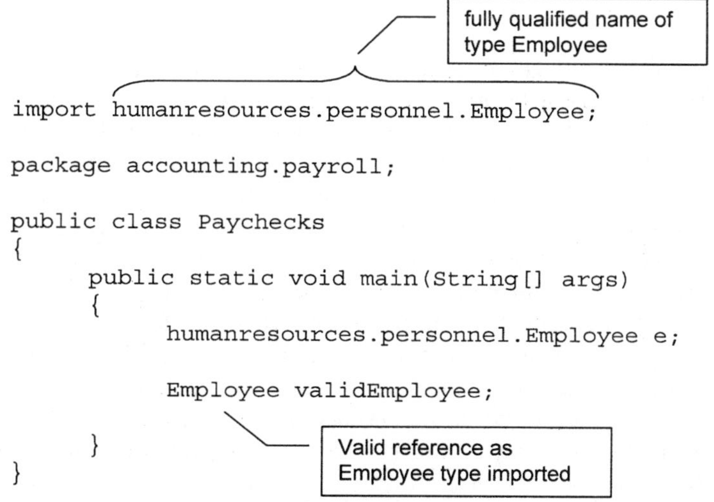

FIGURE 9.4
Type `import` declaration.

type import on demand declaration is formed with the `import` keyword, followed by the fully qualified package name, followed by the "." token, which is followed by an asterisk (*). An example of a type import on demand declaration of the package personnel is shown in Figure 9.5. Types declared in the package personnel are imported as needed.
The statement

```
import humanresources.personnel.*;
```

makes to types declared in the package personnel importable on demand.

> Types declared in `java.lang`, `java.util`, and `java.io` can be referred to by their simple names.

```
package accounting.payroll;

import humanresources.personnel.*;

public class Paychecks
{
      public static void main(String[] args)
      {
            humanresources.personnel.Employee e;

            Employee validEmployeeRef;

            Manager .validManagerRef;
      }
}
```

imports as needed types declared in package personnel

types declared in package personnel

FIGURE 9.5
Type `import` on demand declaration.

The package **`java.lang`** contains the most basic and fundamental Java classes. The package is imported by default. The `Math` class, containing methods commonly used in engineering and science, is included in the `java.lang` package. Textual data are modeled with the `String` class which is also in the `java.lang` package. Classes enabling the object representation of the primitive data types, called wrappers, are also included in the `java.lang` package. The `System` class, another class declared in the `java.lang` package, includes methods for program termination and garbage collection.

The package **`java.util`** contains classes and interfaces modeling objects useful to application programmers. The class `Random`, which can be used to represent random phenomena, is included in the package `java.util`.

The package **`java.io`** contains classes and interfaces enabling input and output. Classes in the package `java.io` are examined in Chapter 14, StreamI/O.

MATH CLASS

The **`Math` class,** included in the package `java.lang`, contains methods commonly used in engineering, science, economics, and business. `Math` class methods are summarized in Table 9.2.

Figure 9.6a lists a program to raise the number 2 to the third power using the `Math` class method `pow`. The output of the program is shown in Figure 9.6b.

TABLE 9.2

`Math` class methods

Method	Function
`abs(`*`number`*`)`	Returns the absolute value of the *number* passed. Valid number types are `double`, `float`, `long`, and `int`. The returned value is the same type as the argument.
`acos(`*`angle`*`)`	Returns the arc cosine of an angle, in the range of 0.0 through pi. The returned value and *angle* are type `double`.
`asin(`*`angle`*`)`	Returns the arc sine of an angle, in the range of –pi/2 through pi/2. The returned value and *angle* are type `double`.
`atan(`*`angle`*`)`	Returns the arc tangent of an angle, in the range of –pi/2 through pi/2. The returned value and *angle* are type `double`.
`atan2(`*`x, y`*`)`	Returns the theta component of the polar coordinate corresponding to the Cartesian coordinates *x* and *y*. The returned value is type *double*. The arguments representing the Cartesian coordinates are type `double`.
`ceil(`*`number`*`)`	Returns the smallest integral value, closest to negative infinity, that is less than *number* and is equal to a mathematical integer. The returned value is type `double`. The argument type is also `double`.
`cos(`*`angle`*`)`	Returns the trigonometric cosine of *angle*. The returned value and *angle* are type `double`.
`exp(`*`number`*`)`	Returns the value of `e` raised to the value of number, that is e^{number}. The returned value and argument *number* are type `double`.
`floor(`*`number`*`)`	Returns the largest number, closest to positive infinity, that is not greater than the argument number and is equal to an integer. The returned value and argument *number* are type `double`.
`IEEEremainder(`*`number1,number2`*`)`	Returns the remainder when *number1* is divided by *number2*. The return value, *number1* and *number2* are type `double`.
`log(`*`number`*`)`	Returns the natural logarithm of *number*, that is `ln` *number*.
`max(`*`number1, number2`*`)`	Returns the greater of the two numbers, *number1* and *number2*. Valid values for *number1* and *number2* are `double`, `float`, and `int`. The returned value is the same as the passed arguments.
`min(`*`number1, number2`*`)`	Returns the smaller of the two numbers, *number1* and *number2*. Valid values for *number1* and *number2* are `double`, `float`, and `int`. The returned value is the same as the passed arguments.
`pow(`*`number1, number2`*`)`	Returns the value of the first number raiser to the power of the second number, that is $number1^{number2}$. The returned value, *number1* and *number2* are type `double`.
`random()`	Returns a random number between 0.0 and 1.0. The returned random number is type `double`.
`rint(`*`number`*`)`	Returns the `double` value equal to the integer value closest to the `double` value *number*.
`round(`*`number`*`)`	Returns *number* rounded. The returned value is type `int` when *number* is type `long`. The returned value is type `long` when *number* is type `double`.
`sin(`*`angle`*`)`	Returns the trigonometric sine of *angle*. The returned value and *angle* are type `double`.
`sqrt(`*`number`*`)`	Returns the square root of number. The returned value and *number* are type `double`.
`tan(`*`angle`*`)`	Returns the trigonometric tangent of *angle*. The returned value and *angle* are type `double`.

```
public class UsingMathMethods
{
	public static void main(String[] args)
	{
		System.out.println("2*2*2 = " + Math.pow(2,3));
	}
}
```

FIGURE 9.6A
Program using the `Math` class `method` `pow`.

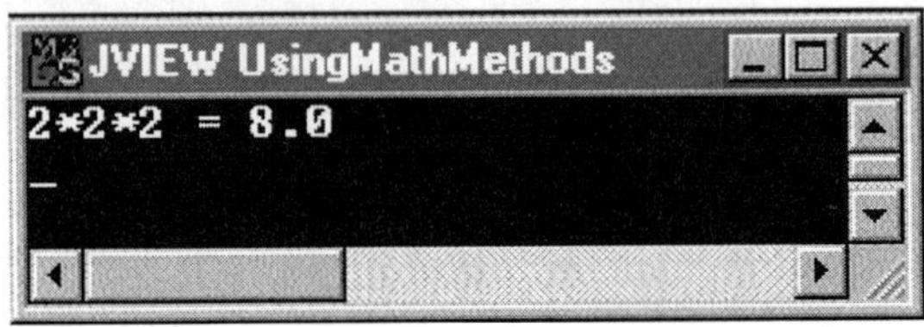

FIGURE 9.6B
Output of a program using the `Math` class method `pow`.

WRAPPER CLASSES

The package `java.lang` contains class types corresponding to the primitive data types. A class corresponding to a primitive type is called a **class wrapper.** The class wrappers are `Boolean`, `Byte`, `Character`, `Double`, `Float`, `Integer`, `Long`, and `Number`. The hierarchy of the `Wrapper` classes is shown in Figure 9.7.

The `Wrapper` classes contain methods for instantiating an object and retrieving the primitive value represented by the wrapper object. An example is provided by the selected methods of the `Double` class summarized in Table 9.3.

A wrapper is so called as it wraps a primitive type with methods for its manipulation. A conceptualization of the `Double` class, a wrapper, is shown in Figure 9.8.

As conceptualized, the `Double` class has a primitive type double field called `value`. The class also has a constructor that converts a textual numerical expression into a number and assigns the number to the field value. The conceptualized class also contains a method `doubleValue` that returns the double field value.

A program utilizing the `Double` class to convert a number expressed in a textual form to a `Double` object is shown in Figure 9.9a. The program also illustrates the retrieval of the double value represented by the object with the method `doubleValue`.

The new instance expression

```
new Double("7.11")
```

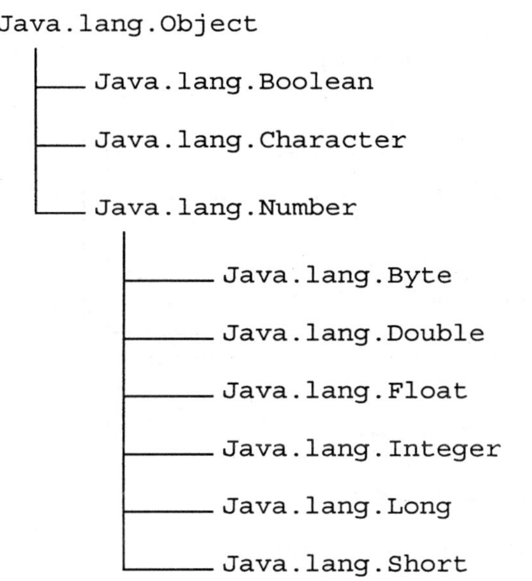

FIGURE 9.7
`Wrapper` Class hierarchy.

TABLE 9.3
Selected `Double` class methods

Method	Function
`Double(`*`number`*`)`	Constructs a `Double` object representing the `double` value *`number`*.
`Double(`*`string`*`)`	Constructs a `Double` object representing the `double` value represented by the `String` argument *`string`*.
`doubleValue( )`	Returns the `double` value of the represented by the `Double` object.
`intValue( )`	Returns the value represented by the `Double` object cast to an `int`.
`toString( )`	Returns a string representation of the `double` value of the `Double` object.

creates a `Double` object representing the numerical equivalent of the text "7.11". The type `Double`, reference `doubleNo`, declared with the expression

```
Double doubleNo
```

references the newly created object.

The `Double` class method, `doubleValue`, is invoked to return the value represented by the object with the statement

```
double No = doubleNo.doubleValue();
```

Program output is shown in Figure 9.9b.

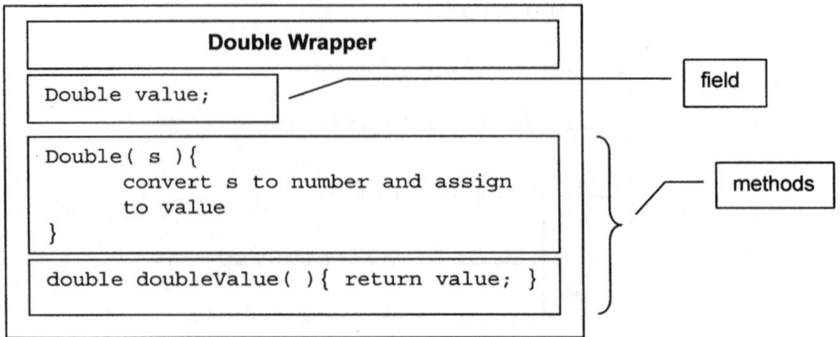

FIGURE 9.8
Conceptualization of `Double` class.

```
public class UsingWrappers
{
      public static void main(String[] args)
      {
            Double doubleNo = new Double("7.11");

            double No = doubleNo.doubleValue();


            System.out.println("double: " + No);
      }
}
```

number expressed as textual value

Double object

method returns double value represented by object

FIGURE 9.9A
Program using `Double` class to convert a double value expressed as text to a real value.

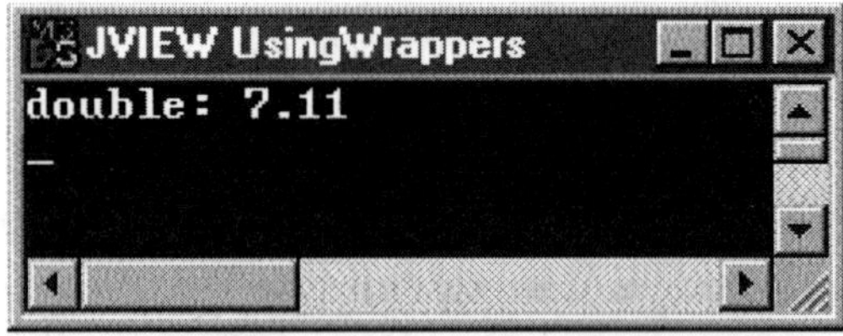

FIGURE 9.9B
Output of program using `Double` class to convert a double value expressed as text to a real value.

STRING CLASS

Textual data, that is character strings, are represented by the **`String`** class. The `String` class is included in the package `java.lang`. String literals are also implemented as instances of the `String` class. A **string literal** is zero or more characters enclosed in double quotes ("..."). For example,

```
"Java is a robust language"
```

is a string literal.

String literals can be joined together with the **concatenation operator** (+). For example,

```
"Java " + "is a robust language"
```

is equivalent to the literal

```
"Java is a robust language"
```

Selected non-graphic characters—the single quote, double quote, and backslash characters—can be represented in a literal string by the character and **string escape sequence.** Table 9.4 lists the escape sequences.

TABLE 9.4

Escape sequences

Sequence	Explanation
\ b	Backslash (\u0008)
\ t	Horizontal tab (\u0009)
\ n	Linefeed (\u000a)
\ f	Form feed (\u000c)
\ r	Carriage return (\u000d)
\ "	Double quote (\u0022)
\ '	Single quote (\u0027)
\ \	Backslash (\u005c)

The backslash is represented in literal strings representing file paths with the " \ \ " sequence

```
"C:\\VJ\\EXERCISES\\HANDSON9.DOC"
```

Also, the statement

```
System.out.println( "Learn Java\r\nnow" );
```

results in the output of

```
Learn Java
now
```

as the sequences " `\r` " and " `\n` " represent carriage return and line feed.

String literals are references to an instance of the class `String`, and thus

```
String moreJave = ".More Java";
```

is simply assigning to the `String` object reference `moreJava` the value of the reference to the string object `"More Java"`.

Strings are constant and cannot be changed once they are created. Note that the concatenation operator (+) creates a new `String` object.

The `String` class includes constructors and methods for searching for strings within the string, extracting substrings, translating all characters to uppercase or lowercase, determining the string length, as well as comparing strings. Selected constructors of the `String` class are summarized in Table 9.5. Selected methods of the `String` class are shown in Table 9.6.

TABLE 9.5

Selected String Class Constructors

Name	Description
`String( )`	Constructs a new `String` containing no bytes.
`String(byte[])`	Constructs a new `String` containing no bytes.
`String(char[])`	Constructs a new `String` by converting the array of characters.
`String(string)`	Creates a new `String` object with the same sequence of characters as the `String` argument.

An application illustrating use of the `String` class is shown in Figure 9.10a. The program output is shown in Figure 9.10b.

SYSTEM CLASS

The **`System`** class, in package `java.lang`, includes useful fields representing the system input, output, and error. The class also contains methods for stopping the Java Virtual Machine and performing garbage collection. The `System` class is declared `final` and cannot be instantiated.

System Input/Output

Operating systems, such as Windows and UNIX, automatically provide for input, output, and error notification. These system-provided input and output provisions are referred to as **standard input, standard output,** and **standard error.** These input and output connections are ready to transmit and receive data when an application starts. Typically, standard input is from the console keyboard. Similarly, standard output is generally on the console monitor but can be directed to another output device. Standard error is typically directed to the console terminal. The relationship of the standard input and output is graphically illustrated in Figure 9.11.

TABLE 9.6

Selected String Class Methods

Name	Description
`char charAt(int i)`	Returns the character at the index value *i*.
`int compareTo(String s)`	Compares two strings lexicographically. The value 0 is returned if the invoking `String` is equal to the `String` argument s. A value less than 0 is returned if the invoking `String` is less than the `String` argument *s*. Similarly, a value greater than 0 is returned if the invoking `String` is greater than the `String` argument *s*.
`int concat(String s)`	Concatenates the `String` argument *s* to the end of the invoking `String`.
`int indexOf(int ch)`	Returns the index within the invoking `String` of the first occurance of the specified character *ch* or –1 if the character *ch* does not exist.
`int indexOf (String s)`	Returns the index within the invoking `String` of the first character of the first occurance of the specified substring *s* or –1 if the substring *s* does not exist.
`int length ( )`	Returns the length of the string, that is, the number of characters in the string.
`String replace(char oldChar, char newChar)`	Returns a new string object in which every occurance of *oldChar* is replaced with *newChar*.
`String substring(int start, int end)`	Returns a new string object in beginning with the character at index *start* and ending with the character at index *end - 1*.
`String toLowerCase( )`	Returns a new string object in which all characters have been converted to their lower case equivalent.
`String toUpperCase( )`	Returns a new string object in which all characters have been converted to their upper case equivalent.
`String valueOf(double d)`	Returns a new string object representation of the `double` *d*.
`String valueOf(float f)`	Returns a new string object representation of the `float` *f*.
`String valueOf(int i)`	Returns a new string object representation of the `int` *i*.
`String valueOf(long l)`	Returns a new string object representation of the `long` *l*.

```
public class UsingStrings
{
        public static void main(String[] args)
        {
                String buf = new String();
                String First = new String("Amy");
                String Last = new String("Anybody");
                buf = Last + ", " + First;
                System.out.println( "buf: " + buf.toUpperCase() );
                System.out.println( "Length: " + buf.length() );
        }
}
```

FIGURE 9.10A
Application illustrating use of `String` class.

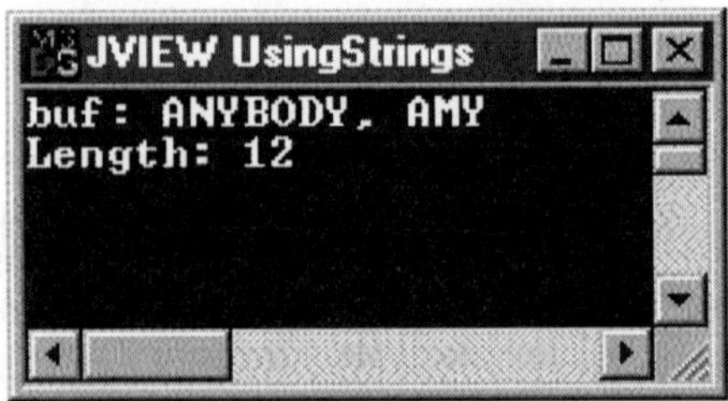

FIGURE 9.10B
Output of application illustrating use of `String` class.

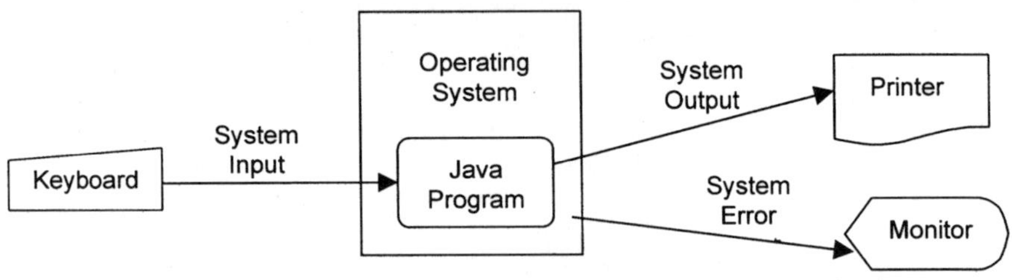

FIGURE 9.11
System input/output.

`System` Class Fields

The `System` class fields represent the system input, output, and error. These fields, summarized in Table 9.7, are stream objects. (Java streams are examined in detail in Chapter 9.)

TABLE 9.7
System class fields.

Name	Purpose
`Printstream err`	Standard error stream.
`InputStream in`	Standard input stream.
`Printstream out`	Standard output stream.

The standard input is represented by the `in` field in the class `System`. The `in` field is an `InputStream` object. Typically the input stream corresponds to the keyboard. The `InputStream` class includes the `read` method, which extracts the next byte of data from a stream—in this case, the next character entered at the keyboard. For example, the statement

```
buf = System.in.read();
```

reads the next keyboard character entered and assigns its numerical value to the `int` type variable `buf`.

```
import java.io.*;

public class StandardIO
{
      public static void main(String[] args)
            throws IOException
      {
            int buf;
            buf = System.in.read();
            System.out.println("buf: " + (char)buf);
      }
}
```


FIGURE 9.12A
Application using standard input and output.

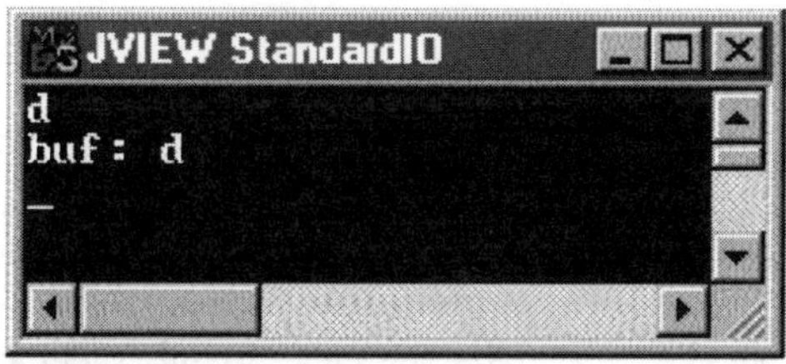

FIGURE 9.12B
Output of an application using standard input and output.

The standard output is represented by the `out` field in the class `System`. The `out` field is a `Printstream` type object. The class `Printstream` includes the method `println`, which can be used to print a line to an output stream. Thus the statement,

```
System.out.println("Java");
```

Prints the data `"Java"` at the standard output device which is typically the console monitor.

Similarly, messages to the standard error are made with the statement

```
System.err.println("Error Message");
```

An application illustrating use of the standard input and output is shown in Figure 9.12a. The application's output is shown in Figure 9.12b.

Garbage Collection

Whenever an object is created, memory is allocated. Although references to these objects go out of scope and the object can no longer be accessed, the memory used by the object remains allocated. This continuation of memory allocation for objects that cannot be accessed is referred to as memory leaks. Over time, as **memory leaks** accumulate, the amount of unallocated or available memory decreases. Eventually, all of the memory will be consumed.

Java reclaims allocated memory that can no longer be referenced and makes it available for use in the creation of other objects. This process is referred to as garbage collection. To effect garbage collection, Java keeps track of references to objects. When an object has no references, it is marked for garbage collection. Java's garbage collection is run periodically at a low priority or when memory is exhausted.

`System` Class Methods

The `System` class includes methods for garbage collection, terminating the Java Virtual Machine and retrieving the current time. Selected `System` class methods are summarized in Table 9.8.

TABLE 9.8

Selected `System` class methods.

Name	Description
`long getTime( )`	Returns the current time in milliseconds between the current time and midnight January 1, 1970.
`void exit(int status)`	Terminates the currently running virtual machine. The *status* value indicates terminating conditions.
`void gc( )`	Runs the garbage collector.

A program is terminated with the statement

```
System.exit( 0 );
```

Traditionally, value 0, as the `exit` method argument, indicates successful execution and 16 catastrophic conditions. Garbage collection is performed with the statement

```
System.gc( );
```

RANDOM CLASS

The **Random** class, in the package `java.util`, generates a stream of pseudorandom numbers. The numbers are referred to as pseudorandom numbers because the sequence repeats after a period of time. The starting point of the sequence generated is determined by a seed used during class instantiation.

Two instances of Random created with the same seed will generate identical sequences of numbers.

The statement

```
Random rn = new Random(7);
```

illustrates instantiation of an instance of class Random object rn with seed value 7. The seed value must be a long type.

The next number in the sequence of random numbers is returned by the method

```
nextDouble()
```

The returned value is type Double and is uniformly distributed between the values 0.0 and 1.0. The program listed in Figure 9.13a displays the first twenty numbers in a random number sequence. The program's output is shown in Figure 9.13b.

```
import java.util.*;
import java.io.*;
public class RandomNumber
{
	public static void main( String args[] )
	{
		Random rn = new Random(7);

		for(int i = 0;i < 20;i++)
			System.out.println( rn.nextDouble() );
	}
}
```

FIGURE 9.13A
Program listing random numbers.

RANDOM NUMBERS AND SIMULATION

Notice that there are as many numbers in the range 0.0 to 5.0 as in the range 5.0 to 1.0. Thus the numbers can be said to be evenly distributed. If the ranges are smaller, the distribution is not as uniform. For example, there are 3 numbers in the range 0.0 to 0.25, 6 numbers in the range 0.25 to 0.5, 5 numbers in the range 0.5 to 0.75, and 6 numbers in the range 0.75 to 1.0. This distribution is illustrated by the histogram shown in Figure 9.14.

The number of random numbers falling within the ranges varies, leading to the conclusion that the distribution is not uniform. Over the short term, that is, over a small list of random numbers, the distribution will not necessarily be uniform. However, over the long term, that is a long list of random numbers the distribution will tend to be more uniform. This is consistent

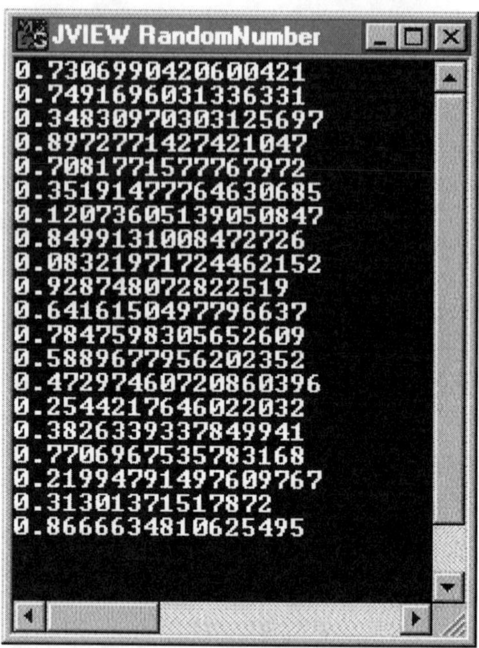

FIGURE 9.13B
Output of program listing random numbers.

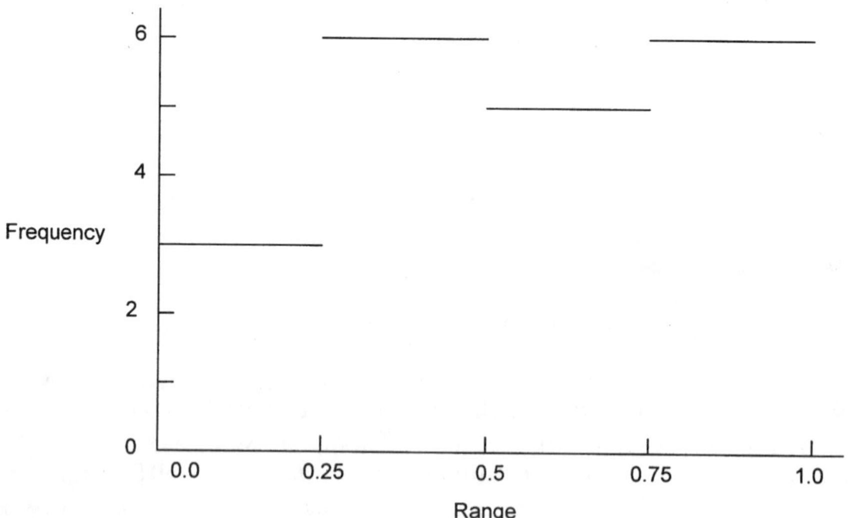

FIGURE 9.14
Histogram showing frequency distribution of random numbers generated.

with our intuitive notion of the results of flipping a coin. It is quite possible that in two flips the coin will either be heads both times or tails both time.

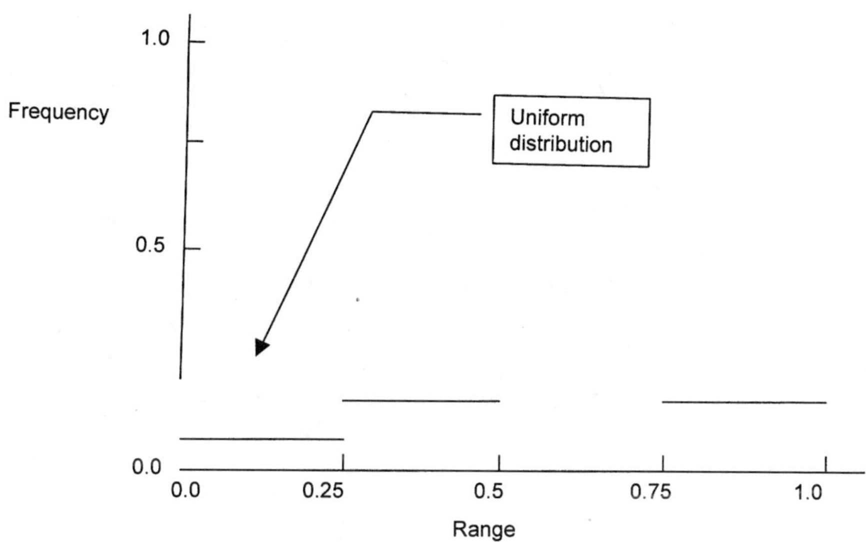

FIGURE 9.15
Frequency diagram of random numbers generated.

However, when the coin is flipped many times, the number of heads will generally be equal to the number of tails.

The number of random numbers falling in a range is referred to as the frequency associated with the range. The frequency of the range is generally normalized by dividing the number in the range by the total number of random numbers all the ranges. A diagram showing normalized frequencies is referred to as a frequency diagram. The frequency diagram associated with the list of random numbers is shown Figure 9.15. The **frequency distribution** shows a nonuniform distribution. As the number of random numbers increases, the distribution will approach the distribution shown by the dashed line.

Many processes have distributions that are not uniform. Consider for instance the arrival of passengers at an elevator. The diagram in Figure 9.16 illustrates the case in which 25% of the passengers arrive 15 seconds after the previous passenger's arrival, 25% of the passengers arrive 45 seconds after the previous passenger's arrival, and 50% of the passengers arrive 60 seconds after the previous passenger's arrival.

Note that the statistics account for 100% of the passengers, and the percentages are measured on the horizontal axes called the range. The range spans the interval 0.0 through 1.0 corresponding to the range of the random numbers. Note that the random number value 0.38 corresponds to the seconds between arrival time of 45 seconds. Note that 100 random numbers will translate very closely to the statistics 25% within 15 seconds, 25% within 45 seconds, and 50% within 60 seconds.

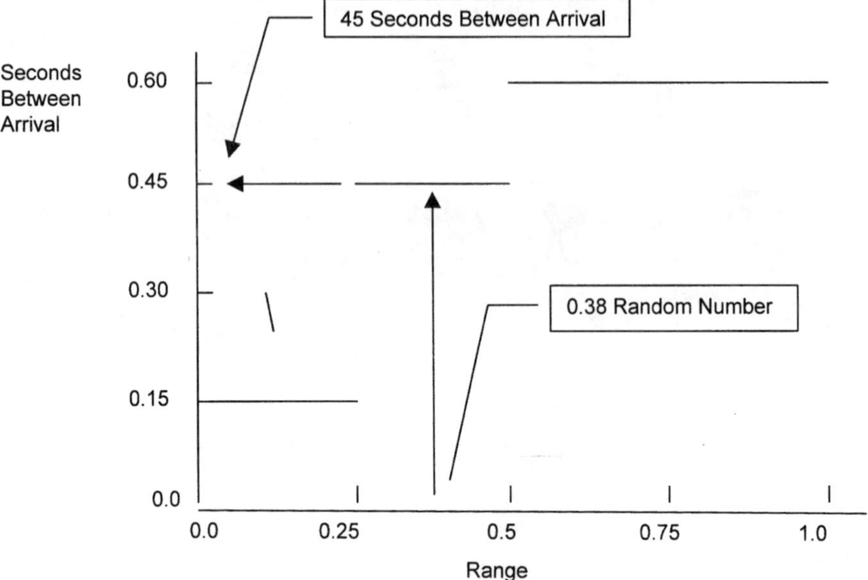

FIGURE 9.16
Seconds between time of arrival of elevator passengers.

HANDS-ON EXERCISE 9: SIMULATING A DIE TOSS

Dice are used in many games. Some games use one die and other games use multiple dice. The game player toss or rolls the die. A die has six faces, one of which faces up after the roll or toss. Each of the six faces contains 1, 2, 3, 4, 5, or 6 spots. The face landing upward varies with each toss or roll depending on the laws of chance. The game progresses depending on the number of spots on the upward face. This exercise develops a program to simulate the tossing of a die. The class `Random` is used to generate random numbers, which are transformed in to the spots facing upward after the toss or roll. The random numbers generated fall within the range 0.0 through 1.0. The range 0.0 through 1.0 is divided into six increments, each corresponding to a die face value. Thus each random number corresponds to a die face value.

Step 1: Create New Project

1. Start Visual J++.
2. Select the New Project ... command in the File menu. The New Project dialog box is displayed Figure HO-9.1).

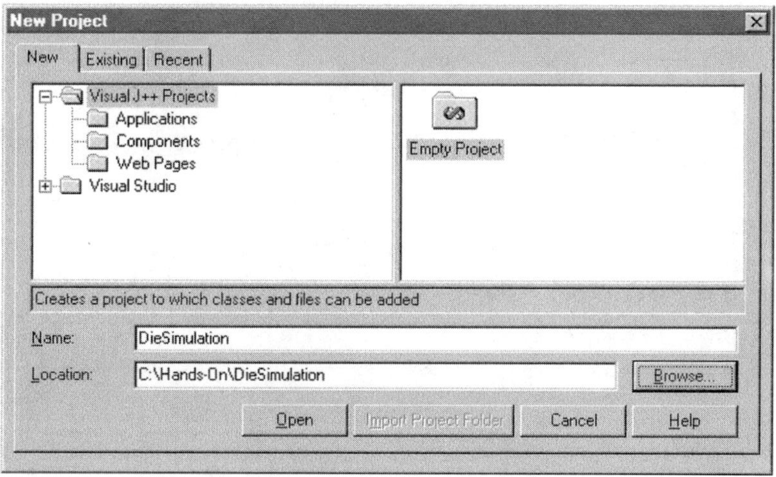

FIGURE HO-9.1
New Project dialog box.

3. If necessary, select the Visual J++ Projects in the Applications Type window, the left window. Select the Empty Project icon in the Options window, the right window.
4. Click the Browse… button and select the Location for the project's files.
5. Enter the project's name in the Name textbox by overtyping the default name `Project1`. Note that the project name is added to the end of the project Location textbox.
6. Click the Open button to create the new project. (Clicking the Cancel button will abort the new projection creation.)

Step 2: Create `.java` File `DieSimulation`

1. Select the Add Class … command in the Project menu. The Add Item dialog box is displayed (Figure HO-9.2).
2. The Class icon is selected in the Item window, the left window. Select the Class Main icon in the Options window, the right window, to create a public class with a main entry point method declaration.
3. Enter the class name in the Name textbox by overtyping the default name `Class1`.
4. Click the Open button to create the new class. (Clicking the Cancel button will abort the new class creation.) The class is opened in the Code Editor (Figure HO-9.3).
5. Note the inclusion of the `DieSimulator.java` file in the Project Explorer.

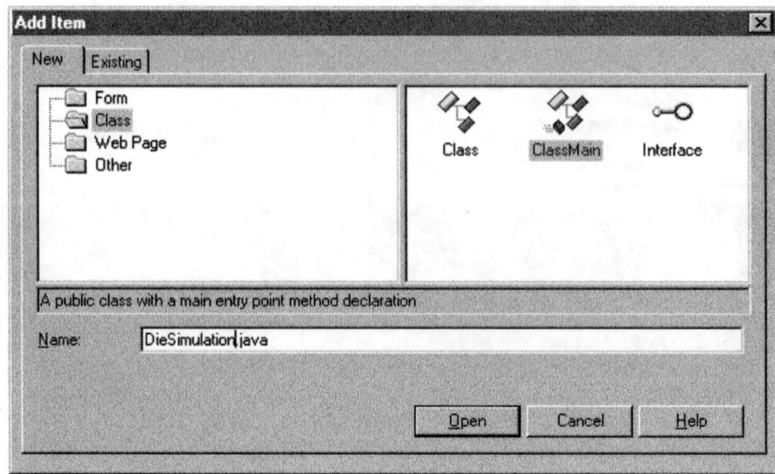

FIGURE HO-9.2
Add Item dialog box.

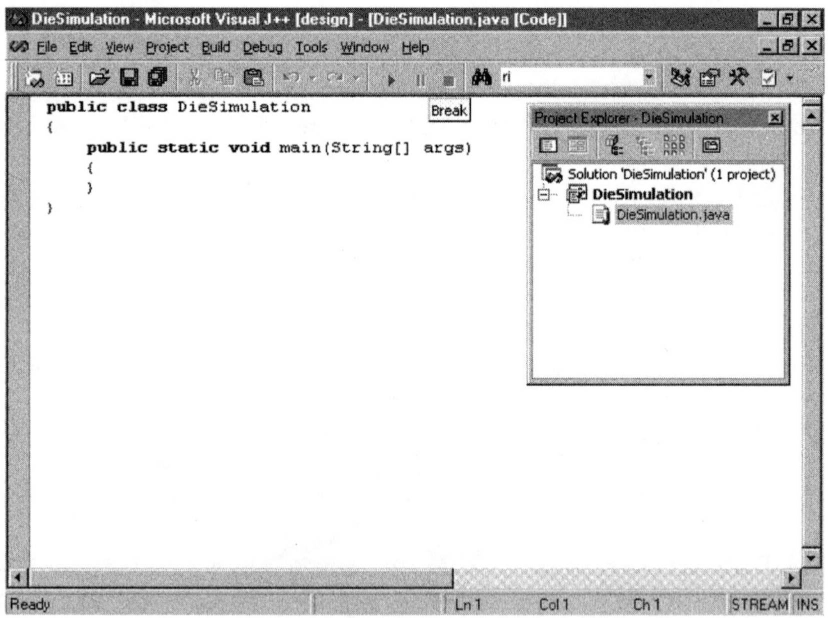

FIGURE HO-9.3
`DieSimulation` class open in the Code Editor.

Step 3: Code the User Response Handler

1. Enter the following statements to import the `java.io` package:

```
import java.io.*;
```

2. The class `UserResponse` models the user's response. The method `TossDie( )` prompts the user to enter **Y** to toss the die. The keyboard input is read into an array `inArray` with the statement

```
System.in.read(inArray);
```

3. The first element of the array is compared with the character `Y` with the expression

```
'Y' == (int) inArray[0]
```

4. If the comparison expression evaluates as true—that is, the first element in the array is the character `Y`—the method returns the value `true`. The method returns the value `false` otherwise.

Step 4: Code the Die Simulation Control Loop

1. Code the `DieSimulation` class control loop.

```
public class DieSimulation
{
    public static void main(String[] args)
    {
        UserResponse ur = new UserResponse();

        while (ur.TossDie()){
            // Simulate die toss here
        }
    }
}
```

2. The statement

```
UserResponse ur = new UserResponse();
```

creates a `UserResponse` object `ur`. The `TossDie( )` method is called in the expression of the `while` loop. If the user enters `Y`, the expression evaluates as `true` and the body of the `while` loop is executed, and the die toss is simulated. If, on the other hand, the user enters other than `Y`, the `while` expression evaluates as `false` and the program ends.

Step 5: Save the Project

Select the Save DieSimulation.java in the File menu.

Step 6: Build the Project

The program code is shown below.

```
import java.io.*;

class UserResponse {

    byte inArray[ ] = new byte[80];

    UserResponse(){ }

    public boolean TossDie( ){
        try {
            System.out.println("Enter Y to toss die: ");
            System.in.read(inArray);
        }
        catch (IOException ioe) {
            System.out.println(ioe.toString());
        }
        if ( 'Y' == (int) inArray[0]   )
            return true;
        else
            return false;
    }
}
public class DieSimulation
{
    public static void main(String[] args)
        {
        UserResponse ur = new UserResponse();

        while (ur.TossDie()){
            // Code die simulation here
        }
    }
}
```

Select the Build command in the Build menu. Correct any programming errors, save, and build the revised project.

Step 7: Run the Project

1. Select the Start command in the Debug menu. The `DieSimulation` dialog box is displayed the first time the project is run. Check the Launch as Console Application checkbox (Figure HO-9.4).
2. Click the OK button.
3. The project is compiled and the results output (Figure HO-9.5).
4. The `while` loop continues as long as the user enters **`Y`**.

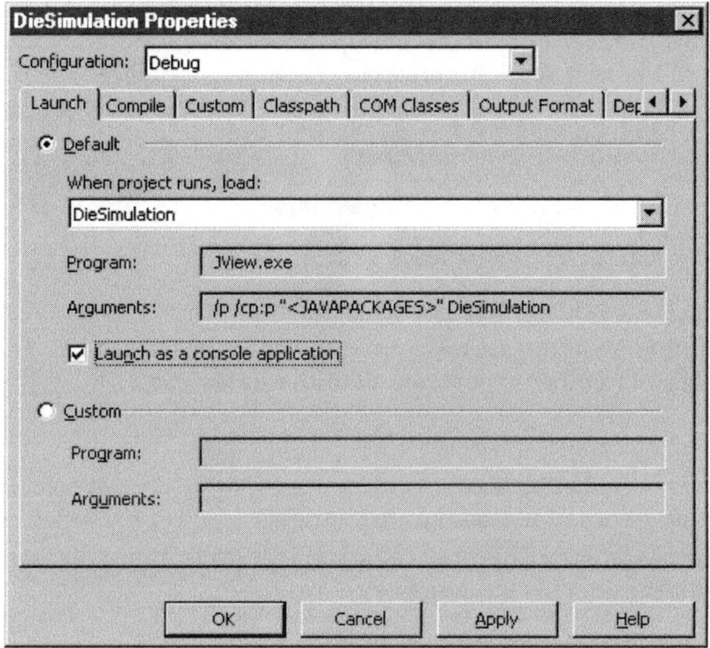

FIGURE HO-9.4
`DieSimulation` dialog box.

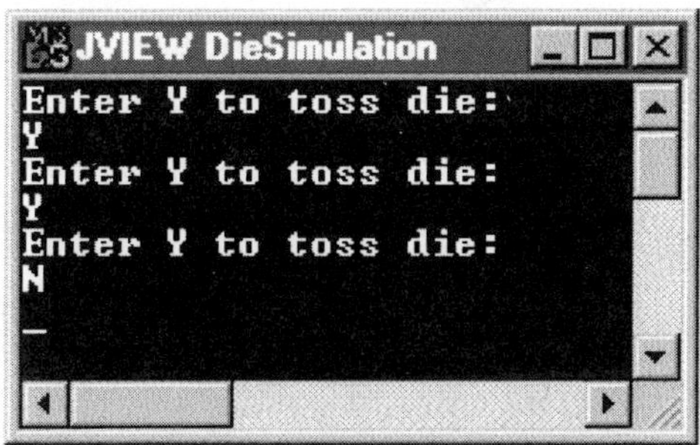

FIGURE HO-9.5
Output

Step 8: Code the Die Simulation

1. Enter the following statements to import the `java.util` package that contains the class `Random`.

```
import java.util.*;
```

2. Create variables to represent the face value of the tossed die, that is the number of spots facing up. In addition, create a variable to store the next random number.

```
int dieFaceValue = 0;
double nextRandomNumber;
```

3. Create an instance of the `Random` class.

```
Random rn = new Random(11);
```

   The seed value in this case is 11.

4. Simulation of the die toss is performed in the `while` loop. The next random number is obtained with the statement

```
nextRandomNumber = rn.nextDouble( );
```

5. The method `nextDouble( )` is invoked for the `Random` object `rn`. The value returned is stored in the variable `nextRandomNumber`.

6. The next random number value is transformed into a die face value with the `if-else` structure.

```
if (nextRandomNumber <= 0.1666)
        dieFaceValue = 1;
else if (nextRandomNumber <= 0.3333)
        dieFaceValue = 2;
else if (nextRandomNumber <= 0.5)
        dieFaceValue = 3;
else if (nextRandomNumber <= 0.6666)
        dieFaceValue = 4;
else if (nextRandomNumber <= 0.8333)
        dieFaceValue = 5;
else
        dieFaceValue = 6;
```

7. The structure divides the range 0.0 through 1.0 into six equal increments. The increments are assigned the numbers 1 through 6 corresponding to the die face values. The value of the `nextRandomNumber` is tested in the `if` expression. The value of `nextRandomNumber` will fall within one of the increments and result in the assignment of the associated `dieFaceValue`. The `dieFaceValue` is displayed with the statement

```
System.out.println(dieFaceValue);
```

Step 9: Save the Project

Select the Save DieSimulation.java in the File menu.

Step 10: Build the Project

The complete program is shown below.

```
import java.io.*;
import java.util.*;
class UserResponse {
   byte inArray[ ] = new byte[80];
   UserResponse(){ }

   public boolean TossDie( ){
      try {
         System.out.println("Enter Y to toss die: ");
         System.in.read(inArray);
      }
      catch (IOException ioe) {
         System.out.println(ioe.toString());
      }
      if ( 'Y' == (int) inArray[0]  )
         return true;
      else
         return false;
}
public class DieSimulation
{
   public static void main(String[] args)
   {
      UserResponse ur = new UserResponse();

      int dieFaceValue = 0;
      double nextRandomNumber;
      Random rn = new Random(11);

      while (ur.TossDie()){

         nextRandomNumber = rn.nextDouble( );
         if (nextRandomNumber <= 0.1666)
            dieFaceValue = 1;
         else if (nextRandomNumber <= 0.3333)
            dieFaceValue = 2;
         else if (nextRandomNumber <= 0.5)
            dieFaceValue = 3;
         else if (nextRandomNumber <= 0.6666)
            dieFaceValue = 4;
         else if (nextRandomNumber <= 0.8333)
            dieFaceValue = 5;
         else
            dieFaceValue = 6;
         System.out.println(dieFaceValue);
      }
}
```

Correct any programming errors, save, and build the revised project.

Step 11: Run the Project

Select the Start command in the Debug menu. The project is executed and the results output.

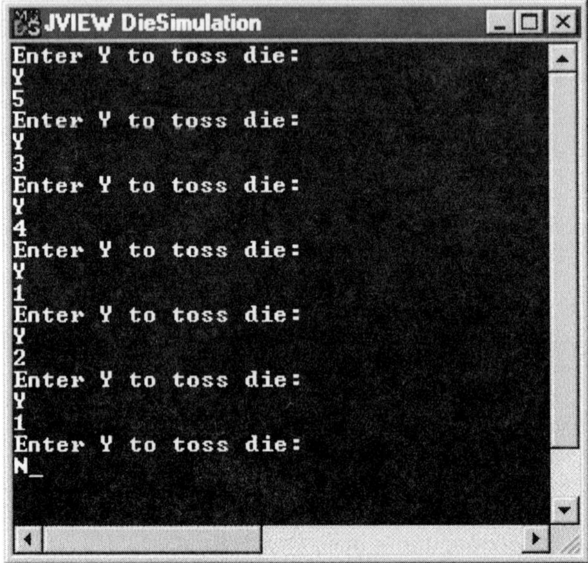

Step 12: Revise the Simulation to Simulate the Toss of Two Dice Simultaneously

Revise the simulation to simulate the tossing of two dice at the same time. The value of both die faces should be displayed.

SUMMARY

1. Classes and interfaces enabling application development are declared in Java API packages.
2. Java packages are organized in a hierarchical organization.
3. Java `import` statements enable access to classes and interfaces declared in other packages.

4. The `Math` class contains methods used in business, scientific, and engineering applications.
5. `String` class objects represent textual data.
6. `System` class methods and fields are used to facilitate input and output to standard devices, terminate operation of the Java Virtual Machine, and garbage collection.
7. Java performs garbage collection automatically.
8. Class `Random` contains methods for generating pseudo-random numbers.
9. Random processes can be simulated with Java applications.

KEY WORDS AND CONCEPTS

(+)
concatenation operator (+)
containing package
frequency diagram
fully qualified name
fully qualified type name
garbage collection
`import` (keyword)
import declaration
`java.lang`
`Math` class
memory leaks
packages
`package` (keyword)
package declaration
`Random` class
simple name
single type import declaration
standard error
standard input
standard output
String class
string escape sequence
string literal
subpackage
`System` class
type import on demand declaration
wrapper
wrapper class

REVIEW QUESTIONS AND EXERCISES

1. (T/F) Interfaces are not included in packages?
2. What package contains classes to implement the following:

Arrays	______
Database access	______
Mathematical functions	______
GUI pushbotton	______

File creation ______________________

3. (T/F) The fully qualified name of the wrapper class `long` in the package `java.lang.math` must always be used.
4. Develop and test a program to output the sin of 7 degrees.
5. Develop and test a program to demonstrate use of the `substring` method of the class `String`.
6. Develop a program to simulate the time between arrivals of an elevator. Develop statistics for an actual elevator in a busy building during the morning rush.
7. (T/F) Garbage collection is automatic in Java.
8. Review documentation on the `finalize` method.

   (T/F) The `finalize` method is explicitly called.

Develop and test a program demonstrating use of the finalize method.

9. Identify the devices generally used:

   standard input ______________________

   standard output ______________________

   standard error ______________________

10. (T/F) Generated random numbers always begin with the same number. Explain your answer.

PRACTICE WITH VISUAL J++

Develop a program to simulate the game of "craps." Craps is played with two dice. A player tosses the dice. If the sum of the faces is 7 or 11 on the first throw, the player wins. If the sum of the faces is 2 (called "dead eyes") or 12 (called "box cars") on the first throw, the player has thrown craps and looses. If the sum of the faces on the first throw is 3, 4, 5, 6, 8, 9, or 10, the sum becomes the player's "point." If the player on subsequent rolls, throws the point value before the player throws 7, the player wins.

Part 3

Abstract Window Toolkit

10

Windows and Events

OBJECTIVES

After reading this chapter, you will be able to:

1. Understand Java's Abstract Window Toolkit and its use in implementing windows.
2. Create a Frame window.
3. Understand the event model and its use in developing event-driven programs.
4. Implement a listener for user interaction.
5. Understand menus and their use in presenting application options.
6. Create a menu.
7. Demonstrate use of the Abstract Window Toolkit in developing application graphical interface windows and menus.

OVERVIEW

The **Abstract Window Toolkit** is a set of classes that enable user interface components and events. Typical components are windows, menus, buttons, and text fields. Typical events include selecting a button with the mouse and pressing the ENTER key after typing in a text field. Using the component classes, applications consisting of windows, menus, buttons, and text fields can be implemented. Events are automatically generated when the user interacts with the components using the mouse or keyboard. These events are modeled in the event classes. Using the Abstract Window Toolkit's event classes and interfaces, the specific actions to be taken when an event occurs can be programmatically selected. Menus provide an effective means for presenting program options. The Abstract Window Toolkit includes classes for implementing menus and using menus.

WINDOWS ENVIRONMENT

Windows was introduced by Microsoft in the early 1980s. The graphical interface, wherein graphical icons and buttons were selected by a pointing device, proved to be user friendly, was quickly accepted by users and has become the most widely used interface today. A typical window is shown in Figure 10.1. The window is a framed display of the ions and controls used to access the application. Controls such as text boxes for displaying and inputting data, buttons enabling the user to indicate a choice, and graphics are displayed in the client area. A title bar contains the window's title, system menu icon, and the minimize, maximize, and close application icons. A menu bar contains pull-down menus enabling the easy selection of application options. In general, user access to an application is through its window.

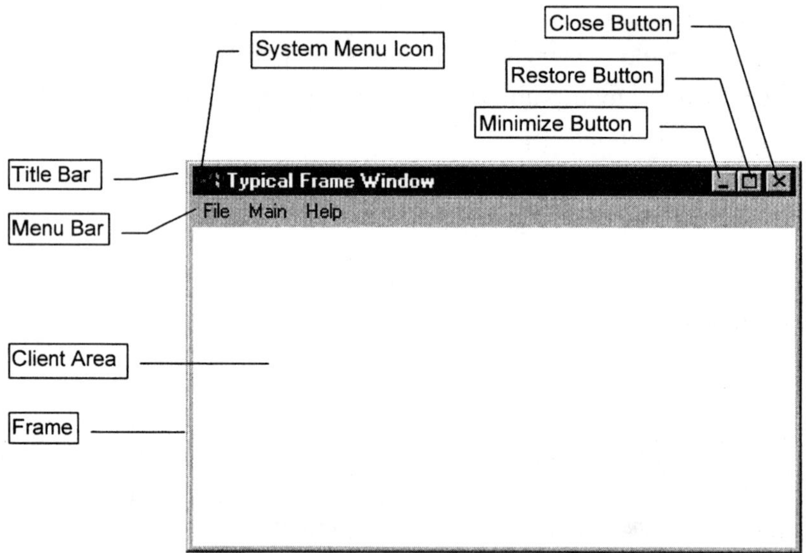

FIGURE 10.1
Typical frame window.

In the Windows operating environment, the user can have several programs executing simultaneously. Each program generally has its own window. The user moves back and forth between the programs by simply clicking the program's window. The selected program can process inputs such as mouse clicks, keystrokes, or window resizing. When the input has been processed, control returns to the operating system, which waits for the next user input. The next user input might be directed at the same program or intended for a different program.

ABSTRACT WINDOW TOOLKIT

The **Abstract Window Toolkit,** or **AWT,** augments the computer's operating system enabling Java applications to approximate the appearance and feel of Windows and Macintosh applications. AWT subsystems enable the construction of user interface components, containers, and events. The components, buttons, checkboxes, labels, and text fields provide the nuts and bolts of user interface. The containers, such as Frames, Panels, and Dialogs, enable the development of windows that hold the components, that can be moved, resized, minimized and restored by the application's user. The event system enables the user to communicate with the program. For example, when the user clicks the mouse over a button or presses a key, an event occurs. When an event occurs, an event object is created and sent as part of the event message to listening components. The AWT layout managers determine the size and position of components whenever a window is drawn. In addition, AWT includes a graphics capability that provides for drawing lines, images, and animation.

This chapter examines AWT components and events.

COMPONENTS

The Object class is extended in the AWT as a hierarchy of classes providing user interface capability, as shown in Figure 10.2.

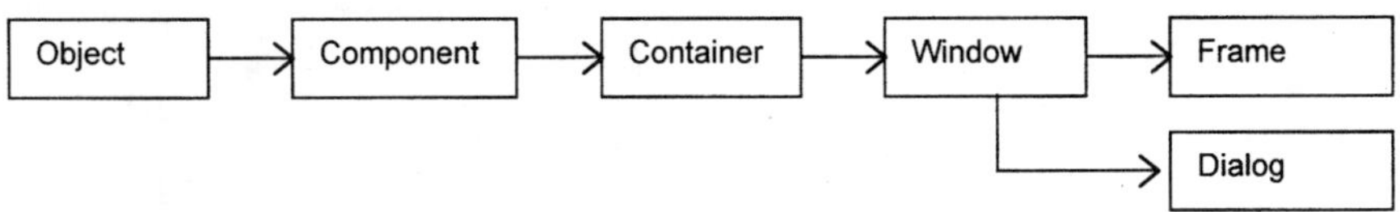

FIGURE 10.2
Component–Container hierarchy.

Component, Window, and Container Objects

A **component** is an object that can (1) represent a graphical image that can be displayed on the screen and (2) enable user interaction. Examples of components are buttons, text fields, labels as well as frames and windows. The Component class is an abstract class containing methods and attributes shared by all components. Selected key methods of the Component class are summarized in Table 10.1. These methods are inherited by all components derived from class Component.

TABLE 10.1

Selected key Component class methods

Name	Description
`Rectangle getBounds( )`	Returns a `Rectangle` object specifying height and width of the rectangle enclosing the component.
`Dimension getSize( )`	Returns a `Dimension` object specifying height and width of the component.
`boolean isEnabled( )`	Returns `true` if the component is able to respond to user input and generate events, `false` otherwise. Components are initially enabled.
`boolean isVisible( )`	Returns `true` if the component is visible, `false` otherwise. Components, with the exception of top-level components such as Frames, are initially visible.
`void paint(Graphic g )`	Called to paint the component.
`void repaint( )`	Request that AWT call `update( )` to repaint the component.
`void setBound(int x, int y, int width, int height)`	Moves and resizes the component.
`void setEnabled(boolean cond )`	Enables the component if the *cond* is `true` and disables the component if the *cond* is `false`.
`void setFont(Font f)`	Sets the components font to the specified `Font` value *f*.
`void setLocation(int x, int y)`	Moves the component to the new location.
`void setsize(int width, int height)`	Resizes the component to the new location. Moves the component to the new location.
`void setVisible(boolean cond )`	Shows component if the *cond* is `true` and hides the component if the *cond* is `false`.
`void update(Graphic g )`	AWT calls the `update` method when requested by the `repaint` method. The update method after clearing the components bounding rectangle and setting the color context calls the components `paint` method.

`Window` class objects are top-level windows without borders and menu bar. A `Window` class object blocks input to other windows when it is shown. Selected key methods of the `Window` class are shown in Table 10.2.

TABLE 10.2

Selected key `Window` class methods inherited by the `Frame` class.

Name	Description
`void dispose( )`	Disposes of the window including releasing resources associated with the window.
`void show( )`	Shows and brings the window to the front.
`void toBack( )`	Sends the window to the back.
`void toFront( )`	Sends the window to the front.

TABLE 10.3

Selected key `container` class methods.

Name	Description
`void add(Component` *`comp`*`)`	Adds the component *`comp`* to the end of the container.
`void paint( )`	Paints the container and passes the `paint` method to children components in the container.
`void remove(Component` *`comp`*`)`	Removes the component *`comp`* from the container.
`void removeAll( )`	Removes all components from the container.
`void setLayout( LayoutManager` *`laymgr`*`)`	Set the container's layout manager to the specified *`laymgr`*.

TABLE 10.4

`Frame` class constructor

Name	Description
`Frame(String` *`title`*`)`	Creates a frame with title specified by the type `String` argument *`title`*.

The `Window` class is derived from the `Container` class. The container object is an object that can contain other objects. Container objects are represented by the `Container` class. Selected `Container` class methods that are passed on to window and component objects are summarized in Table 10.3.

Frame Objects

Frames are top-level windows. Frames have a title bar; icons that are used to minimize, maximize, or restore the frame; a system icon with pull down list; and borders that can be grabbed to size the frame. A constructor is summarized in Table 10.4. Selected key methods of the `Frame` class are shown in Table 10.5.

Note that `Frames` inherit the methods of the `Component`, `Container`, and `Window` classes. The `dispose` method overrides the `dispose` method in the class `Window`.

TABLE 10.5
Selected `Frame` class methods

Name	Description
`Void dispose( )`	Disposes of the frame including the release of resources.
`setTitle(`*`title`*`)`	Sets the frame's title to the specified *`title`*.

Creating a Frame

The code for creating a frame is shown in Figure 10.3a. A class `FrameWindow` that extends the `Frame` class is declared. The main method consists of one line, creating a `FrameWindow` object.

```
new FrameWindow();
```

During creation, the `FrameWindow` constructor is called. The frame is created as invisible. The method `show`, inherited form the class `Frame`, is invoked in the constructor to make the frame visible. In addition, the method `setSize`, also inherited from the class `Frame`, is called to set the size of the frame.

```
        setSize(400,200);
        setVisible(true);


import java.awt.*;
public class FrameWindow extends Frame
{
        public FrameWindow( )
        {
                setSize(400,200);
                setVisible(true);
        }

        public static void main(String[] args)
        {
                new FrameWindow();
        }
}
```

FIGURE 10.3A
Creating a frame.

The frame created by the program is shown in Figure 10.3b. The frame created has a border, and can be sized and moved. The frame also has a menu bar, minimize, maximize, restore and close icons. The frame can be minimized, maximized, and restored. These operations are automatically performed by AWT. The frame cannot be closed, however. Imagine an application that cannot be closed.

FIGURE 10.3B
Creating a frame.

The title of the frame could be set using the `Frame` method `setTitle`. Alternately, the frame constructor `Frame(String title)` could be invoked in the constructor to create a frame with a title.

EVENTS

A user or system action that result in a message is called an **event.** User interaction with AWT objects are signaled by events. For example, every time the user moves the mouse, a message—in this case a mouse, moved message—is directed to the application. Similarly, when the user selects a menu option, a message regarding the option selected is directed to the application. Applications can be programmed to listen for messages and respond when they are applicable. The methods specifying the program actions to be taken for messages are referred to as **message handlers.** Programs designed to respond to events are referred to as **event-oriented programs.**

Events

The upper levels of the AWT event hierarchy are shown in Figure 10.4. The higher level classes of the event hierarchy are summarized in Table 10.6.

All event classes are derived from the `EventObject` class. The `AWTEvent` class, which is derived from `EventObject`, is an abstract class that serves as the root class for all AWT events. The `ActionEvent`, `AdjustementEvent`, `ComponentEvent`, and `ItemEvent` classes are all derived from the `AWTEvent` class. `ActionEvent` objects are emitted for action events. `AdjustmentEvent` objects are emitted when adjustable objects, for example a scroll bar, are adjusted. The `ComponentEvent` events are notification events for all components. Events

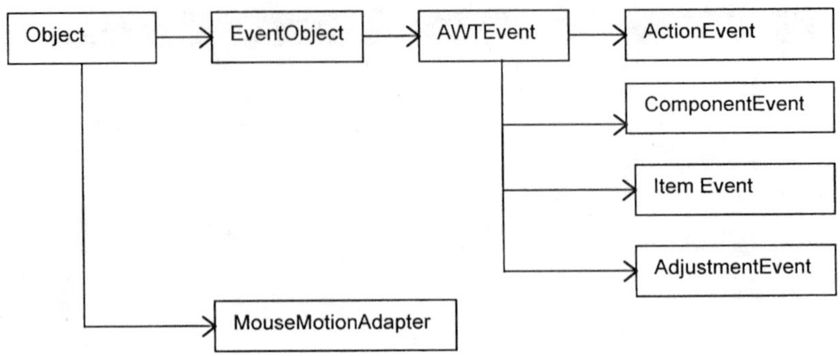

FIGURE 10.4
AWT event hierarchy.

TABLE 10.6
Summary of selected event classes

Class	Events Modeled
`ActionEvent`	Action
`AdjustmentEvent`	Adjustments of adjustable objects
`AWTEvent`	AWT
`ComponentEvent`	Component-level events
`EventObject`	Object events
`ItemEvent`	Item selection or de-selection in an item-selectable object

regarding moves and resizing of component objects are excluded as they are handled automatically by `AWT`. `ItemEvent` objects are emitted by item selectable objects whenever an item is selected or de-selected.

The `ComponentEvent` class hierarchy is shown in Figure 10.5. The `ComponentEvent` subclasses are summarized in Table 10.7. The `ContainerEvent` events are emitted for container-level events. These events are for notification only as container add and remove operations are handled automatically by AWT.

Listeners

Objects that want to receive notification of an event are referred to as listeners. `EventObjects` are sent directly to all listeners. These objects must implement the interface **`EventListener`** or one of its subclasses. The interfaces contain methods that are invoked when the events occur. Table 10.8 summarizes the key window-level events handled by the methods of the **`WindowListener`** interface.

The `WINDOW_CLOSED` event is delivered after the window is closed resulting from a call to hide or destroy the window. The `WINDOW_DESTROY` is delivered

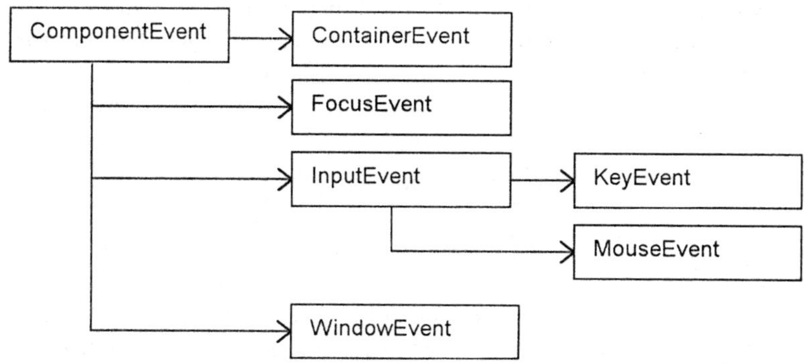

FIGURE 10.5
`ComponentEvent` class hierarchy.

TABLE 10.7

Summary of `ComponentEvent` class subclasses

Class	Events Modeled
`ContainerEvent`	Container-level events
`FocusEvent`	Focus events
`InputEvent`	Input events
`KeyEvent`	Component-level keyboard events
`MouseEvent`	Mouse events
`WindowEvent`	Window-level events

TABLE 10.8

Selected WindowEvent events and WindowListener methods

Event	Method
`WINDOW_CLOSED`	`windowClosed(WindowEvent we)`
`WINDOW_DESTROY`	`windowClosing(WindowEvent we)`

TABLE 10.9

Selected `Component` events and `ComponentListener` methods

Event	Method
`COMPONENT_HIDDEN`	`componentHidden(ComponentEvent ce)`
`COMPONENT_MOVED`	`componentMoved(ComponentEvent ce)`
`COMPONENT_RESIZED`	`componentResized(ComponentEvent ce)`
`COMPONENT_SHOWN`	`componentShown(ComponentEvent ce)`

when the user selects the Close option in the system menu or clicks the X icon in the left corner of the menu bar. Note that if the `hide` or `destroy` methods are not called the `WINDOW_DESTROY` operation is cancelled.

The **`ComponentListener`** has methods that handle component-level events. Table 10.9 summarizes key component-level events and **`ComponentListener`** methods.

The **AdjustmentListener** has methods that handle adjustment events. The interface is defined in the package `java.awt.event`. Table 10.10 summarizes key adjustment events and `AdjustmentListener` methods.

TABLE 10.10

Selected `Adjustment` events and `AdjustmentListener` methods

Event	Method
TRACK	adjustmentValueChanged(AdjustmentEvent ae)
UNIT_INCREMENT	adjustmentValueChanged(AdjustmentEvent ae)
UNIT_DECREMENT	adjustmentValueChanged(AdjustmentEvent ae)
BLOCK_INCREMENT	adjustmentValueChanged(AdjustmentEvent ae)
BLOCK_DECREMENT	adjustmentValueChanged(AdjustmentEvent ae)

The **ActionListener** has the `actionedPerformed` method that handle actions events, for example a mouse click over a button or menu item. The interface is defined in the package `java.awt.event`.

The `ItemListener` has methods that handle item selection and de-selection events. The interface is defined in the package `java.awt.event`. Table 10.11 summarizes key `Item` events and `ItemListener` methods.

TABLE 10.11

Selected `Item` events and `ItemListener` methods

Event	Method
DESELECTED	itemStateChanged(ItemEvent ie)
SELECTED	itemStateChanged(ItemEvent ie)

The **MouseListener** has methods that handle nonmoving mouse events. The interface is defined in the package `java.awt.event`. Table 10.12 summarizes nonmovement mouse events and `MouseListener` methods.

TABLE 10.12

Selected `mouse` events and `MouseListener` methods

Event	Method
MOUSE_CLICKED	mouseClicked(MouseEvent *me*)
MOUSE_ENTERED	mouseEntered(MouseEvent *me*)
MOUSE EXITED	mouseExited(MouseEvent *me*)
MOUSE_PRESSED	mousePressed(MouseEvent *me*)
MOUSE_RELEASED	mouseReleased(MouseEvent *me*)

The **MouseMotionListener** has methods that handle moving mouse events. The interface is defined in the package `java.awt.event`. Table 10.13 summarizes movement mouse events and `MouseMotionListener` methods.

The `MOUSE_DRAGGED` event is emitted when the mouse button is pressed over a component and then moved without releasing the button. The events con-

TABLE 10.13

Selected mouse events and `MouseMotionListener` methods

Event	Method
`MOUSE_DRAGGED`	`mouseDragged(MouseEvent` *`me`*`)`
`MOUSE_ MOVED`	`mouseMoved (MouseEvent` *`me`*`)`

tinue until the button is released. The `MOUSE_MOVED` event is emitted when the mouse moves over a component with no buttons pressed.

The **`KeyListener`** has methods that handle keyboard events. The interface is defined in the package `java.awt.event`. Table 10.14 summarizes key events and `KeyListener` methods.

TABLE 10.14

Selected key events and `KeyMotionListener` methods

Event	Method
`KEY_PRESSED`	`keyPressed(MouseEvent` *`me`*`)`
`KEY_RELEASED`	`keyReleased(MouseEvent` *`me`*`)`
`KEY_TYPED`	`keyTyped(MouseEvent` *`me`*`)`

The `KEY_PRESSED` event is emitted when a key is pressed. Similarly, the `KEY_RELEASED` event is emitted when a key is released. The `KEY_TYPED` event is emitted when a key is pressed followed by its release.

The **`FocusListener`** has methods that focus events. The interface is defined in the package `java.awt.event`. Table 10.15 summarizes focus events and `FocusListener` methods.

TABLE 10.15

Selected focus events and FocusMotionListener methods

Event	Method
`FOCUS_GAINED`	`focusGained(FocusEvent` *`fe)`*
`FOCUS_LOST`	`focusLost(FocusEvent` *`fe`*`)`

The `FOCUS_GAINED` event is emitted when a component gains keyboard focus. Similarly, the `FOCUS_LOST` event is emitted when a component loses keyboard focus.

Implementing Listening

A listener is implemented in a four-step process:

1. Import the package `java.awt.event` enabling the component to receive and handle events.
2. Use the `implements` *`listenerInterface`* to specify a listener for the events in the class definition.

3. Invoke the `add listener` method enabling the component to listen for events.
4. Implement handlers for the methods of the listener that specify what actions are to be taken when the events occur.

An example illustrating handling the window closing event is shown in Figure 10.6a. First, the interface `WindowListener` is implemented with the statement

```
public class FrameWindow extends Frame
implements WindowListener
```

Then, the `addWindowListener` method of the `Window` class is implemented with the statement

```
addWindowListener(this);
```

Finally, the `windowClosing` method of the `WindowListening` interface is overridden.

```
public void windowClosing( WindowEvent we )
{
    setVisible(false );
    dispose( );
    System.exit( 0 );
}
```

A window closing method results in invocation of the `windowClosing` method. First, the frame is hidden by invoking the `setVisible` method with parameter value `false`. Resources used by the frame object are released for garbage collection with the `dispose` method. Finally, the program is terminated by invoking the `System.exit` method.

Note that all of the methods of the `WindowListening` interface must also be overridden.

```
public void windowActivated( WindowEvent we ){}
public void windowClosed( WindowEvent we ){}
public void windowDeactivated( WindowEvent we ){}
public void windowDeiconified( WindowEvent we ){}
public void windowIconified( WindowEvent we ){}
public void windowOpened( WindowEvent we ){}
```

The program's output is shown in Figure 10.3b. The user can quit the application by selecting the Close option in the system menu or clicking the exit icon in the menu bar.

```
import java.awt.*;
import java.awt.event.*;
public class FrameWindow extends Frame
        implements WindowListener
{
        public FrameWindow( )
        {
                addWindowListener(this);

                setSize(400,200);
                setVisible(true);
        }

        public static void main(String[] args)
        {
                new FrameWindow();
        }
        public void windowClosing( WindowEvent we )
        {
                setVisible(false );
                dispose( );
                System.exit( 0 );
        }
        public void windowActivated( WindowEvent we ){}
        public void windowClosed( WindowEvent we ){}
        public void windowDeactivated( WindowEvent we ){}
        public void windowDeiconified( WindowEvent we ){}
        public void windowIconified( WindowEvent we ){}
        public void windowOpened( WindowEvent we ){}
}
```

FIGURE 10.6
Implementing the window closing event.

MENUS

Menus provide an effective means for user selection of various program options. A typical menu is shown in Figure 10.7. A **menu bar** is attached to the window frame. Menus are contained in the bar. The menu bar shown contains three menus: File, Main, and Help. When clicked, a **menu** displays a list of options and as such the menu is often referred to as a **pull-down menu.** The options displayed in a pull-down menu are called **menu items.** Typically, selection of a menu item initiates a program action. For example, the File menu contains an Exit option in its pull-down menu which when clicked will result in termination of program operation. The pull-down menu can contain a menu as well as menu items. A menu item in a pull-down will also display another pull-down called a **submenu.** In the example, the Main menu contains a menu captioned `Submenu` that has a pull-down submenu containing two menu items: `Submenu1` and `Submenu2`. Additionally, a menu item can be a separator or a checkbox menu item. Selecting the checkbox menu item toggles the checkmark shown to the left of the caption on and off.

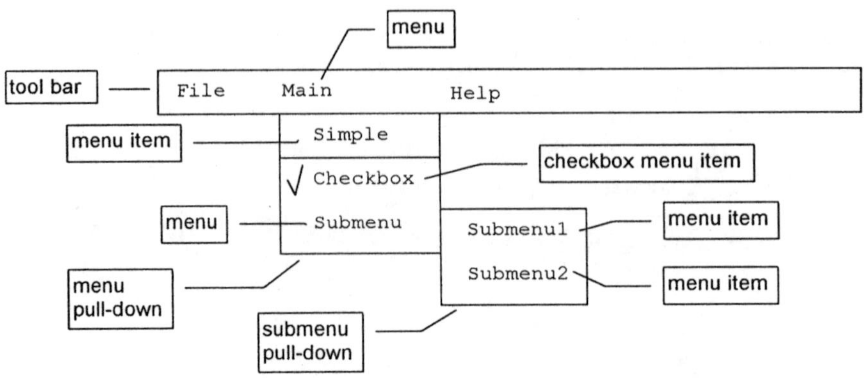

FIGURE 10.7
Typical menu and its components.

The menu provides the primary user application interface. In general, users have come to expect a consistent menu layout. As such, the File menu, containing Save, Print, and Exit program options, is generally placed in the leftmost menu bar position. When editing is required in the application, editing options are placed in the Edit menu which follows the File menu in the menu bar. The rightmost menu is the Help menu. Application-related menus are placed between the Edit and Help menus in the toolbar.

Menu Bar

The **`MenuBar` class** models menu bars. The class has one constructor that does not take any parameters. Methods of the class are summarized in Table 10.16.

TABLE 10.16

Selected members of the **`Menu`** class

Name	Description
`Menu add (Menu m)`	Adds the menu *m* to the toolbar. The menu added is returned.
`void remove (int index)`	Removes the menu at the position specified by *`index`*.

A menu bar is added to a frame with the `Frame` method `setMenuBar`. For example, the menu `mb` is added to the frame `f` with the statement

```
f.setMenuBar( mb )
```

Menu

Menus are components of the menu bar. Each menu contains a list of items that are displayed when the menu is clicked; menus often are referred to as

TABLE 10.17

Menu class constructors

Name	Description
`Menu( )`	Constructs a new menu with no label.
`Menu(String label)`	Constructs a new menu with text *label*.

TABLE 10.18

Menu items

Type	Description
`MenuItem`	Simple menu item.
`CheckboxMenuItem`	Checkbox menu item.
`Menu`	Menu item with a pull-down submenu.

TABLE 10.19

Selected Menu class methods

Name	Description
`MenuItem add()`	Adds the menu item *mi* to the menu and returns the menu item added.
`void add(String label)`	Adds a menu items to the menu with the text *label*.
`void addSeparator( )`	Adds a separator line to the menu.
`void insert(MenuItem mi, int index)`	Inserts the menu item *mi* into the menu at the position specified by *index*.
`void insert(String label, int index)`	Inserts the menu item with text *label* into the menu at the position specified by *index*.
`void insertSeparator(int index)`	Inserts a separator into the menu at the position specified by *index*.
`void remove(int index)`	Removes the menu item at the position *index*.
`void remove(MenuComponent item)`	Removes the menu item *item*.
`void removeAll( )`	Removes all the menu items.

pull-down menus. Menus are modeled by the **Menu class.** Menu class constructors are summarized in Table 10.17.

For example, the statement

```
Menu menuMain = new Menu("Main");
```

creates a Menu object with the caption "Main" referenced by menuMain.

Items in the menu must belong to the MenuItem, CheckboxMenuItem, or Menu classes as described in Table 10.18. The items are added and manipulated with the members of the Menu class. Selected methods of the Menu class are shown in Table 10.19.

For example, a MenuItem type itemSimple is added to the menu menuMain with the statement

```
menuMain.add(itemSimple);
```

Note that if the added menu item is part of another menu it is removed from the previous menu. Note also that a separator line is added at the current position in the formation of the menu's items.

Menu Item

The **`MenuItem` class** models menu items that are not checkbox or submenu items. Constructors of the `MenuItem` class are summarized in Table 10.20.

TABLE 10.20

MenuItem class constructors

Name	Description
`MenuItem( )`	Constructs a new `MenuItem` with no label and no keyboard shortcut.
`MenuItem(String` *`label`*`)`	Constructs a new `MenuItem` with text *`label`* and no keyboard shortcut.
`MenuItem(String` *`label`* `,) KeyboardShortcut` *`sc)`*	Constructs a new `MenuItem` with text *`label`* and keyboard shortcut *`sc`*.

A menu item with the caption `"Simple"` and reference `itemSimple` is created with the statement

```
MenuItem itemSimple = new MenuItem("Simple");
```

The use of a dash (-) as a label is used to create a separator menu item.

Menu items are manipulated with the `MenuItem` class methods. Selected `MenuItem` class methods are summarized in Table 10.21.

TABLE 10.21

Selected `MenuItem` class methods

Name	Description
`void addActionListener( ActionListener` *`al`* `)`	Adds the action listener *`al`* to receive the menu item's events.
`void setEnabled(boolean` *`status`*`)`	Enables the menu item if *`status`* is `true` and disables the menu item if *`status`* is `false`.
`void setLabel(String` *`label`*`)`	Sets the menu item's label to the value *`label`*.
`void setShortcut( MenuShortcut` *`sc`* `)`	Sets or replaces the item's keyboard shortcut to *`sc`*.

For example, a menu item is enabled or disabled with the `setEnabled` method. An enabled menu item is user-responsive, such as mouse clicks over the menu item, whereas a disabled menu item is not responsive. Disabled menu items are displayed with a faint caption and are said to be grayed. For example, the menu item `itemSubmenu2` is disabled with the statement

```
itemSubmenu2.setEnabled(false);
```

Checkbox Menu Items

A checkbox menu item contains a checkbox that can be toggled between the "on" state and the "off" state by clicking over the menu item. When in the "on" state a small check appears to the left of the caption. The checkbox menu item is modeled with the **CheckboxMenuItem** class. Constructors of the class are shown in Table 10.22.

TABLE 10.22

CheckboxMenuItem class constructors

Name	Description
`CheckboxMenuItem( )`	Constructs a checkbox menu item with no caption and the checkbox initially set to "off."
`CheckboxMenuItem(String label)`	Constructs a checkbox menu item with caption *label* and the checkbox initially set to "off."
`CheckboxMenuItem(String label,) boolean state)`	Constructs a checkbox menu item with caption *label* and the checkbox initially set to "off" if *state* is false and "on" if *state* is true.

For example, a checkbox item in the "on" state—that is, checked—and reference itemCheckbox is created with the statement

```
CheckboxMenuItem itemCheckbox = new
                    CheckboxMenuItem("Checkbox",true);
```

Selected methods of the CheckboxMenuItem class are summarized in Table 10.23.

TABLE 10.23

Selected CheckboxMenuItem class methods

Name	Description
`void addItemListener( ItemListener il )`	Adds the item listener *il* to receive events associated with the checkbox.
`boolean getState( )`	Returns the value true if the checkbox is true and false if unchecked.
`void setState(boolean state )`	Sets the checkbox to "on" if *state* is true and "off" if *state* is false.

Creating the Menu

A program illustrating creation of a menu bar is shown in Figure 10.8a. The program's output is shown in Figure 10.8b. The menu bar object is created with the statement

```
MenuBar menuBar = new MenuBar();
```

and referenced with menuBar. The menu bar contains two menus: File and Main. These menus are created and reference with the statements

```
Menu menuFile = new Menu("File");
Menu menuMain = new Menu("Main");
```

```
import java.awt.*;
import java.awt.event.*;
public class UsingMenus extends Frame
      implements WindowListener
{
      MenuBar menuBar = new MenuBar();
      Menu menuFile = new Menu("File");
      Menu menuMain = new Menu("Main");
      Menu menuSubmenu = new Menu("Submenu");

      MenuItem itemSimple = new MenuItem("Simple");
      MenuItem itemSeparator = new MenuItem("-");
      CheckboxMenuItem itemCheckbox =
                    new CheckboxMenuItem("Checkbox",true);
      MenuItem itemSample = new MenuItem("Sample");
      MenuItem itemSubmenu1 = new MenuItem("Submenu 1");
      MenuItem itemSubmenu2 = new MenuItem("Submenu 2");



      public UsingMenus( )
      {
           addWindowListener(this);



           setMenuBar(menuBar);

           mb.add(menuFile);
           mb.add(menuMain);

           menuFile.add("Exit");
           menuMain.add(itemSimple);
           menuMain.add(itemSeparator);
           menuMain.add(itemCheckbox);
           menuMain.add(menuSubmenu);

           menuSubmenu.add(itemSubmenu1);
           menuSubmenu.add(itemSubmenu2);

           itemSubmenu2.setEnabled(false);

           setTitle("Menus");
           setSize(600,300);
           setVisible(true);
      }

      public static void main(String[] args)
      {
           new UsingMenus();
      }

      public void windowClosing( WindowEvent we )
      {
           setVisible(false );
           dispose( );
           System.exit( 0 ) ;
      }
```

```
        public void windowActivated( WindowEvent we ){}
        public void windowClosed( WindowEvent we ){}
        public void windowDeactivated( WindowEvent we ){}
        public void windowDeiconified( WindowEvent we ){}
        public void windowIconified( WindowEvent we ){}
        public void windowOpened( WindowEvent we ){}
}
```

FIGURE 10.8A
Program illustrating creation of a menu.

FIGURE 10.8B
Output of a program illustrating creation of a menu.

Another menu, which will serve as a submenu, is created with the statement

```
Menu menuSubmenu = new Menu("Submenu");
```

Note that menus and submenus are referenced and created the same.

Menu items for the menu `Main` are created with the statements

```
MenuItem itemSimple = new MenuItem("Simple");
MenuItem itemSeparator = new MenuItem("-");
CheckboxMenuItem itemCheckbox =
        new CheckboxMenuItem("Checkbox",true);
```

Two additional menu items are created with the statements

```
MenuItem itemSubmenu1 = new MenuItem("Submenu 1");
MenuItem itemSubmenu2 = new MenuItem("Submenu 2");
```

These menu items are intended as submenu menu items.

The menu bar, menu, and menu item components are connected together in the constructor of the `UsingMenus` class that is called when an instance of the class is created in the program's main method. The statement

```
setMenuBar(menuBar);
```

attaches the menubar, `menuBar`, to the Frame object `UsingMenus`. The menus are added to the menu bar with the `MenuBar` add method

```
menuBar.add(menuFile);
menuBar.add(menuMain);
```

The menuFile menu has the caption "File" and the menuMain menu has the caption "Main." The menu method add with a String type parameter is used to add a menu item with "Exit" caption to the menu menuFile.

```
menuFile.add("Exit");
```

The menu add method with a MenuItem parameter is used to add the menu items itemSimple, ItemSeparator, and itemCheckbox to the menu menuMain.

```
menuMain.add(itemSimple);
menuMain.add(itemSeparator);
menuMain.add(itemCheckbox);
```

Similarly, the menu add method with a Menu parameter is used to add the menu item menuSubmenu to the menu menuMain.

```
menuMain.add(menuSubmenu);
```

Note that menuSubmenu is a Menu type. The menu items itemSubmenu1 and itemSubmenu2 are added to the menu menuSubmenu with the statements

```
menuSubmenu.add(itemSubmenu1);
menuSubmenu.add(itemSubmenu2);
```

Finally, the menu item itemSubmenu2 is disabled with the statement

```
itemSubmenu2.setEnabled(false);
```

Using the Menu

The usefulness of menus is contained in their ability to direct program action; for example, program termination when the Exit option in the File menu is selected. Again, implementation is a four-step process:

1. Import the package java.awt.event containing event classes and listener interfaces.
2. Implement the ActionInterface, thereby declaring the actionPerformed method.
3. Invoke the addActionListener method for the menu item object.
4. Implement handlers for the actions in the actionPerformed override.

Implementation of these steps is shown in Figure 10.9.

Note presence of the java.awt.event import. This package was previously imported to enable handling of the window closing event.

Listening for action events is implemented for the class by adding the ActionListener interface to the comma-separated list of implemented interfaces in the class declaration.

```
public class UsingMenus extends Frame
implements WindowListener,
              ActionListener
```

The addActionListener method is invoked for the menu item itemExit with the statement

```
itemExit.addActionListener(this);
```

Finally, a handler for the action event is implemented with the actionPerformed.

```
public void actionPerformed(ActionEvent ae)
{
    if(ae.getSource()==itemExit)
    {
        setVisible(false );
        dispose( );
        System.exit( 0 ) ;
    }
}
```

The getSource method, inherited by the ActionEvent class from the Event class, is invoked to ascertain the source object of the event. If the source is the menu item itemExit the program is terminated.

When the program is executed, selecting the Exit option of the File menu will terminate program execution. Handlers for action events associated with the other menu items are implemented similarly.

```
import java.awt.*;
import java.awt.event.*;          ——— event classes and interfaces

public class UsingMenus extends Frame
     implements WindowListener,
                     ActionListener   ——— ActionListener interface
{

     MenuBar mb = new MenuBar();
     Menu menuFile = new Menu("File");
     Menu menuMain = new Menu("Main");
     Menu menuSubmenu = new Menu("Submenu");

     MenuItem itemExit = new MenuItem("Exit");
     MenuItem itemSimple = new MenuItem("Simple");
     MenuItem itemSeparator = new MenuItem("-");
     CheckboxMenuItem itemCheckbox = new
          CheckboxMenuItem("Checkbox",true);
     MenuItem itemSample = new MenuItem("Sample");
     MenuItem itemSubmenu1 = new MenuItem("Submenu 1");
     MenuItem itemSubmenu2 = new MenuItem("Submenu 2");
```

FIGURE 10.9
Handling selection of the Exit option in the File menu. (Continued on the next page.)

```
public UsingMenus( )
{
      addWindowListener(this);

      setMenuBar(mb);

      mb.add(menuFile);
      mb.add(menuMain);

      menuFile.add(itemExit);
      menuMain.add(itemSimple);
      menuMain.add("-");
      menuMain.add(itemCheckbox);
      menuMain.add(menuSubmenu);

      menuSubmenu.add(itemSubmenu1);
      menuSubmenu.add(itemSubmenu2);

      itemSubmenu2.setEnabled(false);

      itemExit.addActionListener(this);

      setTitle("Menus");
      setSize(600,300);
      setVisible(true);
}
public static void main(String[] args)
{
      new UsingMenus();
}
public void windowClosing( WindowEvent we )
{
      setVisible(false );
      dispose( );
      System.exit( 0 ) ;
}
public void windowActivated( WindowEvent we ){}
public void windowClosed( WindowEvent we ){}
public void windowDeactivated( WindowEvent we ){}
public void windowDeiconified( WindowEvent we ){}
public void windowIconified( WindowEvent we ){}
public void windowOpened( WindowEvent we ){}

public void actionPerformed(ActionEvent ae)
{
      if(ae.getSource()==itemExit)
      {
            setVisible(false );
            dispose( );


                  System.exit( 0 ) ;
            }
      }
}
```

item listener method invoked

action handler

action event source

> Menu action events are handled in the `actionPerformed` method.

HANDS-ON EXERCISE 10: CUSTOMER WINDOW APPLICATION

Menus provide an efficient method for navigating through and selecting application options. Consider a customer application in which the customer information displayed can be retrieved in two ways: by customer number or customer name. Further, assume information regarding the customer displayed can be printed in a customized report. The menu shown schematically in Figure HO-10.1 provides the required functionally.

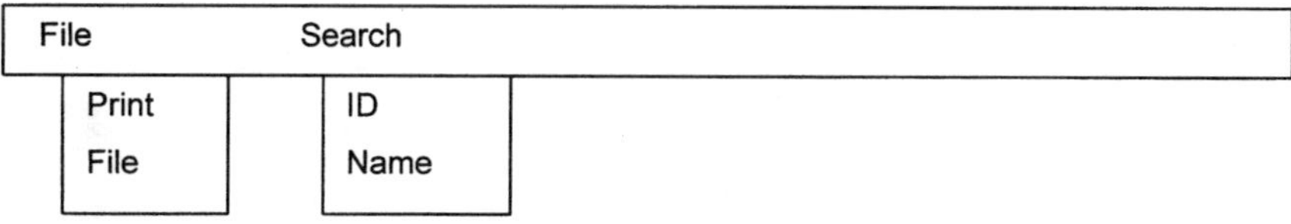

FIGURE HO-10.1
Customer application menu.

One of the advantages of Windows is commonality. For example, the print and exit options are typically located in the File menu, which is typically located in the leftmost position on the menu bar. Similarly, the Help menu is positioned at the rightmost position on the menu bar. Every effort should be made to retain the Windows feel to which users have become accustomed.

This exercise implements a typical Customer application menu.

Step 1: Create a New Project

1. Start Visual J++.
2. Select the New Project... command in the File menu. The New Project dialog box is displayed.
3. Enter **`Customer Window`** in the Name text field.
4. Use the Browse button to enable navigation to the Location `C:\Hands-On\Customer Window`.
5. Click the Open button to create the new project.

Step 2: Create `Customer.java` File

1. Press the Ctrl+D keys. The Add Item dialog box is displayed.
2. Select the Class folder.
3. Select the ClassMain icon.
4. Enter **`Customer.java`** in the Name text field.
5. Click the Open button to create the new class.
6. The Code Editor is opened with the file `Customer.java`.

Step 3: Code the Application

Enter code to create a frame window and handle window events, specifically `windowClosing`.

```
import java.awt.*;
import java.awt.event.*;
public class Customer extends Frame
      implements WindowListener
{
      public Customer( ) {
           addWindowListener(this);
           setSize(400,200);
           setVisible(true);
      }

      public static void main(String[] args)
      {
                    new Customer();
      }

      public void windowClosing( WindowEvent we ) {
           setVisible(false );
           dispose( );
           System.exit( 0 );
      }
      public void windowActivated( WindowEvent we ){}
      public void windowClosed( WindowEvent we ){}
      public void windowDeactivated( WindowEvent we ){}
      public void windowDeiconified( WindowEvent we ){}
      public void windowIconified( WindowEvent we ){}
      public void windowOpened( WindowEvent we ){}
}
```

Note: Completed code is shown in Figure HO-10.2.

Step 4: Build the Basic Application

1. Press the CTRL+SHIFT+B keys.
2. Debug the code as required.

Step 5: Add Code to Declare and Create Menu and Menu Items

Enter the following code

```
MenuBar menuBar = new MenuBar();

Menu menuFile = new Menu("File");
MenuItem itemPrint = new MenuItem("Print");
MenuItem itemExit = new MenuItem("Exit");

Menu menuSearch = new Menu("Search");
MenuItem itemID = new MenuItem("Id");
MenuItem itemName = new MenuItem("Name");
```

Step 6: Build the Basic Application

1. Press the CTRL+SHIFT+B keys.
2. Debug the code as required.

Step 7: Enter Code to Implement Menu in Frame

Enter the following code:

```
setMenuBar(menuBar);

menuBar.add(menuFile);
menuBar.add(menuSearch);

menuFile.add(itemExit);
menuFile.add(itemPrint);
menuSearch.add(itemID);
menuSearch.add(itemName);
```

Step 8: Build the Basic Application

1. Press the CTRL+SHIFT+B keys.
2. Debug the code as required.

```
import java.awt.*;
import java.awt.event.*;
public class Customer extends Frame
      implements WindowListener
{
      MenuBar menuBar = new MenuBar();

      Menu menuFile = new Menu("File");
      MenuItem itemPrint = new MenuItem("Print");
      MenuItem itemExit = new MenuItem("Exit");
```

```
        Menu menuSearch = new Menu("Search");
        MenuItem itemID = new MenuItem("Id");
        MenuItem itemName = new MenuItem("Name");

        public Customer( ) {
             setMenuBar(menuBar);

             menuBar.add(menuFile);
             menuBar.add(menuSearch);

             menuFile.add(itemExit);
             menuFile.add(itemPrint);
             menuSearch.add(itemID);
             menuSearch.add(itemName);

             addWindowListener(this);
             setSize(400,200);
             setVisible(true);
        }

        public static void main(String[] args)
        {
                                    new Customer();
        }

        public void windowClosing( WindowEvent we ) {
             setVisible(false );
             dispose( );
             System.exit( 0 );
        }
        public void windowActivated( WindowEvent we ){}
        public void windowClosed( WindowEvent we ){}
        public void windowDeactivated( WindowEvent we ){}
        public void windowDeiconified( WindowEvent we ){}
        public void windowIconified( WindowEvent we ){}
        public void windowOpened( WindowEvent we ){}
}
```

FIGURE HO-10.2
File `Customer.java`

Step 9: Set the Project Properties

1. Select the Customer Properties… option in the Project menu. The Properties dialog box is displayed.
2. Click the OK button to accept the default properties.

Step 10: Step through Execution of the Application

1. Press the F5 key. The application starts and the Customer window is displayed (Figure HO-10.3)
2. Click the Search menu. The Search menu items are displayed.

FIGURE HO-10.3
Customer application window.

3. Click either the ID or Name items to close the menu.

Step 11: Save All the Project Files

Press the CTRL+SHIFT+5 keys.

Step 12: Close the Project

Select the Close All option in the File menu.

Step 13: Exit Visual J++ the Project

Select the Exit option in the File menu.
Challenge: Add code to implement a Help menu with Help About menu item.

> *Note:* Help is generally located at the rightmost position on the menu bar.

SUMMARY

1. Java's Abstract Window Toolkit provides the classes and interfaces for developing event-driven applications.
2. Frame windows are the foundation of the graphical user interface.

3. Menus are the primary means for user selection of program options.
4. Menu classes enable the creation of menus.
5. Selection of a menu option results in an event.
6. User and system actions, selection of a menu option, or clicking a title bar icon, for example, result in messages called `events`.
7. Event classes and interfaces enable programmatic response to events.

KEY WORDS AND CONCEPTS

Abstract Window Toolkit (AWT)
`ActionListener`
`AdjustmentListener`
`CheckboxMenuItem`
component
`Component` class
`ComponentListeners`
container
event
`EventListener`
event-oriented programs
frames
`ItemListener`
`KeyListener`
menu
menu bar
`Menu` class
`Menubar` class
`MenuItem` class
`MouseListener`
pull-down menu
submenu
`Window` class
`WindowListener`

REVIEW QUESTIONS AND EXERCISES

1. Mark the following statements true of false:

   (T/F) The user friendliness of the window is the primary reason for its wide acceptance.

   (T/F) The Abstract Window Toolkit has classes for creating windows.

   (T/F) A frame window object is modeled by the `FrameWindow` class.

   (T/F) Only users initiate events.

   (T/F) Component type objects cannot be associated with events.

(T/F) Selecting a menu option results in an event.

(T/F) Action events are handled in the `ActionHandler` method.

2. What is the difference between a frame and a dialog type window?
3. How is the `setVisible` method used?
4. Why is use of the `dispose` method essential?
5. Describe several user events.
6. What is the purpose of a listener?
7. How is listening enabled?
8. Is it necessary to implement all of the methods of a listener?
9. Add a Help menu to the application developed in the text.
10. Add code to enable the menu item `Submenu2` when the `Simple` menu item is selected in the application developed in the text.
11. Add items to the application's Main menu that will insert and delete menu items in another menu when selected by the user.

PRACTICE WITH VISUAL J++

Create a Frame window with menu for use by a book store in responding to customer requests for specific books. The menu should include a Search menu that has three search options: Title, Author, and ISBN number. Another menu, Range, should provide for searching (1) only in the store or (2) in the store and warehouse. The current Range mode should be checked.

11

Graphics

OBJECTIVES

After reading this chapter, you will be able to:

1. Understand the concept of graphic context.
2. Recognize windows and controls of graphical interfaces as graphically representable components.
3. Specify the color of graphic objects.
4. Stipulate the font and use font metrics in arranging textual graphics.
5. Use graphic methods for rendering text and figures.
6. Demonstrate the rendering of text and figures.

OVERVIEW

The java.awt package includes classes enabling drawing on an output device. Figure 11.1 shows the hierarchical structure of these classes. Class `Color` encapsulates the management of color. Class `Fonts` and `FontMetrics` contains the attributes and methods for management of text. Class `Graphics` contains the attributes and methods for rendering text, figures, and images.

This section presents the Java fundamentals for rendering lines, figures, and text.

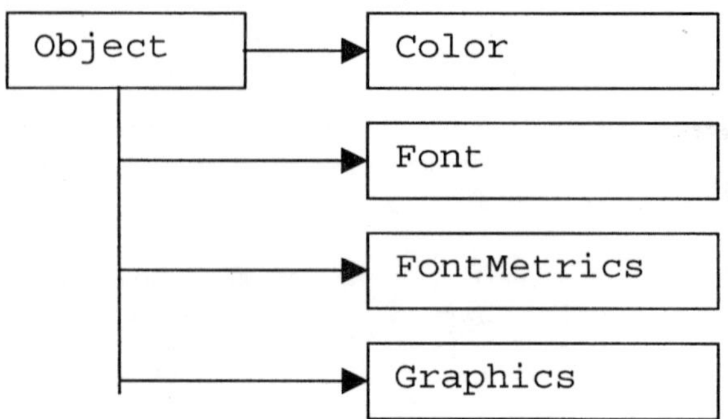

FIGURE 11.1
Selected members of the **`java.awt`** hierarchy.

GRAPHIC CONTEXT

Java applications must be able to render displays on platforms varying widely in terms of software and hardware. For example, the specifics required to draw a line on a PC running Windows vary greatly from those required to draw the same line on a Macintosh. The specifics of rendering displays, both on-screen and off-screen, on a particular platform are characterized in a **graphic context.** The **`Graphics` class** encapsulates graphical context specifics required to enable Java applications to render displays including the following:

- The device on which the rendering is to be performed.
- Current color.
- Current font for text.
- Position of the rendering.
- Clipping coordinates for components that are not fully shown.

The graphic context includes a **coordinate system** illustrated in Figure 11.2. Every point on the output medium is represented by a **coordinate point.** Each coordinate point consists of a x coordinate and a y coordinate. The point (0,0) is located in the upper left corner. The x coordinate is measured along the horizontal axes with increasing values to the left. The y coordinate is measured along the vertical axes with increasing values downward. The coordinates are measured in pixels. A **pixel,** contraction of picture element, is the smallest unit of resolution of the output medium.

A derived version of the `Graphics` class is implemented for every platform. The `Graphics` class includes methods for rendering figures, text, and images.

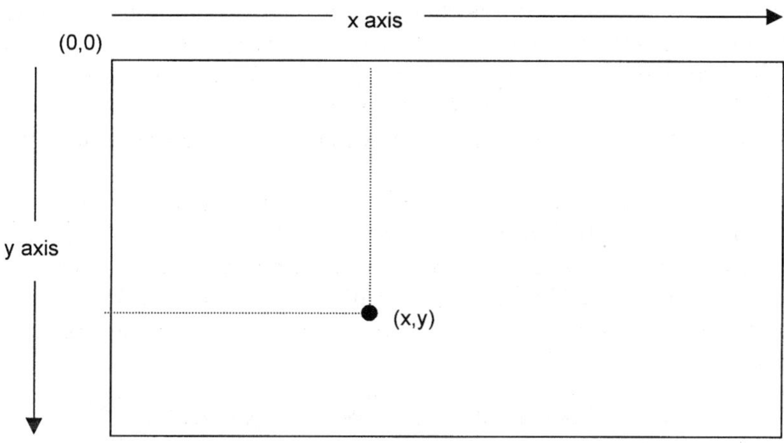

FIGURE 11.2
Java's coordinate system.

COMPONENTS

Components have a graphical representation. Frame and dialog windows are components. Controls, such as buttons, textboxes, scrollbars, and checkboxes, used with windows to provide user interfaces, are also components. Graphic context provides the mechanism for subsuming the requirements of diverse software and hardware platforms in rendering these components. Methods of the `Component` class enable the rendering of graphical representations of components. Graphical methods of the `Component` class are summarized in Table 11.1.

TABLE 11.1
Graphic methods of the `Component` class

Name	Description
`Graphics getGraphics( )`	Returns a graphic context for the component. A `null` value is returned if the component is not visible.
`void paint(Graphics g)`	Paints the component using the graphic context *g*.
`void repaint( )`	Paints the component.
`void update(Graphics g)`	Updates the component including painting.

Method **`paint( )`** is called whenever the component's graphical representation needs to be painted including when the component is first shown, is uncovered or altered. AWT passes the graphical context of the component to the method `paint( )` when it is called. Painting describing the lines, figures, and text to be painted is placed in the `paint( )` method.

Method **`repaint( )`** calls the `update( )` method as soon as possible. Variations in the `repaint( )` method provide for calling the update `method( )` within a specific period of time or specifying a particular rectangle of the component to be painted.

AWT responds to a `repaint( )` call by calling **`update( )`**. The `update( )` method first clears the component by filling it with the background color, then sets the color of the graphic context to the color of the component being updated, and finally, calls the component's `paint( )` method.

The graphic context of a component is returned by the method **`getGraphics( )`**.

> Java statements describing the lines, figures, and text to be painted are placed in the `paint( )` method.

COLOR

All Java **colors** are viewed as a combination of three basic colors—red, blue and green—mixed in a specified proportion. The resulting color depends on the proportion of the basic colors specified. This approach to color, often referred to as the **three-component RBG** model, is modeled by the **`Color` class** in the package `java.awt`. The proportion of each basic color can be specified using an `int` value in the range 0 through 255. The value 255 implies the color is represented with full intensity and 0 implies the color is not represented. The three-tuple 1, 0, 0 implies the color red is represented at full intensity whereas blue and green are not represented. The three-tuple thus represents the color red. The three-tuple 122, 0, 122 represents a mixture of red and green approximating a yellow–brown color. Selected `Color` class constructors are summarized in Table 11.2.

TABLE 11.2

Selected Color class constructors

Name	Description
`Color(int red, int blue, int green)`	Creates a color based on mixing red, blue, and green colors in the proportion specified by the `int` values *`red, blue`* and *`green`*.
`Color(float red, float blue,float green)`	Creates a color based on mixing red, blue, and green colors in the proportion specified by the `float` values *`red, blue`* and *`green`*.

The color white is represented by the three-tuple 255, 255, 255 and the color black is represented by the three-tuple 0, 0, 0.

A program creating a frame window with light blue background is shown in Figure 11.3a.

```
import java.awt.*;
import java.awt.event.*;

public class FrameWindow extends Frame
      implements WindowListener
{
      public FrameWindow( )
      {
           addWindowListener(this);

           setTitle("Blue Window");
           setSize(400,200);
           setVisible(true);

           Color bg = new Color (0, 255, 255);
           setBackground(bg);
      }

      public static void main(String[] args)
      {
           new FrameWindow();
      }

      public void windowClosing( WindowEvent we )
      {
           setVisible(false ); // hide frame
           dispose( );         // free resources
           System.exit( 0 ) ;  // terminate program
      }
      public void windowActivated( WindowEvent we ){}
      public void windowClosed( WindowEvent we ){}
      public void windowDeactivated( WindowEvent we ){}
      public void windowDeiconified( WindowEvent we ){}
      public void windowIconified( WindowEvent we ){}
      public void windowOpened( WindowEvent we ){}
}
```

FIGURE 11.3A
Painting the component background color.

The statement

```
Color bg  = new Color(0, 255, 255);
```

creates a `Color` object referenced by `bg` representing a color consisting of a mixture of blue and green. The statement sets the background color of the frame to the color represented by the `Color` object `bg`.

```
setBackground(bg);
```

FIGURE 11.3B
Output of program painting the component background color.

The program displays a frame window with a light blue or aqua background as shown in Figure 11.3b.

> Color class encapsulates the red-blue-green three component model for representing colors.

FONT AND FONT METRICS

Typeface, size, and style of rendered text are characterized by the **`Font`** class. Renderings of typical fonts are shown in Figure 11.4A.

Typeface	Style	Size
TimesRoman	Plain	12
`Courier`	`Bold`	`36`
TimesRoman	*Italic*	*12*
TimesRoman	***Bold + Italic***	***12***

FIGURE 11.4A
Text renderings illustrating typefaces, styles, and size

Typical font types include Courier, Helvetica, and TimesRoman. The font styles are plain, bold, and italic. Font size is specified in points. A point is 1/72 of an inch. Selected key methods of the `Font` class is shown in Table 11.3.

A `Font fontTRBI` object representing TimesRoman typeface thirty-six points high with bold and italic styling is created with the statement

TABLE 11.3

Selected methods of the `Font` class

Method	Description
`void Font(String fontName, int style, int size)`	Creates a font object with typeface *`fontName`*, style *`style`* and point size *`size`*.
`String getName( )`	Returns the name of the font.
`int getSize( )`	Returns the size of the font in points.
`int getStyle( )`	Returns the integer value of the style.

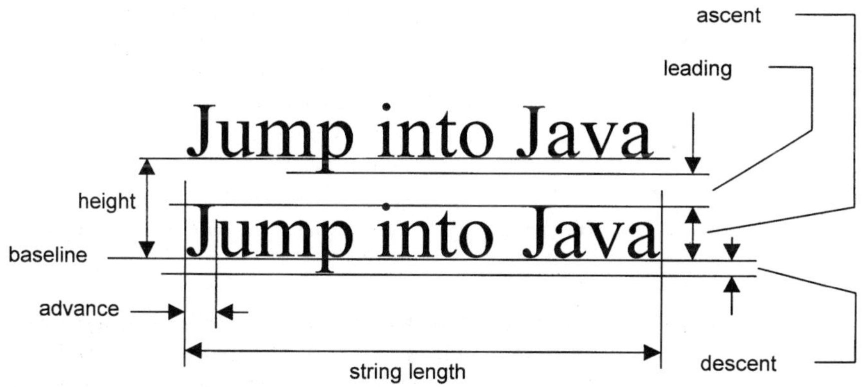

FIGURE 11.4B
Font measurements

```
Font fontTRBI = new Font("TimesRoman",
                         Font.BOLD + Font.ITALIC,36);
```

The typeface is specified with the name of the type. The style is specified as an integer. The Font class specifies constants for plain, bold, and italic styles. Note that the style constants are accessed as `Font.PLAIN`, `Font.BOLD`, and `Font.ITALIC`. The plus sign (+) can be used to joint the style constants to represent compound styles. For example, the compound style of bold and italic is specified with the clause

```
Font.BOLD + Font.ITALIC
```

Size is specified as 36 points.

Font metrics are illustrated in Figure 11.4B. Characters are positioned on a horizontal line called the **baseline.** The distance between the baseline of one line of text and the baseline of the next line of text is the font's **height.** The font's **ascent** is the distance from the baseline to the top of most of the font's capital characters. The font's **descent** is the distance characters extend below the baseline. Every character has an **advance** that specifies the relative position of subsequent characters. Note that the advance can vary from character to character. Fonts like TimesRoman have a varying advance, whereas Courier has a constant advance. **Leading** is the distance between the descent of one line of text and the ascent of the next line of text. Leading can be viewed as the interline spacing.

Measurement of fonts is encapsulated in the **`FontMetrics`** class. A constructor and selected methods of the `FontMetrics` class are shown in Table 11.4.

TABLE 11.4

Constructor and selected methods of the `FontMetrics` class

Method	Description
`void FontMetrics(Font` *`font`*	Creates a FontMetric object characterizing the measurements of the specified *`font`*.
`int charWidth(Char` *`ch`*`)`	Returns the advance measurement of the character *`ch`* in the font.
`int getAscent( )`	Returns the ascent for the font.
`int getDescent()`	Returns the descent for the font.
`int getHeight(.)`	Returns the font's height.
`int getLeading(.)`	Returns the font's leading.
`int stringWidth(String` str)	Returns the string length of the specified string *`str`*.

The Toolkit **class** provides additional methods useful in handling fonts and font metrics. `Toolkit` is an abstract class and is the basis for the actual implementation of the Abstract Window Toolkit. The method `getDefaultToolkit()` returns the default toolkit, which is a system property named `awt.toolkit`. The default toolkit contains the methods of the `Toolkit` class and can be used to retrieve information regarding the graphic context, fonts, and font metrics. Selected members of the `Toolkit` class are summarized in Table 11.5.

TABLE 11.5

Selected methods of the `Toolkit` class

Method	Description
`TString[] getFontList()`	Returns the names of the available fonts as elements of a string array.
`FontMetrics getFontMetrics(Font` *`fnt`*`)`	Returns the font metrics of the specified font *`fnt`*.
`int getScreenResolution()`	Returns the screen's resolution in dot-per-inch.
`Dimension getScreenSize()`	Returns the screen's size in pixels.

METHODS FOR RENDERING TEXT

Text is output to a specific position in the coordinate system of the graphical context. The position is specified by an x and y coordinate. The text baseline is positioned at the y coordinate. Note that the character advances from the x position and that some characters may descend below the y position baseline, as shown in Figure 11.5.

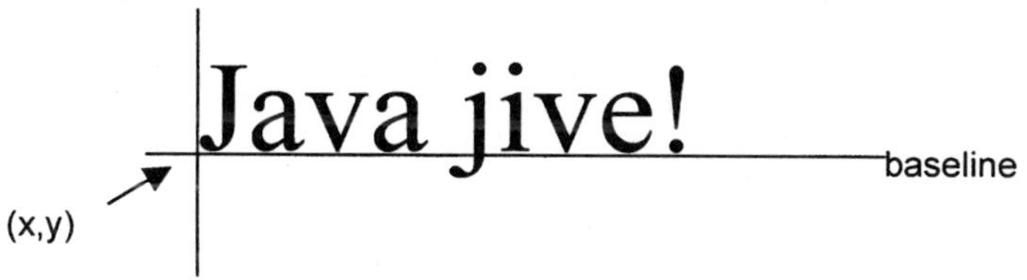

FIGURE 11.5
Positioning of text within the graphic context.

The **`Graphics` class** provides methods for rendering text. Methods of the `Graphics` class for rendering text are summarized in Table 11.6.

TABLE 11.6

Selected methods of the `Graphics` class used for rendering text

Method	Description
`void drawBytes(byte` *`data[]`*`, int` *`offset`*`, int` *`length`*`, int` *`x`*`, int` *`y`*`)`	Draws *`length`* textual characters from the array `data[]` starting with the *`offset`* element at the point (*x*, *y*)
`void drawChars(char` *`data[]`*`, int` *`offset`*`, int` *`length`*`, int` *`x`*`, int` *`y`*`)`	Draws *`length`* textual characters from the array `data[]` starting with the *`offset`* element at the point (*x*, *y*)
`void drawString(String` *`string`*`, int` *`x`*`, int` *`y`*`)`	Draws textual characters comprising *`string`* at the point (*x*, *y*)

When the textual data are specified in an array, the first element drawn is specified by the *`offset`* argument and the number of characters drawn is specified by the `length` argument. The first character is positioned on a baseline specified by the `x` and `y` arguments. The text is rendered in the current color and font.

Text are displayed by the program listed in Figure 11.6a. The statement

```
Font fontSP = new Font("Serif",Font.PLAIN,12);
```

creates a `Font` object `fontSB` representing 12-point plain serif type. Similarly, the statements

```
Font fontSSB = new Font("SansSerif",Font.BOLD,18);
Font fontMBI = new Font("Monospaced",Font.BOLD +
                                        Font.ITALIC,24);
```

create font objects representing 18-point bold sans-serif type and 24-point monospaced bold and italic type.

```
import java.awt.*;
import java.awt.event.*;

public class FrameWindow extends Frame
      implements WindowListener
{
      Font fontSP = new Font("Serif",Font.PLAIN,12);
      Font fontSSB = new Font("SansSerif",Font.BOLD,18);
      Font fontMBI = new Font("Monospaced",Font.BOLD +
                                        Font.ITALIC,24);

      public FrameWindow()
      {
       addWindowListener(this);

       setTitle("Painting Text");
       setSize(600,200);
       setVisible(true);

       Color bg = new Color (0, 255, 255);
       setBackground(bg);
      }
      public void paint(Graphics g)
      {
       Color r = new Color(255,0,0);

       g.setFont(fontSP);
       g.drawString("Serif 12 point plain ", 30, 40);
       g.setFont(fontSSB);
       g.drawString("SansSerif 24 point bold", 30, 68);
       g.setFont(fontMBI);
       g.drawString("Monospaced 36 point bold & italic",
                                        30, 102);
      }

      public static void main(String[] args)
      {
       new FrameWindow();
      }

      public void windowClosing( WindowEvent we )
      {
       setVisible(false );
       dispose( );
       System.exit( 0 );
      }
      public void windowActivated( WindowEvent we ){}
      public void windowClosed( WindowEvent we ){}
      public void windowDeactivated( WindowEvent we ){}
      public void windowDeiconified( WindowEvent we ){}
```

```
        public void windowIconified( WindowEvent we ){}
        public void windowOpened( WindowEvent we ){}
}
```

FIGURE 11.6A
Rendering text in a frame

The statement

```
g.setFont(fontSP);
```

sets the font of the graphic context `g` to font type represented by the `Font` object `fontSB`. The string `"Serif 12 point plain "` is output to the position (`30,40`) in the graphic context's coordinate system by the statement

```
g.drawString("Serif 12 point plain ", 30, 40);
```

Similarly, the statements

```
g.drawString("SansSerif 24 point bold", 30, 68);
g.drawString("Monospaced 36 point bold & italic",
                                              30, 102);
```

paint the strings `"SansSerif 24 point bold"` and `"Monospaced 36 point bold & italic"` at positions `(30,68)` and `(30,102)` in the fonts `fontSSB` and `fontMB`, respectively. The programs output is shown in of Figure 11.6b.

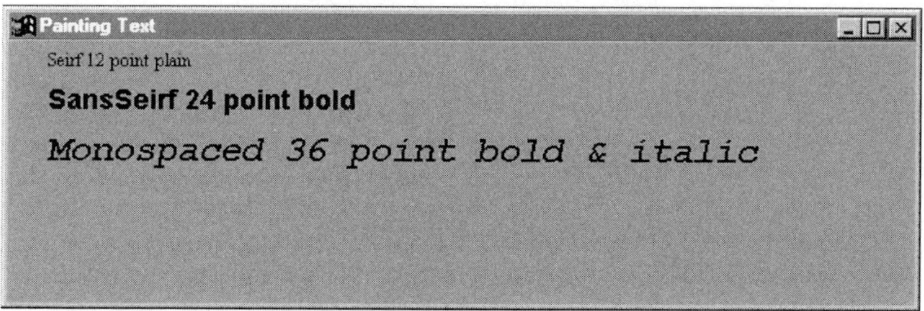

FIGURE 11.6B
Output of program rendering text in a frame

METHODS FOR RENDERING FIGURES

Methods of the `Graphics` class for rendering figures are summarized in Table 11.7.

TABLE 11.7

Selected methods of the `Graphics` class

Method	Description
`void drawArc(int x, int y, int width, int height, int beginningAngle, int endingAngle)`	Draws an arc in a rectangle with upper-left corner at the point (*x*, *y*) with `width` wide and `height` high in the current color where `beginningAngle` is the starting angle and `endingAngle` is the ending angle of the arc.
`void drawLine(int x1, int y1, int x2, int y2)`	Draws a line connecting the point (x1, y1) with the point (x2, y2) in the current color.
`void drawOval(int x, int y, int width, int height)`	Draws outline of an oval in a rectangle with upper-left corner at the point (*x*, *y*) with `width` wide and `height` high in the current color.
`void drawRect(int x, int y, int width, int height)`	Draws outline of a rectangle with upper-left corner at the point (*x*, *y*) with `width` wide and `height` high in the current color.
`void drawRoundrect(int x, int y, int width, int height, int horizArcWidth, int vertArcHeight)`	Draws a rounded rectangle with upper-left corner at the point (*x*, *y*) with `width` wide and `height` high in the current color where `horizArcWidth` is the horizontal diameter of the corner rounding arc and `vertArcHeight` is the vertical diameter of the corner rounding arc.
`void drawPolygon(int xPoints[], int yPoints[], int nPoints)`	Draws an `nPoint` polygon defined by `nPoints` line segments connecting the points `(xPoint[0],yPoint[0])` through `(xPoint[nPoints-1], yPoint[nPoints-1])`.
`void fillArc(int x, int y, int width, int height, int beginningAngle, int endingAngle)`	Draws a filled arc in a rectangle with upper-left corner at the point (*x*, *y*) with `width` wide and `height` high in the current color where `beginningAngle` is the starting angle and `endingAngle` is the ending angle of the arc.
`void fillOval(int x, int y, int width, int height)`	Draws a filled oval in a rectangle with upper-left corner at the point (*x*, *y*) with `width` wide and `height` high in the current color.

TABLE 11.7

Selected methods of the Graphics class (continued)

Method	Description
void fillRect(int *x*, int *y*, int *width*, int *height*)	Draws a filled rectangle with upper-left corner at the point (*x*, *y*) with *width* wide and *height* high in the current color.
void drawRoundrect(int *x*, int *y*, int *width*, int *height*, int *horizArcWidth*, int *vertArcHeight*)	Draws a filled rounded rectangle with upper-left corner at the point (*x*, *y*) with *width* wide and *height* high in the current color where *horizArcWidth* is the horizontal diameter of the corner rounding arc and *vertArcHeight* is the vertical diameter of the corner rounding arc.
void drawPolygon(int *xPoints[]*, int *yPoints[]*, int *nPoints*)	Draws a filled *nPoint* polygon defined by *nPoints* line segments connecting the points *(xPoint[0],yPoint[0])* through *(xPoint[nPoints-1], yPoint[nPoints-1])*.

Drawing Lines

Lines are rendered with the drawLine method. The method takes two coordinates. The first coordinate specifies the line's beginning point or origin and the second coordinated specifies the line's end point or terminus. For example, the statement

```
g.drawLine(20,40,360,140);
```

draws a line one pixel wide beginning at point (20,40) and ending at point (360,140), as illustrated in Figure 11.7.

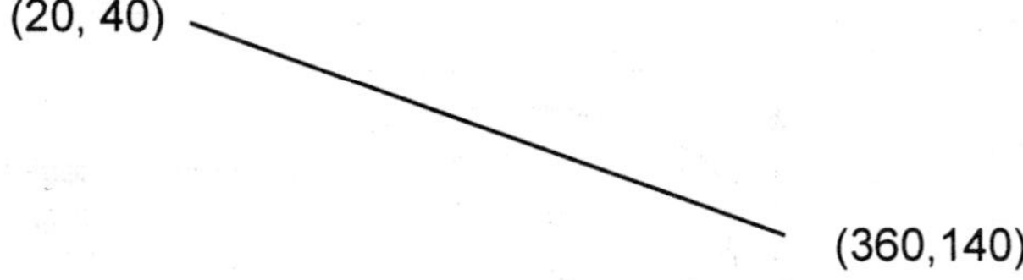

FIGURE 11.7
Line drawn with method drawLine(20,40,360,140);.

Drawing Rectangles

A rectangular outline is drawn by the drawRect method. For example, the statement

```
g.drawRect(20,40,360,140);
```

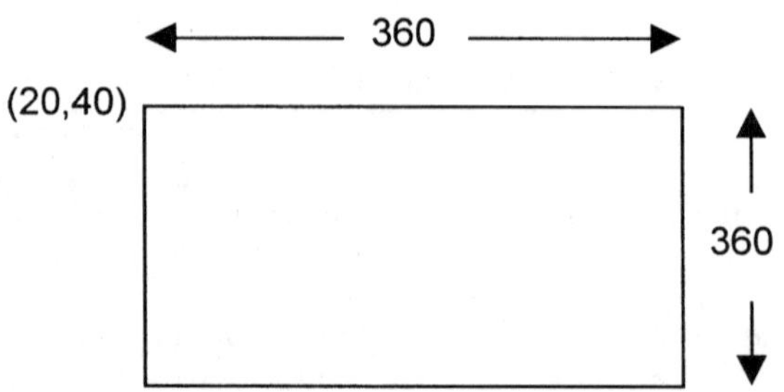

FIGURE 11.8
Rectangle drawn with the method `drawRect(20,40,360,140);`.

draws a rectangle with the upper-left corner at the point (20,40), 360 wide and 140 high as shown in Figure 11.8. The outline is one pixel wide.

The method `fillRect( )` draws a filled rectangle.

Drawing Arcs

The method `drawArc( )` draws a circular or elliptical arc. The point of origin of the angles is the center of the rectangle, 0 degrees is at the 3 o'clock position, and positive angles correspond to counter-clockwise rotation. For example, the statement

```
drawArc(20,40, 360,140,45,225);
```

draws an elliptical arc where the ellipse is contained within a box with upper-left corner at (20,40), height and width of 140 and 360, and where the beginning angle is 45 degrees and ending angle of 225 degrees, as shown in Figure 11.9.

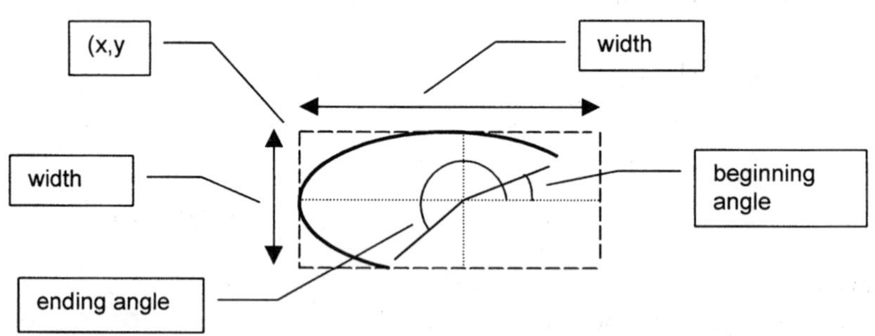

FIGURE 11.9
Arc drawn with method `drawArc(20,40, 360,140,45,225);`.

If the bounding rectangle is square the arc is circular, otherwise the arc is elliptical. In the case shown in Figure 11.9, the arc is elliptical.

The method `fillArc( )` will draw and fill the arc with the current color.

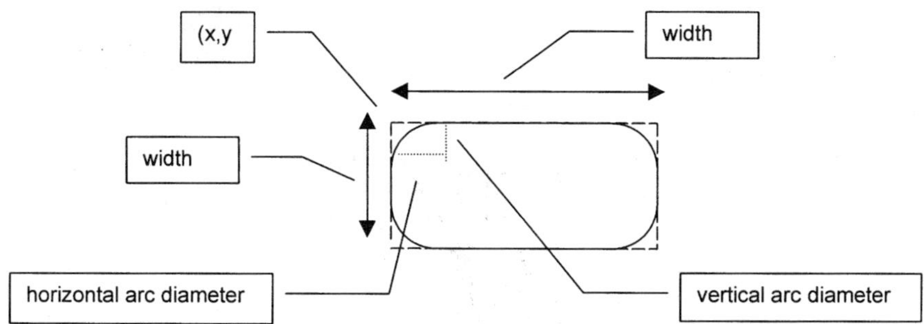

FIGURE 11.10
Rounded corner rectangle drawn with method `drawRoundRect(20,40,360,140,80,40);`.

Drawing Rounded Rectangles

The method `drawRoundRect( )` draws the outline of a rectangle with rounded corners. For example, the statement

```
g.drawRoundRect(20,40,360,140,80,40);
```

draws a rectangle with rounded corners. The horizontal and vertical diameters of the corner rounding arc are specified by the fifth and sixth arguments, 80 and 40 in this case. An example is shown in Figure 11.10.

The method `fillRoundRect( )` will draw and fill a rounded corner rectangle drawn with the current color.

Drawing Polygons

The method `drawPolygon` draws a closed polygon. For example, given the arrays of x and y coordinates

```
int yPoints[ ] = {260,190,200,130,140,60,50,120,110,
                        180,170};
int xPoints[ ] = {20,50,30,60,40,70,130,80,100,60,80};
```

the statement

```
g.drawPolygon(xPoints,yPoints,11);
```

draws the polygon shown in Figure 11.11.

The method takes three arguments. The first argument is an array containing the x coordinates of the polygon's corners. The second array contains the corresponding y coordinates. The third array contains the number of points in the polygon. The first points in each array correspond to the coordinates of the polygon's first corner. Subsequent points in each array correspond to sub-

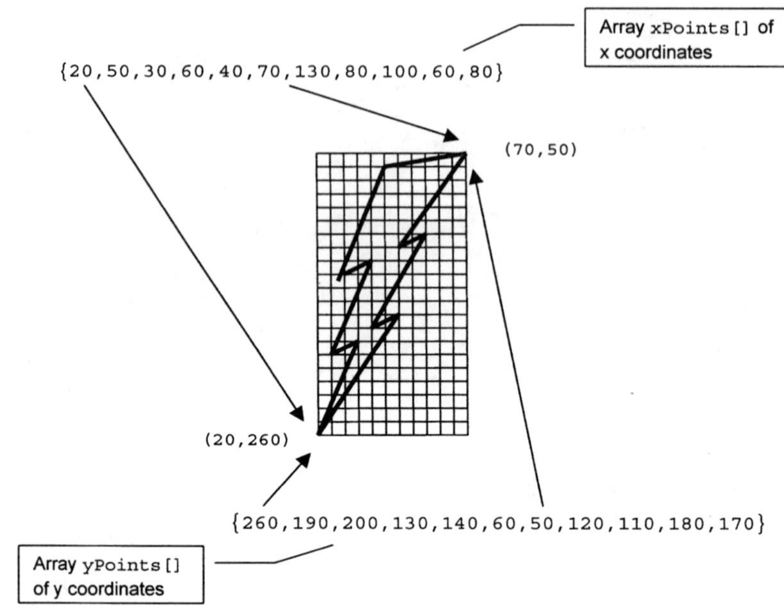

FIGURE 11.11
Polygon drawn with the statement `drawPolygon(xPoints,yPoints,11);`.

sequent corners of the polygon. The polygon is formed by connecting the corners with lines.

A filled polygon is drawn with the `fillPolygon` method.

Drawing Figures

The Java program in Figure 11.12a illustrates the drawing of a line, rounded rectangle, and arc using the methods of the `Graphics` class. The program's output is shown in Figure 11.12b.

```
import java.awt.*;
import java.awt.event.*;
public class FrameWindow extends Frame
      implements WindowListener
{
      public FrameWindow( )
      {
           addWindowListener(this);

           setTitle("Painting Figures");
           setSize(600,300);
           setVisible(true);

           Color bg = new Color (0, 255, 255);
```

```
            setBackground(bg);
      }
      public void paint(Graphics g)
      {
            g.setColor(Color.black);
            g.drawLine(20,40,460,240);
            g.setColor(Color.red);
            g.fillRoundRect(190,90,360,120,80,40);
            g.setColor(Color.blue);
            g.drawArc(20,40,360,140,45,180);
      }

      public static void main(String[] args)
      {
            new FrameWindow();
      }

      public void windowClosing( WindowEvent we )
      {
            setVisible(false );
            dispose( );
            System.exit( 0 );
      }
      public void windowActivated( WindowEvent we ){}
      public void windowClosed( WindowEvent we ){}
      public void windowDeactivated( WindowEvent we ){}
      public void windowDeiconified( WindowEvent we ){}
      public void windowIconified( WindowEvent we ){}
      public void windowOpened( WindowEvent we ){}
}
```

FIGURE 11.12A
Drawing a line, rounded rectangle, and arc with `Graphics` class methods.

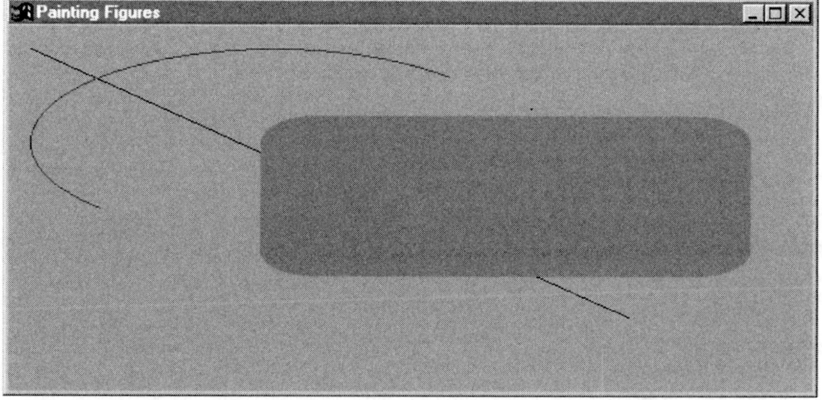

FIGURE 11.20B
Drawing a line, rounded rectangle, and arc with `Graphics` class methods.

HANDS-ON EXERCISE 11: RESPONDING TO KEYBOARD AND MOUSE ACTIONS

The mouse and keyboard are primary user interface devices. This exercise implements user keyboard and mouse interaction.

Step 1: Create a New Project

1. Start Visual J++.
2. Select the New Project... command in the File menu. The New Project dialog box is displayed.
3. Enter **`Mouse And Keyboard`** in the Name text field.
4. Use the Browse button to enable navigation to the Location `C:\Hands-On\Mouse And Keyboard`.
5. Click the Open button to create the new project.

STEP 2: Create `UserAction.java` File

1. Press the CTRL+D keys. The Add Item dialog box is displayed.
2. Select the Class folder.
3. Select the ClassMain icon.
4. Enter **`UserAction.java`** in the Name text field.
5. Click the Open button to create the new class.
6. The Code Editor is opened with the file `UserAction.java`.

Step 3: Code the Application

Enter code to create a frame window and handle window events, specifically `windowClosing`.

```
import java.awt.*;
import java.awt.event.*;
public class UserAction extends Frame
      implements WindowListener
{
      public static void main(String[] args)
      {
            new UserAction();
      }
```

```
        public UserAction() {
              addWindowListener(this);
              setSize(400,200);
              setVisible(true);
        }

        public void windowClosing(WindowEvent we) {
              setVisible(false);
              dispose();
              System.exit(0);
        }
        public void windowActivated(WindowEvent we){}
        public void windowClosed(WindowEvent we){}
        public void windowDeactivated(WindowEvent we){}
        public void windowDeiconified(WindowEvent we){}
        public void windowIconified(WindowEvent we){}
        public void windowOpened(WindowEvent we){}
}
```

Note: Completed code is shown in Figure HO-11.1.

Step 4: Build the Basic Application

1. Press the CTRL+SHIFT+B keys.
2. Debug the code as required.

Step 5: Add Code to Implement a Mouse Listener

1. Modify the class declaration to implement a mouse listener.

```
public class UserAction extends Frame
    implements WindowListener, MouseListener
```

2. Add a mouse listener to the frame window in the constructor

```
addMouseListener(this);
```

3. Enter the following code to implement the mouse listener event handlers

```
public void mouseClicked(MouseEvent me){ }
public void mouseEntered(MouseEvent me){ }
public void mouseExited(MouseEvent me){ }
public void mousePressed(MouseEvent me){
    setBackground(new Color(0,255,255));
    repaint();
}
public void mouseReleased(MouseEvent me){ }
```

4. Note that the background color is changed in the `mousePressed` event handler with the statement

```
setBackground(new Color(0,255,255));
```

Step 6: Build the Basic Application

1. Press the CTRL+SHIFT+B keys.
2. Debug the code as required.

Step 7: Set the Project Properties

1. Select the Mouse And Keyboard Properties… option in the Project menu. The properties dialog box is displayed.
2. Click the OK button to accept the default properties.

Step 8: Execute the Application

1. Press the F5 key. The application starts and the application window is displayed.
2. Position the cursor over the window.
3. Click a mouse key. The window background color changes.

Step 9: Enter Code to Implement a Keyboard Listener

1. Modify the class declaration to implement a key event listener.

```
public class UserAction extends Frame
    implements WindowListener, MouseListener,
                              KeyListener
```

2. Add a key event listener to the frame window in the constructor

```
addKeyListener(this);
```

3. Enter the following code to implement the key listener event handlers

```
public void keyPressed(KeyEvent ke){ }
public void keyReleased(KeyEvent ke){ }
public void keyTyped(KeyEvent ke){
    setBackground(new Color(255,0,0));
}
```

4. Note that the background color is changed in the `keytyped` event handler with the statement

```
setBackground(new Color(255,0,0));
```

Step 10: Build the Basic Application

1. Press the CTRL+SHIFT+B keys.
2. Debug the code as required.

```
import java.awt.*;
import java.awt.event.*;
public class UserAction extends Frame
      implements WindowListener, MouseListener,
              KeyListener
{
      public static void main(String[] args)
      {
       new UserAction();
      }

      public UserAction() {
       addKeyListener(this);
       addMouseListener(this);
       addWindowListener(this);
       setSize(400,200);
       setVisible(true);
      }
      public void keyPressed(KeyEvent ke){}
      public void keyReleased(KeyEvent ke){}
      public void keyTyped(KeyEvent ke){
       setBackground(new Color(255,0,0));
      }

      public void mouseClicked(MouseEvent me){}
      public void mouseEntered(MouseEvent me){}
      public void mouseExited(MouseEvent me){}
      public void mousePressed(MouseEvent me){
       setBackground(new Color(0,255,255));
      }
      public void mouseReleased(MouseEvent me){}


      public void windowClosing(WindowEvent we) {
       setVisible(false);
       dispose();
       System.exit(0);
      }
      public void windowActivated(WindowEvent we){}
      public void windowClosed(WindowEvent we){}
      public void windowDeactivated(WindowEvent we){}
      public void windowDeiconified(WindowEvent we){}
      public void windowIconified(WindowEvent we){}
      public void windowOpened(WindowEvent we){}
}
```

FIGURE HO-11.1
File `UserAction.java`

Step 11: Execute the Application

1. Press the F5 key. The application starts and the UserAction window is displayed
2. Press a key. The background color changes.
3. Position the cursor over the window.
4. Press a mouse key, left or right. The window background color changes.
5. Press a key. The background color changes.

Step 12: Save All the Project Files

Press the CTRL+SHIFT+5 keys.

Step 13: Close the Project

Select the Close All option in the File menu.

Step 14: Exit Visual J++ the Project

Select the Exit option in the File menu.
Challenge 1: Implement a mouse motion listener.
Challenge 2: Develop and test an application that will draw a line from the position where the left mouse button was pressed to the position where the right mouse button is pressed when the F1 key is pressed.

SUMMARY

Drawing is performed in an application's window. Drawings can be sent to different devices: screens, printers and plotters, and different operating systems. A graphic context provides the link necessary to draw to a specific platform—that is, operating system and hardware. The graphic context contains the information necessary to render lines, figures, and text to the platform's output devices. The `Graphic` class encapsulates the attributes and methods of the graphic context. Class `Color` provides a three-component color model for rendering color in the graphic context. Fonts are modeled by the `Font` class and font metrics are encapsulated in the `FontMetrics` class. System parameters such as available fonts, and hardware parameters such as display screen size, are accessed with methods contained in the `Toolkit` class.

KEY WORDS AND CONCEPTS

ascent
baseline
BOLD
colors
Color class
components
coordinate system
descent
Font
Font class
FontMetric class
getGraphics()
graphic context
Graphics
Graphics class
height
ITALIC
height
leading
paint()
pixel
PLAIN
Point
repaint()
three-component RBG
Toolkit class
update()
x-coordinate
y-coordinate

REVIEW QUESTIONS AND EXERCISES

1. Mark the following statements true or false:

   (T/F) A frame window is a component.

   (T/F) The paint method is invoked only to render filled figures.

   (T/F) Statements to draw lines, figures, and text are placed in the paint method.

   (T/F) SCRIPT is a Font class style.

   (T/F) The method drawLine(String *string*) will render the parameter *string* as a line of text.
2. What is graphic context?
3. What methods in the Component class are used to render graphical representations?
4. Show the RBG tuples required to render the following colors:

Green	Brown	Gray
Orange	Violet	Chartreuse
5. What is the value of the Font.BOLD + Font.ITALIC style. (*Hint:* Use the getStyle() method.)

6. Show a statement to render the string "Way to go!" in a boldface TimesRoman font 24 points high.
7. What are the advance values for the characters "m" and "n" in the Courier font? Will the values change if the characters are rendered in an italic style?
8. Show a statement to draw a line between the point (10,100) and point (100,50).
9. Show a statement to draw a circle centered at point (100,50) with radius 75.
10. Show a statement to draw a right triangle with height 75 and base 100.
11. Show a statement to draw a filled trapezoid. Test the statement in a program.

PRACTICE WITH VISUAL J++

1. Develop an application to render a chart presenting the high, low, and closing values of a stock for a week.
2. Render an idealized representation of your state's boundary. Show and label the capital, major cities, and your home town.

12

Graphical User Interface: Buttons, Labels, and Text Components

OBJECTIVES

After reading this chapter, you will be able to:

1. Understand how components of the Abstract Window Toolkit are used to build a Graphical User Interface.
2. Use buttons to initiate program action.
3. Insert labels in a user interface.
4. Use text fields and text areas to input and display information.
5. Demonstrate the ability to construct user interfaces with labels, buttons, text fields, and text areas.

OVERVIEW

The **Graphical User Interface (GUI)** enables the user to input data and actions and receive data output. Components used to construct interfaces include the following:

- Buttons that enable the user to initiate program actions
- Check boxes that provide for selection of multiple options
- Labels that provide for describing the interface and the purpose of the interface's components
- Lists that enable the user to communicate the selection of one or multiple items

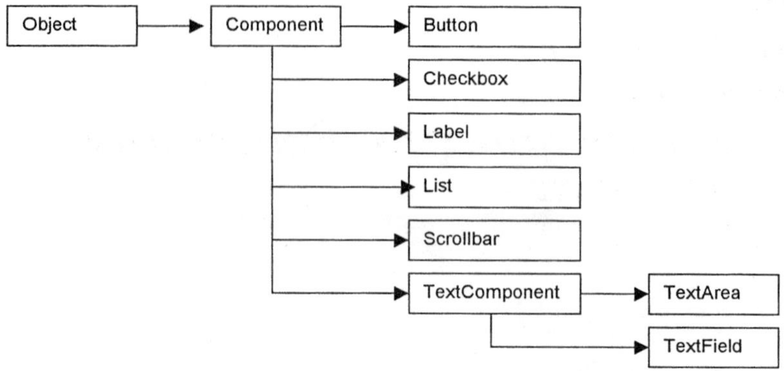

FIGURE 12.1
`Component Class` hierarchy of selected components.

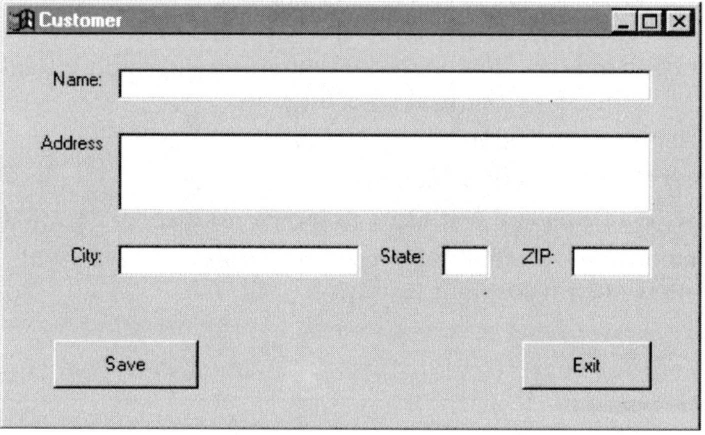

FIGURE 12.2
Simplified customer application window.

- Scrollbars that enable the navigation of text and lists
- Text fields and text areas that provide for output display and keyboard input of data.

Classes modeling components are shown in the `Component` class hierarchy diagram, Figure 12.1. These components are added to the application's window to form a graphical user interface. A simplified window for a customer application is shown in Figure 12.2.

LAYOUTS

Layout is the arrangement of components in a container. Java provides several classes that provide for standard arrangements. In addition, Java provides for placing components in a container according to absolute coordinates. `Container` class methods used to manage layout are summarized in Table 12.1. The `Container` class is an abstract class that models components that can hold other components.

TABLE 12.1

Selected Container methods used to manage layout

Method	Description
`LayoutManager getLayout()`	Returns the layout manager for the container.
`void setLayout( LayoutManger` *`laymgr`*`)`	Sets the layout manger to the specified layout manager *`laymgr`*.

`LayoutManager` is an interface for classes that know how to layout components in a container. The interface contains methods for adding and removing components. Java provides several layouts that automatically locate components in a container as well as the ability to add a component in a specific location.

`null` Layout

The **`null` layout** provides for adding components to absolute positions. This provides for a user interface that is efficient and effective in that it can follow the workflow and patterns of the user. This is a primary concern of applications developed for business enterprises. However, this can result in a component arrangement that does not size appropriately for displays of differing size or resolution. This problem is less likely in applications developed for a business enterprise in which the user displays tend to have similar size or resolution.

The null layout is established in a container by invoking the method `setLayout()` with a null argument.

```
setLayout(null);
```

After adding a component, the component is positioned in the container and sized by invoking the `setBounds()` method. The `setBounds()` method is invoked by the object to be positioned. The `setBounds()` method takes four parameters: two representing the component's x and y coordinates in the container, a third representing the component's width, and a fourth representing the component's height.

The `null` layout will be used in the development of user interfaces in the remainder of this chapter.

> Built-in layout managers are an advantage in user interfaces that will be used on displays with differing sizes or resolutions—for example, applications developed for the Internet—but the `null` layout is better suited for business applications.

BUTTON COMPONENTS

A **button** is a clickable component. A button's appearance changes when it has the focus or is clicked. Various button appearances, as rendered in a Windows environment, are shown in Figure 12.3.

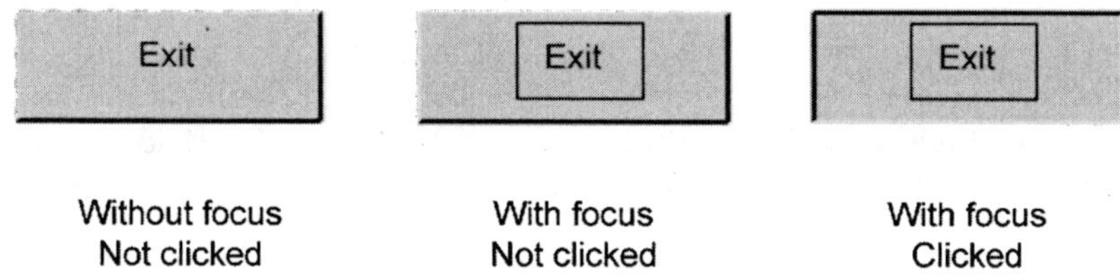

FIGURE 12.3
The appearance of a button changes when it has focus or is clicked.

The button has a centered label and is rendered with 3-D shading. When the button has the focus the labeled is outlined with a dashed line. A button is given focus by clicking on it with the mouse. When the button is clicked the heavy shading moves to the right and bottom sides. The button emits action events when subjected to user interaction.

The **`Button`** class models the graphical interface button. Button class constructors are summarized in Table 12.2. The constructors provide for creating button objects with and without labels, which are often referred to as the button's *caption*.

TABLE 12.2
Constructors of the `Button` class

Method	Function
`Button()`	Creates a button object with no centered label.
`Button(String lbl)`	Creates a button with a centered label *`lbl`*.

TABLE 12.3
Selected methods of the `Button` class

Method	Function
`Void addActionListener( ActionListener al)`	Adds the specified action listener *al* to receive actions from the invoking button.
`string getLabel( )`	Returns the button's label. `null` is returned if the button has no label.
`void setLabel(String lbl)`	Sets the button's label to the specified string *lbl*.

Selected methods of the `Button` class are shown in Table 12.3. `Button` class methods enable the implementation of listening for action events regarding the button as well as management of the label.

ACTION EVENTS

When a button is pressed—that is clicked—an `ActionEvent` is created. To respond to an action event an application must implement `ActionListener` and register the listener to receive events from the button by calling the button's `addActionListener` method. Response to actions regarding the button is coded in the **`actionPerformed`** method, which is called when an `ActionEvent` is emitted. The `actionPerformed` method is a member of the `ActionListener` interface.

The `ActionEvent` class hierarchy is shown in Figure 12.4.

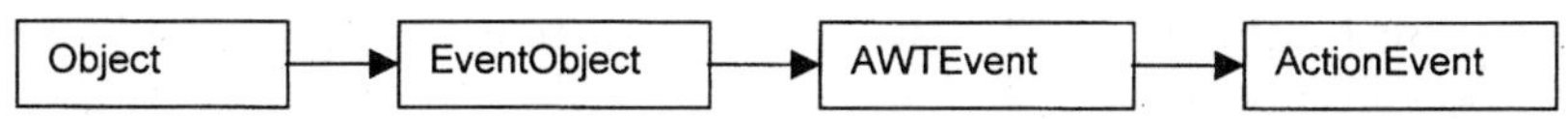

FIGURE 12.4
Action event hierarchy.

The `EventObject` class contains the `getSource` method that returns the `ActionEvent` object's source.

Responding to action events emitted by pressing or clicking buttons is illustrated in Figure 12.5a. In this application, the background color of the frame is changed to red by clicking the Red button and to blue by clicking the Blue button. Methods of the `ActionListener` interface are inherited in the application's main frame with the statement

```
public class GUI extends Frame
                 implements WindowListener,
                                        ActionListener
```

Button objects with the labels "Red" and "Blue" are created with the statements

```
Button bRed = new Button("Red");
Button bBlue = new Button("Blue");
```

The buttons are added to the container with the statements

```
add( bRed );
add( bBlue );
```

Note that the null layout is employed with the statement

```
setLayout(null);
```

enabling positioning and sizing the buttons.

The size and location of the buttons are set with the `setBounds` method inherited from the `Control` class. The method is invoked by the button to be bounded in the window.

```
bRed.setBounds(250, 200, 100, 50);
bBlue.setBounds(50, 200, 100, 50);
```

The `bRed` button's x coordinate, y coordinate, width and height are specified as 250, 200, 100, and 50, respectively. The `bBlue` button's position and size are similarly specified.

The application's `ActionListener` is registered to listen for `ActionEvents` from the buttons by invoking the button's `addActionListener`:

```
bRed.addActionListener(this);
bBlue. addActionListener (this);
```

Note that the button object is identified as the listener by use of the `this` keyword to specify the listening component.

Whenever the buttons are clicked an `ActionEvent` is emitted. The `actionPerformed` method is called automatically. The `ActionEvent` is passed as a parameter. The statement

```
if(ae.getSource() == bRed)
```

is used to determine if the source of the `ActionEvent` object `ae` is the button `bRed`. The source of the `ActionEvent` object is returned by the method `getSource` which is inherited by the `ActionEvent` class from the `EventObject` class. If the source of the `ActionEvent` object `ae` is the button `bRed` the frame background is turned red with the statement

```
setBackground(new Color(255,0,0));
```

Similarly, if the source of the `ActionEvent` is button `bBlue` the frame background is turned blue:

```
if(ae.getSource() == bBlue)
```

```
        {
            setBackground(new Color(0,0,255));
        }
```

```
import java.awt.*;
import java.awt.event.*;
public class GUI extends Frame
      implements WindowListener,
                   ActionListener
{
      Button bRed = new Button("Red");
      Button bBlue = new Button("Blue");

      public GUI()
      {
           addWindowListener(this);

           setTitle("Using Buttons");
           setLayout(null);

           setSize(400,300);

           add(bRed);
           add(bBlue);

           bRed.setBounds(250, 200, 100, 50);
           bBlue.setBounds(50, 200, 100, 50);
           bRed.addActionListener(this);
           bBlue.addActionListener(this);
           setVisible(true);
      }

      public static void main(String[] args)
      {
           GUI gui = new GUI();
      }
      public void actionPerformed(ActionEvent ae)
      {
           if(ae.getSource() == bRed)
           {
                setBackground(new Color(255,0,0));
           }
           if(ae.getSource() == bBlue)
           {
                setBackground(new Color(0,0,255));
           }
      }

      public void windowClosing(WindowEvent we)
```

```
        {
            setVisible(false);
            dispose();
            System.exit(0);
        }
        public void windowActivated(WindowEvent we){}
        public void windowClosed(WindowEvent we){}
        public void windowDeactivated(WindowEvent we){}
        public void windowDeiconified(WindowEvent we){}
        public void windowIconified(WindowEvent we){}
        public void windowOpened(WindowEvent we){}
}
```

FIGURE 12.5A
Program responding to `ActionEvents` emitted by buttons.

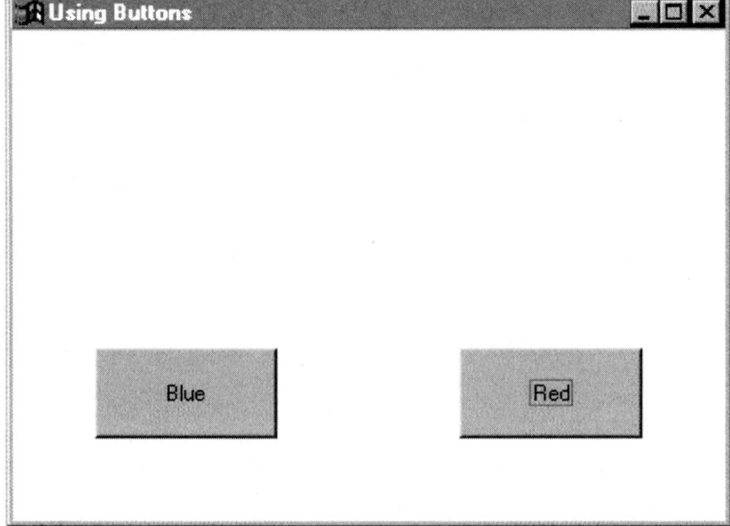

FIGURE 12.5B
Output of program responding to `ActionEvents` emitted by buttons.

The application (shown in Figure 12.5a) displays the frame window shown in Figure 12.5b. Initially the frame background is white. When the Red button is clicked the frame background turns red. When the Blue button is clicked the frame background turns blue.

LABEL COMPONENTS

A **label** is a component that places a single line of text in a container. Labels are modeled by the **`Label` class.** A label can be changed by the application

TABLE 12.4

`Label` class constructors

Method	Function
`Label()`	Creates a label object with no text.
`Label(String text)`	Creates a label object with centered string *text*.
`Label(String text, int alignment)`	Creates a label with a centered string *text* with alignment `Label.LEFT`, `Label.RIGHT`, or `Label.CENTER`.

TABLE 12.5

Selected methods of the `Label` class

Method	Function
`int getAlignment()`	Returns the label's text alignment: `Label.LEFT`, `Label.RIGHT`, or `Label.CENTER`.
`String getText()`	Returns the label's text.
`void setAlignment{ int alignment}`	Aligns the label's text; *alignment*: values are `Label.LEFT`, `Label.RIGHT`, or `Label.CENTER`.
`void setText(String text)`	Sets the label's text to the specified string *text*.

but not by the user and is said to be **read-only.** `Label` class constructors are summarized in Table 12.4. Selected methods of the `Label` class are summarized in Table 12.5.

The program shown in Figure 12.6a illustrates the creation and manipulation of labels. Two label objects are created with the statements

```
Label lShowAlign = new Label("Showing
                              Alignment",Label.RIGHT);
Label lLeft = new Label("Left Alignment");
```

The first creates a label object `lShowAlign` with the text "Showing Alignment" aligned to the right side of the label's bounding rectangle. The second creates a label object `lLeft` with centered text "Left Alignment." The label objects are added to the Frame component with the add method in the Frames constructor.

```
add(lShowAlign);
add(lLeft);
```

A label's position and size are set with the `setBounds` method inherited form the `Component` class. For example,

```
lShowAlign.setBounds(40, 40, 200, 50);
lLeft.setBounds(40, 100, 200, 50);
```

Similarly, the label's font is set with the `setFont` method and the background color is set with the `setBackground` method:

```
lLeft.setFont(new Font("Serif",Font.ITALIC,22));
lLeft.setBackground(new Color(0,255,255));
```

Although a label is read-only, in Windows it can receive focus, which is indicated by a visible rectangle surrounding the text, when the mouse is clicked in the label's bounding rectangle. This property can be disabled with the `setEnabled` method, also inherited from the `Component` class.

```
lLeft.setEnabled(false);
```

Methods of the `Component` class can be used to manipulate labels.

```
import java.awt.*;
import java.awt.event.*;
public class Labels extends Frame
      implements WindowListener,
                 ActionListener
{
      Button bRight = new Button("Right");
      Button bLeft = new Button("Left");
      Label lShowAlign = new Label("Showing
                                                      Align-
                                           ment",Label.RIGHT);
      Label lLeft = new Label("Left Alignment");

      public Labels()
      {
          addWindowListener(this);

          setTitle("Labels");
          setLayout(null);

          setSize(400,300);

          add(bRight);
          add(bLeft);

          add(lShowAlign);
          add(lLeft);

          bRight.setBounds(250, 200, 100, 50);
          bLeft.setBounds(50, 200, 100, 50);

          lShowAlign.setBounds(40, 40, 200, 50);
          lLeft.setBounds(40, 100, 200, 50);

          lLeft.setFont(new Font("Serif",Font.ITALIC,22));
```

```
        lLeft.setBackground(new Color(0,255,255));
        lLeft.setEnabled(false);

        bRight.addActionListener(this);
        bLeft.addActionListener(this);

        setVisible(true);
    }

    public static void main(String[] args)
    {
        Labels lbs = new Labels();
    }
    public void actionPerformed(ActionEvent ae)
    {
        if(ae.getSource() == bRight)
        {
            lShowAlign.setAlignment(Label.RIGHT);
        }
        if(ae.getSource() == bLeft)
           {
            lShowAlign.setAlignment(Label.LEFT);
        }
    }

    public void windowClosing(WindowEvent we)
    {
        setVisible(false);
        dispose();
        System.exit(0);
    }
    public void windowActivated(WindowEvent we){}
    public void windowClosed(WindowEvent we){}
    public void windowDeactivated(WindowEvent we){}
    public void windowDeiconified(WindowEvent we){}
    public void windowIconified(WindowEvent we){}
    public void windowOpened(WindowEvent we){}
}
```

FIGURE 12.6A
Program illustrating creation and manipulation of label objects.

The output of the application's shown in Figure 12.6a is illustrated in Figure 12.6b. Note the size of the bounding rectangle is made clear by the contrasting background color of the `lLeft` label. Note also the font type, style and size of the `lLeft` label. In addition, note the right alignment of the `lShowAlign` label's text in its bounding rectangle.

Realignment of the top label is performed when the Left and Right buttons are pressed. For example, Figure 12.6c shows the alignment of the "Showing Alignment" label text when the Left button is pressed. The `setAlignment`

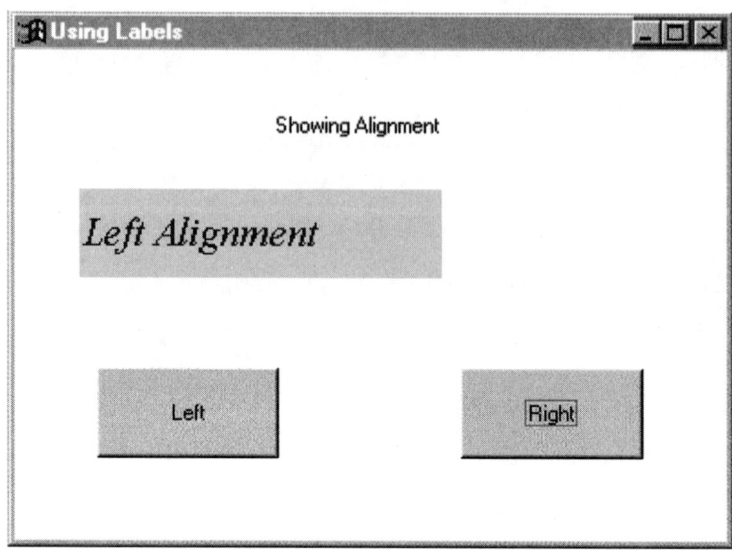

FIGURE 12.6B
Output of program illustrating creation and manipulation of label objects.

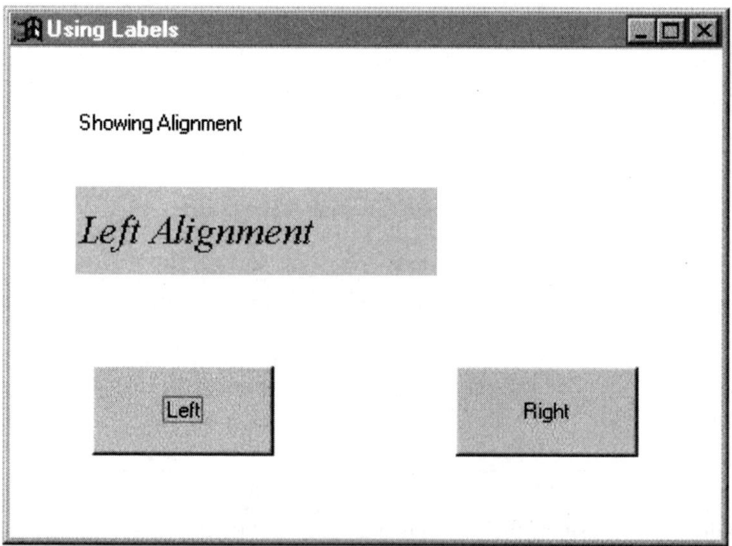

FIGURE 12.6C
Output of program illustrating creation and manipulation of label objects.

method with `Label.LEFT` argument is invoked for the `lShowAlign` label when the Left button is pressed.

```
lShowAlign.setAlignment(Label.LEFT);
```

The result of pressing the Left button is illustrated in Figure 12.6c.

Similarly, the `setAlignment` method with `Label.RIGHT` argument

```
lShowAlign.setAlignment(Label.RIGHT);
```

is invoked for the `lShowAlign` label when the Right button is pressed.

> Labels are said to be read-only but can be manipulated programmatically.

TEXT COMPONENTS

Text can be displayed and entered in a text component. Text components are modeled by the `TextComponent` class. Methods of the `TextComponent` class are shown in Table 12.6.

TABLE 12.6

Selected methods of the `TextComponent` class

Method	Function
`String getSelectedText()`	Returns the selected text.
`int getSelectionEnd()`	Returns the position of the end of the selected text.
`int getSelectionStart()`	Returns the position of the start of the selected text.
`String getText{`	Returns the text presented in the component.
`boolean isEditable()`	Returns true if the component is editable, false otherwise.
`void select(int` *`start`*`, int` *`end`*`)`	Selects a portion of the text starting with position *`start`* and ending at position *`end`*.
`void selectAll()`	Selects the entire text.
`void setCaretPosition( int` *`position`*`)`	Sets the insertion caret at *`position`* in the text.
`void setEditable(boolean` *`flag`*`)`	Enables the component for user editing of text when *`flag`* is `true` and disables the component when *`flag`* is `false`.
`void setSelectionEnd(int` *`end`*`)`	Sets the end of the selected text to position *`end`*.
`void setSelectionStart( int` *`start`*`)`	Sets the start of the selected text to position *`start`*.
`void setText(String` *`text`*`)`	Sets the text to the specified string *`text`*.

A typical text component is shown in Figure 12.7. Text can be entered in the component if the component is **editable;** otherwise the user cannot enter or remove text. Portions of the text in the component can be **selected.**

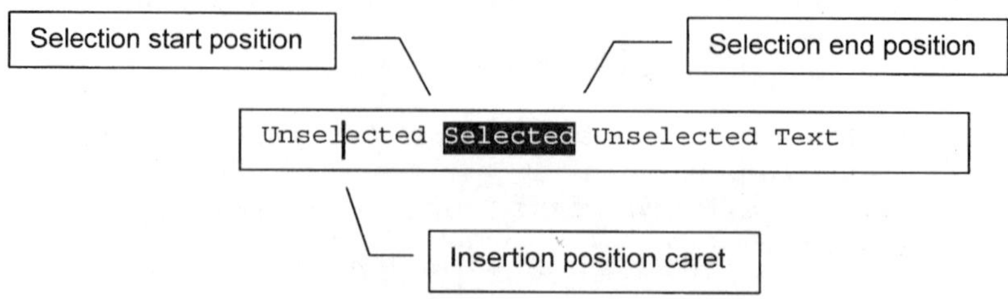

FIGURE 12.7
Characteristics of a text component.

The selection position are denoted by a **selection start position** and **selection end position.** The position of the first character in a text component is zero. All of the text or only a selected portion of the text can be retrieved. Text is entered at the insertion caret.

`TextComponent` serves as the superclass for user-editable text components.

TEXT FIELDS

A **text field** holds a single row or line of text. When the text field has the focus, a character of text is entered at the insert caret position whenever a key is pressed. Text fields are modeled by the **`TextField` class.** `TextField` inherits all of the method of the `TextComponent` class and adds additional methods. `TextField` constructors are summarized in Table 12.7. Selected methods of the `TextField` class are summarized in Table 12.8.

TABLE 12.7

`TextField` class constructors

Method	Function
`TextField()`	Creates a text field object without text.
`TextField(String` *`text`*`)`	Creates a text field object with string *`text`* aligned left.

TABLE 12.8

Selected key methods of the `TextField` class

Method	Function
`boolean echoCharIsSet()`	Returns `true` if an echo character is set, `false` otherwise.
`Char getEchoChar()`	Returns the echo character.
`void setEchoChar(char` *`c`*`)`	Sets the echo character to the character *`c`*.

An **echo character** is the character that is echoed to the text field when the user presses a key. Echo characters are commonly used to mask PINs, passwords, and other sensitive information. A application illustrating use of text fields is listed in Figure 12.8a. The application's window contains two text fields. When the left button, captioned "`Selected`", is clicked, the selected text in the left window is transferred to the right window. When the right button, captioned "`All`", is clicked, the selected text in the left window is transferred to the right window. Note that the transferred text overwrites the existing text.

```
import java.awt.*;
import java.awt.event.*;
public class TextFields extends Frame
      implements WindowListener,
                  ActionListener
{
      Button bAll = new Button("All");
      Button bSelection = new Button("Selection");
      TextField tSource = new TextField("Initial Text",30);
      TextField tTarget = new TextField(30);
      Label lMove = new Label("->");

      public TextFields()
      {
            addWindowListener(this);

            setTitle("Using Text Fields");
            setLayout(null);

            setSize(400,300);

            add(bAll);
            add(bSelection);
            add(tSource);
            add(tTarget);
            add(lMove);

            bAll.setBounds(250, 200, 100, 50);
            bSelection.setBounds(50, 200, 100, 50);
            tSource.setBounds(50, 50, 100, 80);
            tTarget.setBounds(250, 50, 100, 80);
            lMove.setBounds(200, 50, 10, 10);

            bAll.addActionListener(this);
            bSelection.addActionListener(this);

            setVisible(true);
      }
```

```
    public static void main(String[] args)
    {
        TextFields tx = new TextFields();
    }
    public void actionPerformed(ActionEvent ae)
    {
        if(ae.getSource() == bAll)
        {
            String buf = tSource.getText();
            tTarget.setText(buf);
        }
        if(ae.getSource() == bSelection)
        {
            String buf = tSource.getSelectedText();
            tTarget.setText(buf);
        }
    }

    public void windowClosing(WindowEvent we)
    {
        setVisible(false);
        dispose();
        System.exit(0);
    }
    public void windowActivated(WindowEvent we){}
    public void windowClosed(WindowEvent we){}
    public void windowDeactivated(WindowEvent we){}
    public void windowDeiconified(WindowEvent we){}
    public void windowIconified(WindowEvent we){}
    public void windowOpened(WindowEvent we){}
}
```

FIGURE 12.8A
Application illustrating use of text fields.

The application initial window of Figure 12.8b is shown in Figure 12.8b. Two text fields are created with the statements

```
TextField tSource = new TextField("Initial Text",30);
TextField tTarget = new TextField(30);
```

The first, `tSource`, creates a text field 30 columns wide containing the text "`Initial Text`" when first displayed. The second, `tTarget`, creates a text field 30 columns wide without text when first displayed. The text fields are added to the window with the statements

```
add(tSource);
add(tTarget);
```

Bounding rectangles for the text fields are set with the statements

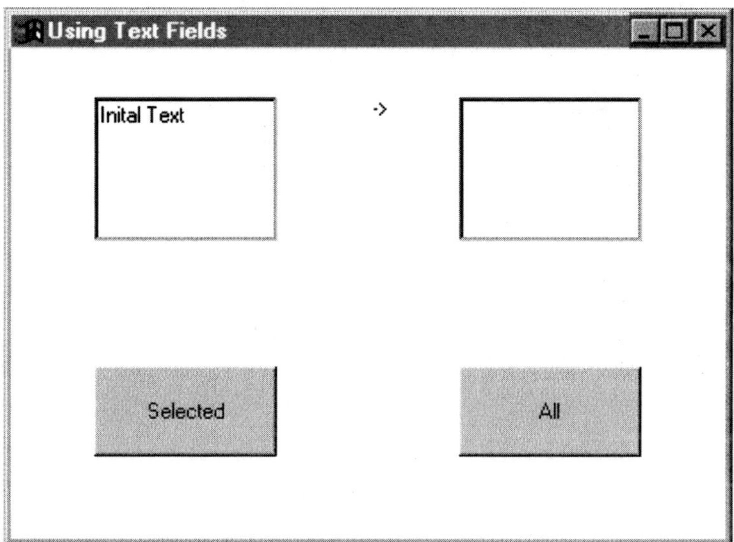

FIGURE 12.8B
Output of application illustrating text fields.

```
tSource.setBounds(50, 50, 100, 80);
tTarget.setBounds(250, 50, 100, 80);
```

The size attribute of the text field set by the constructor parameter is overridden by the width and height parameters of the `setBounds` method.

When the "`All`" button is clicked, the entire text of the left text field is assigned to the `String` variable `buf`.

```
String buf = tSource.getText();
```

The contents of `buf` are then inserted in the right text field

```
tTarget.setText(buf);
```

Similarly, when the "`Selected`" button is clicked the selected text in the left text field is transferred to the right text field.

```
String buf = tSource.getSelectedText();
tTarget.setText(buf);
```

Figure 12.8c illustrates the application's response when a portion of the text in the left text field is selected and the "`Selected`" button is clicked.

All text is inserted in a single row. The text is automatically scrolled when the text length exceeds the width of the window. The left and right arrow, END and HOME keys can be used to navigate the text field.

> `A text field is limited to a single row of text.`

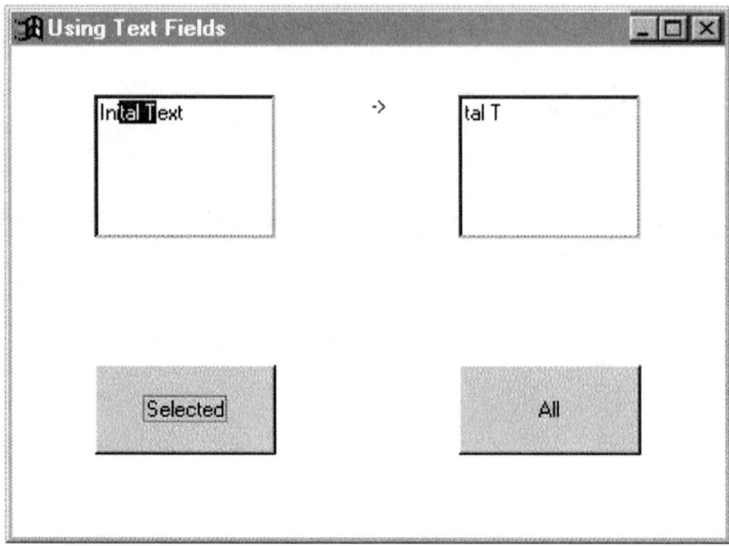

FIGURE 12.8C
Output of application illustrating text fields.

TEXT AREAS

Multiple rows of text are displayed in text areas. Users start a new row of text by pressing the enter key. Scrollbars can be displayed when the columns or rows of text exceed the number of columns or rows displayed. These scrollbars can be used to navigate through the text. Additionally, the left and right arrow, HOME and ENTER keys can be used to navigate through the text. Text areas are modeled by the **`TextArea`** class. Methods of the class `TextComponent` are inherited by the class `TextArea`. `TextArea` constructors are summarized in Table 12.9. Some of the methods, defined in the class `TextArea`, are summarized in Table 12.10.

Scrollbars become visible automatically when the text exceeds the text area, except if not specified when the text area is created with the constructor `TextArea(String` *`text`*`, int` *`noRows`*`, int` *`noCols`*`, int` *`scrollbars`*`)`. Values of *`scrollbars`* are attributes of the class `TextArea` and are `SCROLLBARS_BOTH`, `SCROLLBARS_HORIZONTAL_ONLY`, `SCROLLBARS_NONE`, and `SCROLLBARS_VERTICAL_ONLY`.

An application illustrating text areas is listed in Figure 12.9a. The application's window is shown in Figure 12.9b.

TABLE 12.9

TextArea class constructors

Method	Function
`TextArea()`	Creates a text area object without text.
`TextArea(String text)`	Creates a text area object with string *text* aligned left.
`TextArea(String text, int noRows, int noCols, int scrollbars)`	Creates a text area object with string *text*, number of rows and columns specified as *noRows* and *noCols* and either none, both, vertical-only or horizontal-only scrollbars specified *scrollbars*.

TABLE 12.10

Selected methods of the TextArea class

Method	Function
`void append(String text)`	Appends the string *text* to the text already in the text area.
`void insert(String text, int pos)`	Inserts the string *text* at position *pos*.
`void replaceRange(String text, int start, int end)`	Replaces the text starting at position *start* and position *end* with the string *text*.

```
import java.awt.*;
import java.awt.event.*;
public class TextAreas extends Frame
      implements WindowListener,
                 ActionListener
{
      String text =
"line1\nline2\nline3\nline4\nline5\nline6\nline7";
      Button bAll = new Button("All");
      Button bSelected = new Button("Selected");
      TextArea tSource = new TextArea(text);
      TextArea tTarget = new TextArea();
      Label lMove = new Label("->");

      public TextAreas()
      {
           addWindowListener(this);

           setTitle("Using Text Areas");
           setLayout(null);

           setSize(400,300);

           add(bAll);
           add(bSelected);
```

```
            add(tSource);
            add(tTarget);
            add(lMove);

            bAll.setBounds(250, 200, 100, 50);
            bSelected.setBounds(50, 200, 100, 50);
            tSource.setBounds(50, 50, 100, 80);
            tTarget.setBounds(250, 50, 100, 80);
            lMove.setBounds(200, 50, 10, 10);

            bAll.addActionListener(this);
            bSelected.addActionListener(this);

            setVisible(true);
      }

      public static void main(String[] args)
      {
            TextAreas tx = new TextAreas();
      }
      public void actionPerformed(ActionEvent ae)
      {
            if(ae.getSource() == bAll)
            {
                  String buf = tSource.getText();
                  tTarget.setText(buf);
            }
            if(ae.getSource() == bSelected)
            {
                  String buf = tSource.getSelectedText();
                  tTarget.setText(buf);
            }
      }

      public void windowClosing(WindowEvent we)
      {
            setVisible(false);
            dispose();
            System.exit(0);
      }
      public void windowActivated(WindowEvent we){}
      public void windowClosed(WindowEvent we){}
      public void windowDeactivated(WindowEvent we){}
      public void windowDeiconified(WindowEvent we){}
      public void windowIconified(WindowEvent we){}
      public void windowOpened(WindowEvent we){}
}
```

FIGURE 12.9A
Application illustrating use of text areas.

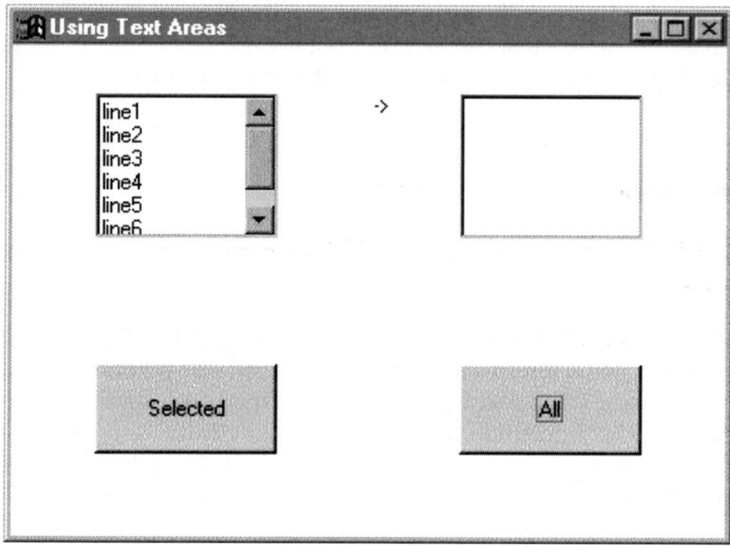

FIGURE 12.9B
Output of application illustrating use of text areas.

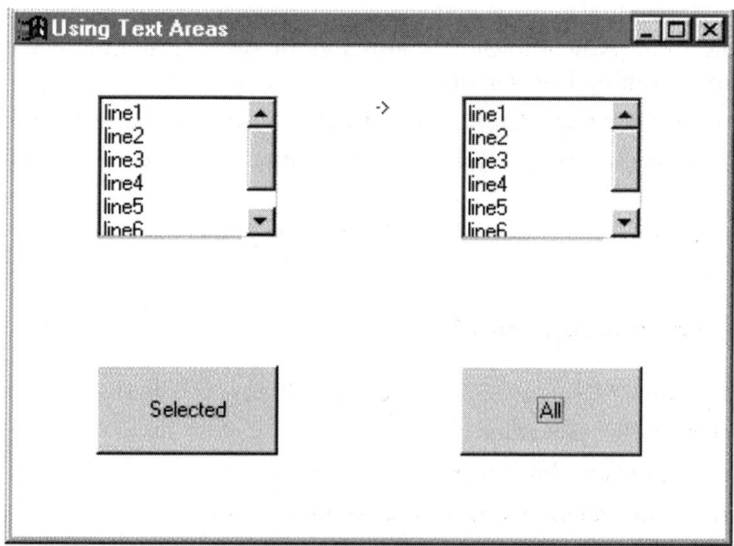

FIGURE 12.9C
Output of application illustrating use of text areas.

The application contains two text areas. The vertical scrollbar is displayed in the left text area to provide easy access to rows initially displayed that are not visible. The application also contains two buttons. Pressing the "Selected" button transfers the selected text in the left text area to the right text area. Pressing the "All" button transfers the entire text in the left text area to the right text area (Figure 12.9c).

Note that a vertical scrollbar is automatically displayed in the right text area when all of the rows of text are not visible. Similarly, a horizontal scrollbar will be displayed when all the columns of text cannot be displayed.

HANDS-ON EXERCISE 12: EDITING USER INPUT FOR VALIDITY

User-entered codes for a customer's region are edited for validity when the user presses the ENTER button. If the code is invalid, the text field background color is set to red and the text field is given the focus. When the user changes the text field the background color is returned to white.

Information entered by users must be test for validity where possible. This application illustrates the testing of user information.

Step 1: Create a New Project

1. Start Visual J++.
2. Select the New Project... command in the File menu. The New Project dialog box is displayed.
3. Enter **`Customer Input Validity`** in the Name text field.
4. Use the Browse button to enable navigation to the Location `C:\Hands-On\Customer Input Validity`.
5. Click the Open button to create the new project.

Step 2: Create `Customer.java` File

1. Press the CTRL+D keys. The Add Item dialog box is displayed.
2. Select the Class folder.
3. Select the ClassMain icon.
4. Enter **`Customer.java`** in the Name text field.
5. Click the Open button to create the new class.
6. The Code Editor is opened with the file `Customer.java` (Figure HO-12).

Step 3: Code the Application

Enter code to create a frame window and handle window events, specifically `windowClosing`.

```
import java.awt.*;
import java.awt.event.*;

public class Customer extends Frame
      implements WindowListener
{
      MenuBar menuBar = new MenuBar();

      Menu menuFile = new Menu("File");
      MenuItem itemSave = new MenuItem("Save");
      MenuItem itemPrint = new MenuItem("Print");
      MenuItem itemExit = new MenuItem("Exit");

      Menu menuSearch = new Menu("Search");
      MenuItem itemID = new MenuItem("Id");
      MenuItem itemName = new MenuItem("Name");

      public Customer( ) {
           setMenuBar(menuBar);

           menuBar.add(menuFile);
           menuBar.add(menuSearch);

           menuFile.add(itemSave);
           menuFile.add(itemExit);
           menuFile.add(itemPrint);
           menuSearch.add(itemID);
           menuSearch.add(itemName);

           addWindowListener(this);
           setSize(400,200);
           setVisible(true);
      }

      public static void main(String[] args)
      {
                    new Customer();
      }

      public void windowClosing( WindowEvent we ) {
           setVisible(false );
           dispose( );
           System.exit( 0 );
      }
      public void windowActivated( WindowEvent we ){}
      public void windowClosed( WindowEvent we ){}
      public void windowDeactivated( WindowEvent we ){}
      public void windowDeiconified( WindowEvent we ){}
      public void windowIconified( WindowEvent we ){}
      public void windowOpened( WindowEvent we ){}
}
```

Note 1: The code is identical to the code of Hands-On Exercise 10 with the addition of the Save menu item statements shown in bold type.

Note 2: Completed code is shown in Figure HO-12.1.

Step 4: Build the Basic Application

1. Press the CTRL+SHIFT+B keys.
2. Debug the code as required.

Step 5: Add Code to Implement an Action Listener

1. Modify the class declaration to implement an action listener.

```
public class UserAction extends Frame
    implements WindowListener, ActionListener
```

2. Add an action listener to the Save menu item window in the constructor

```
itemSave.addActionListener(this);
```

3. Enter the following code to implement an action event handler:

```
public void actionPerformed(ActionEvent ae) {
    if (ae.getSource() == itemSave)
    {
        // Edit user input
    }
}
```

4. Note that the editing of user input will be performed when the Save menu option is selected.

Step 6: Build the Basic Application

1. Press the CTRL+SHIFT+B keys.
2. Debug the code as required.

Step 7: Set the Project Properties

1. Select the Customer Input Validity Properties... option in the Project menu. The properties dialog box is displayed.
2. Click the OK button to accept the default properties.

Step 8: Execute the Application

1. Press the F5 key. The application starts and the application window is displayed.
2. Click the File menu.
3. Select the Save menu. The menu closes.
4. Close the application.

Step 9: Add a Region Label and Text Field to the Window

1. Create a Region label and text field with the statements

```
Label labelRegion = new Label("Region:");
TextField tieldRegion = new TextField(10);
```

2. Add the label and text field to the frame window in the constructor

```
    setLayout(null);
    add(labelRegion);
    add(textRegion);
    labelRegion.setBounds(20,100,40,20);
    textRegion.setBounds(80,100,30,20);
```

Step 10: Build the Basic Application

1. Press the CTRL+SHIFT+B keys.
2. Debug the code as required.

Step 11: Edit Region Text Field

1. Enter the following method:

```
private boolean editRegion() {
     String stringNE = new String("NE");
     if (textRegion.getText().equals(stringNE)) {
          textRegion.setBackground(Color.white);
          return true;
     }
     else {
          textRegion.setBackground(Color.red);
          return false;
     }
}
```

2. The method retrieves the region text field and compares it with the `stringNE`. If the region text field content equals the `StringNE` the method returns `true` and sets the text field background color to white. If the Region text field content does not compare equally to `stringNE` the method returns `false` and sets the text field background to red.
3. Modify the `actionPerformed` method to save the data if the edit method returns `true` and nothing otherwise.

```
public void actionPerformed(ActionEvent ae) {
     if (ae.getSource() == itemSave) {
          if (editRegion()) {
               // save Customer information
          }
          else {
               // do not save Customer information
          }
     }
}
```

```
import java.awt.*;
import java.awt.event.*;

public class Customer extends Frame
      implements WindowListener, ActionListener
{
      MenuBar menuBar = new MenuBar();

      Menu menuFile = new Menu("File");
      MenuItem itemSave = new MenuItem("Save");
      MenuItem itemPrint = new MenuItem("Print");
      MenuItem itemExit = new MenuItem("Exit");

      Menu menuSearch = new Menu("Search");
      MenuItem itemID = new MenuItem("Id");
      MenuItem itemName = new MenuItem("Name");

      Label labelRegion = new Label("Region:");
      TextField textRegion = new TextField(10);

      public Customer( ) {
           setMenuBar(menuBar);

           menuBar.add(menuFile);
           menuBar.add(menuSearch);

           menuFile.add(itemSave);
           menuFile.add(itemExit);
           menuFile.add(itemPrint);
```

```
        menuSearch.add(itemID);
        menuSearch.add(itemName);

        itemSave.addActionListener(this);

        setLayout( null );
        add(labelRegion);
        add(textRegion);
        labelRegion.setBounds(20,100,40,20);
        textRegion.setBounds(80,100,30,20);

        addWindowListener(this);
        setSize(400,200);
        setVisible(true);
    }

    public static void main(String[] args)
    {
        new Customer();
    }

    public void actionPerformed(ActionEvent ae) {
        if ( ae.getSource( ) == itemSave) {
            if (editRegion( )) {
                // save Customer information
            }
            else {
                // do not save Customer information
            }
        }
    }

    private boolean editRegion( ) {
        String stringNE = new String("NE");
        if (textRegion.getText( ).equals(stringNE)) {
            textRegion.setBackground(Color.white);
            return true;
        }
        else {
            textRegion.setBackground(Color.red);
            return false;
        }
    }

    public void windowClosing( WindowEvent we ) {
        setVisible(false );
        dispose( );
        System.exit( 0 );
    }
```

```
        public void windowActivated( WindowEvent we ){}
        public void windowClosed( WindowEvent we ){}
        public void windowDeactivated( WindowEvent we ){}
        public void windowDeiconified( WindowEvent we ){}
        public void windowIconified( WindowEvent we ){}
        public void windowOpened( WindowEvent we ){}
}
```

FIGURE HO-12.1
File Customer.java.

Step 12: Build the Application

1. Press the CTRL+SHIFT+B keys.
2. Debug the code as required.

Step 13: Execute the Application

1. Press the F5 key. The application starts and the Customer window is displayed.

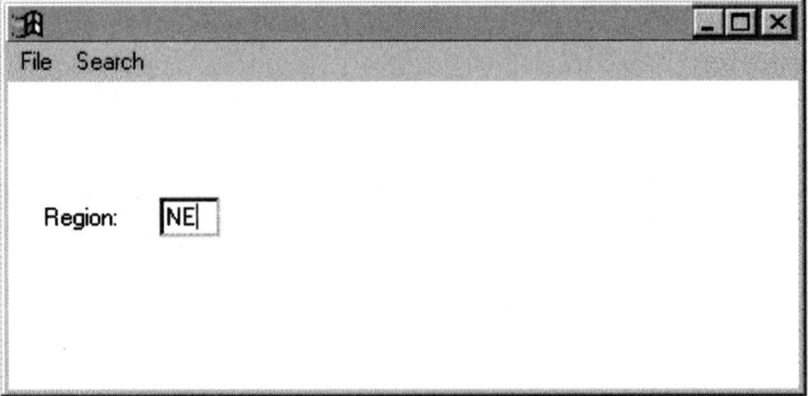

FIGURE HO-12.2
Customer Window.

2. Enter "**NE**" in the region window (Figure HO-12.2).
3. Select the Save option in the File menu. The region text field background color does not change.
4. Enter "`SW`" in the region window.
5. Select the Save option in the File menu. The region text field background color changes to red.

6. Enter "NE" in the region window.
7. Select the Save option in the File menu. The region text field background color changes to white.

Step 14: Save All the Project Files

Press the CTRL+SHIFT+F5 keys.

Step 15: Close the Project

Select the Close All option in the File menu.

Step 16: Exit Visual J++ the Project

Select the Exit option in the File menu.

Challenge: Add a Customer ID text field. Edit to ensure the text field contains a string that starts with an capitalized alphabetic letter that is followed by four digits, for example, A1234.

SUMMARY

1. Java's AWT components, used to construct Graphical User Interfaces, include buttons, labels, text fields, and text areas.
2. Components can be arranged in the application's window using the `null` layout.
3. Components are added to the interface using the `add` method derived from the `Container` class.
4. Components are positioned and sized in the container with the `setBounds` method.
5. Action events, such as clicking a button, are handled in an override of the `Component` class `actionPerformed` method.

KEY WORDS AND CONCEPTS

`actionPerformed`
button
`Button` class
echo character
editable
`getSource`
Graphical User Interface (GUI)
insertion caret
interface
label
`Label` class
layout
`LayoutManager`
`null` layout
read-only
selected
selection end position
selection start position
`setBounds`
`setLayout`
text area
text component
`TextArea` class
`TextField` class

REVIEW QUESTIONS AND EXERCISES

1. Mark the following statements true or false.
   a. A `null` value is not a valid `setLayout` method parameter.
   b. Buttons can only have a gray background.
   c. Labels can be aligned right.
   d. Text fields can be multi-line.
   e. Textual content exceeding the length of a text field is lost.
   f. Text area scrollbars are automatic unless specified otherwise.
2. When is the `null` layout not appropriate?
3. How are the `setBounds` method parameters related to the text shown in a text field?
4. Can two lines of text can be displayed in a text field?
5. Are scrollbars shown automatically for all text areas?
6. Add a button to a frame with colored background to turn the background white.
7. Create a red label.
8. Create a text box with blue background.
9. Create a text area with red text on blue background.

PRACTICE WITH VISUAL J++

1. Create and test an application for input or output of information regarding a customer.
2. Create and test an application for input or output of information regarding a product.
3. Create and test an application for input or output of information regarding a student.
4. Create and test an application for input or output of information regarding a customer's order.

13

Graphical User Interface: Lists, Scrollbars, Checkboxes, Checkbox Groups, and Choice Components

OBJECTIVES

After reading this chapter, you will be able to:

1. Understand and use list, scrollbar, checkbox, checkbox group, and choice components.
2. Understand and use adjustment and item events.
3. Demonstrate the use of these components in building graphical user interfaces.

OVERVIEW

Abstract Window Toolkit also enables the user to select from lists, check selections, choose one option in a group, and select a value from a continuum. Classes modeling these user interfaces are shown in the hierarchy diagram in Figure 13.1.

Figure 13.2 illustrates the use of a list box to enable the user to select a customer's region from a fixed list. Additionally, the user can select a customer type from the radio button group. Only one type in the group can be selected. Thus selection of Platinum results in deselection of the default. Note also the use of a choice box, State, which restricts the user to entry of a valid state abbreviation, in this case NJ for New Jersey. Although not used in the Customer application Graphical User Interface, scrollbars are very useful in enabling the user to select a value from a continuum.

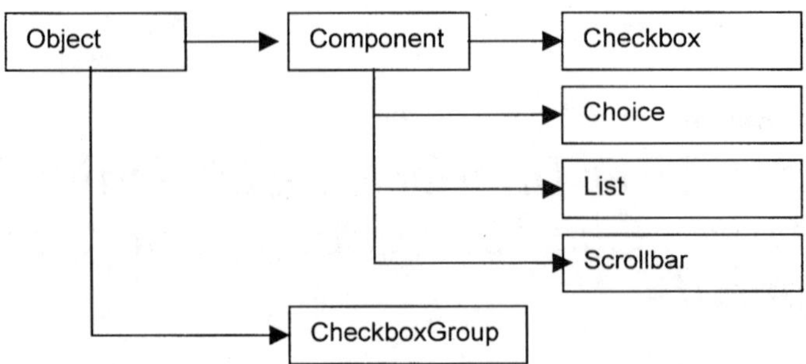

FIGURE 13.1
Selected `Component` Class hierarchy.

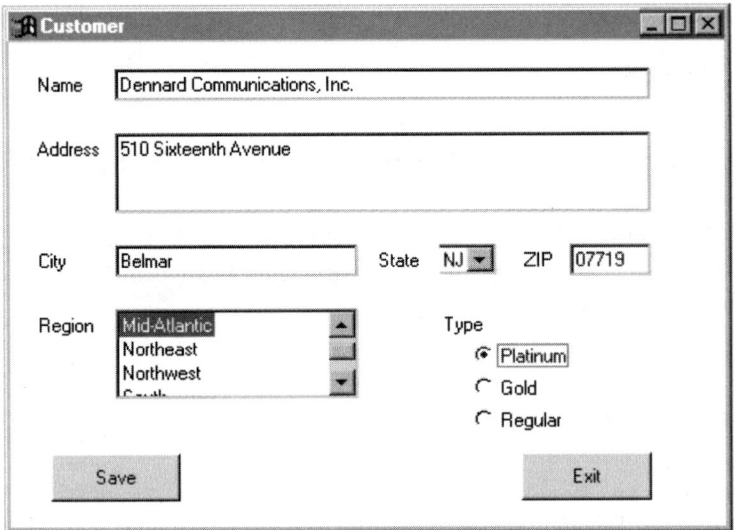

FIGURE 13.2
Enhanced customer application window with list box, choice, and radio button group.

LIST COMPONENTS

A list allows the user to select one or many items. A list, as rendered in a Windows environment, is shown in Figure 13.3. A selected item in a list is highlighted.

Items are added to the list programmatically. A scrollbar is automatically added to the list box when the number of items in the list exceeds the number of rows displayed. List boxes can provide for user selection of only one item or user selection of multiple items. In the case of list boxes enabling multiple

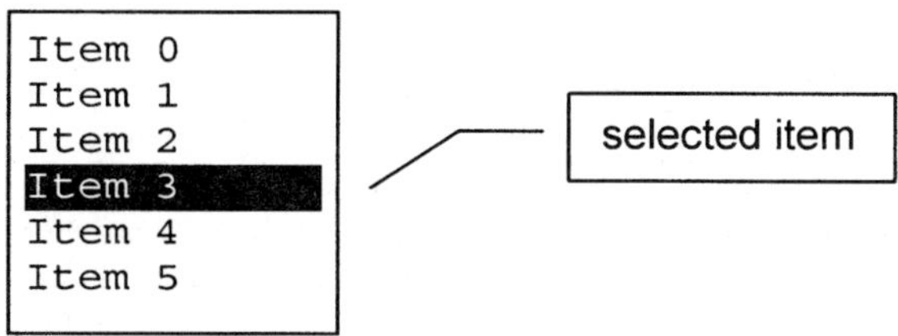

FIGURE 13.3
List box with selected item.

selection, clicking an item selects it. Clicking a selected item deselects it. In the case of a list box enabling single selection only, an item is selected by clicking it. Selecting an item, in a single selection list box, deselects any other selection. When an item is selected or deselected an item event is created.

The **`List` class** models the graphical interface list. Constructors of the `List` class are presented in Table 13.1.

TABLE 13.1
`List` class constructors

Method	Function
`List( )`	Creates a list object enabling selection of a single item.
`List(int` *`initCapacity`*`)`	Creates a list object enabling selection of a single item with *`initCapacity`* items.
`List(Object [ ]` *`items`*`)`	Creates a list object enabling selection of a single item initially populated with elements of the array *`item`*.

A list object created with the default constructor has ten items. List size doubles whenever space for items is exhausted. Selected methods of the `List` class are shown in Table 13.2.

ITEM EVENTS

When a list item is selected or deselected an **`ItemEvent`** is created. The `ItemEvent` class hierarchy is shown in Figure 13.4. To respond to an item event an application must implement **`ItemListener`** and register the listener to receive item events by calling the list's `addItemListener` method. The response is coded in the `itemStateChanged` method which is called when an `ItemEvent` is emitted. The `itemStateChanged` method is a member of the `ItemListener` interface.

TABLE 13.2

Selected methods of the `List` class

Method	Function
`void add(String item)`	Adds the string *item* at the end of the list.
`void add(String item, int index)`	Adds the string *item* to the list at the position specified *index*.
`void delete(int index)`	Deletes the list item at the position *index*.
`string getItem(int index )`	Returns the list item at position *index*.
`String[ ] getItems( )`	Returns all the items in the list.
`int getSelectedIndex( )`	Returns the index of the selected item.
`Int[ ] getSelectedIndexs( )`	Returns the indexes of the selected items.
`String getSelectedItem( )`	Returns the selected item.
`String[ ] getSelectedItems( )`	Returns all the selected items.
`void remove(int index)`	Removes the item at position *index*.
`void remove(String item)`	Removes the first occurrence of an item matching *item*.
`void removeAll( )`	Removes all the items in the list.
`void select(int index)`	Selects the item at position *index*.
`void setMultipleMode(Boolean mode)`	Enables multiple selections if mode is `true` and single selection if `false`.

Object → EventObject → AWTEvent → ItemEvent

FIGURE 13.4
Action event hierarchy.

TABLE 13.3

Selected methods of the `ItemEvent` class

Method	Function
`Object getItem( )`	Returns the item where the event occurred.
`int getStateChanged( )`	Returns the state change type generating the event.

The `ItemEvent` class inherits the `getSource` method that returns the `ItemEvent` object's source. Additional `ItemEvent` members are summarized in Table 13.3.

The state-changed types are defined as attributes of the `ItemEvent` class: **`SELECTED`** and **`DESELECTED`**.

USING LIST BOXES

Responding to item events emitted by selecting or deselecting list items is illustrated in Figure 13.5. In this application, the background color of the frame is changed to red when an item is selected and to blue when an item is deselected. Methods of the `ItemListener` interface are inherited in the application's main frame with the statement

```
public class Lists extends Frame
    implements WindowListener,
               ActionListener,
               ItemListener
```

List object `lSource` is created with the statement

```
List lSource = new List(4, true);
```

List items are added to the list by invoking the `add` method of the `List` class:

```
lSource.add("Item 0");
lSource.add("Item 1");
lSource.add("Item 2");
lSource.add("Item 3");
lSource.add("Item 4");
lSource.add("Item 5");
lSource.add("Item 6");
```

These items are displayed when the list is initially shown.

The list is added to the frame container with the statement

```
add(lSource);
```

its bounds set

```
lSource.setBounds(50, 50, 100, 80);
```

and a listener implemented

```
lSource.addItemListener(this);
```

Note that the application is identified as the listener by use of the `this` keyword to specify the listening component.

Selecting or deselecting an item generates an `ItemEvent`. The `itemStateChanged` method is invoked automatically with the `ItemEvent` as an argument. State change type is an attribute of the `ItemEvent` object and is returned with the `getStateChange` method. When the state change type is `SELECTED` the frame background color is set red. When the state change type is `DESELECTED` the frame background color is set blue.

```
public void itemStateChanged(ItemEvent ie)
{
    if(ie.getStateChange() == ie.SELECTED)
        setBackground(new Color(255,0,0));
    if(ie.getStateChange() == ie.DESELECTED)
        setBackground(new Color(0,0,255));
}
```

```
import java.awt.*;
import java.awt.event.*;
public class Lists extends Frame
      implements WindowListener,
                     ActionListener,
                     ItemListener
{
      Button bClear = new Button("Clear");
      Button bSelected = new Button("Selected");
      List lSource = new List(4, true);
      TextArea tTarget = new TextArea( );
      Label lMove = new Label("->");

      public Lists( )
      {
           addWindowListener(this);

           setTitle("Using Text Fields");
           setLayout( null );

           setSize(400,300);

           add( bClear );
           add( bSelected );
           lSource.add("Item 0");
           lSource.add("Item 1");
           lSource.add("Item 2");
           lSource.add("Item 3");
           lSource.add("Item 4");
           lSource.add("Item 5");
           lSource.add("Item 6");
           add( lSource );
           add( tTarget );
           add( lMove );

           bClear.setBounds( 250, 200, 100, 50 );
           bSelected.setBounds( 50, 200, 100, 50 );
           lSource.setBounds( 50, 50, 100, 80 );
           tTarget.setBounds( 250, 50, 100, 80 );
           lMove.setBounds( 200, 50, 10, 10 );

           bClear.addActionListener(this);
           bSelected.addActionListener(this);
           lSource.addItemListener(this);

           setVisible(true);
      }

      public static void main(String[] args)
      {
```

```
            Lists list = new Lists();
        }
        public void actionPerformed( ActionEvent ae )
        {
            if( ae.getSource( ) == bClear )
            {
                tTarget.setText(
                tTarget.setBackground(new Color(255,255,255));
            }
            if( ae.getSource( ) == bSelected )
            {
                String buf[ ] = lSource.getSelectedItems();
                for(int i = 0; i < buf.length; i++ )
                {
                    tTarget.append(buf[i]);
                    tTarget.append("\n");
                }
            }
        }

        public void itemStateChanged( ItemEvent ie )
        {
            if( ie.getStateChange( ) == ie.SELECTED )
                setBackground(new Color(255,0,0));
            if( ie.getStateChange( ) == ie.DESELECTED )
                setBackground(new Color(0,0,255));
        }

        public void windowClosing( WindowEvent we )
        {
            setVisible(false );
            dispose( );
            System.exit( 0 );
        }
        public void windowActivated( WindowEvent we ){}
        public void windowClosed( WindowEvent we ){}
        public void windowDeactivated( WindowEvent we ){}
        public void windowDeiconified( WindowEvent we ){}
        public void windowIconified( WindowEvent we ){}
        public void windowOpened( WindowEvent we ){}
}
```

FIGURE 13.5
Program using `Lists` and responding to `ItemEvents`.

The application shown in Figure 13.5 displays the window shown in Figure 13.6. List items are displayed in the list. Because the number of rows of list items exceeds the number of visible rows in the list, a scrollbar is automatically added.

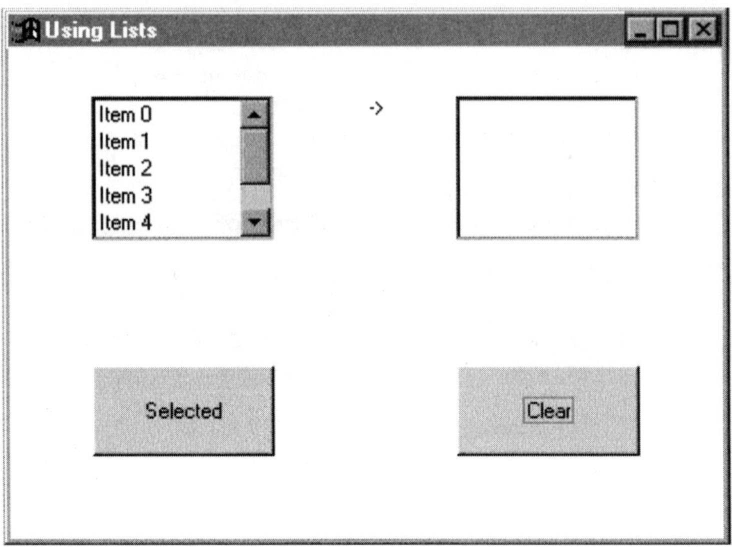

FIGURE 13.6A
Window from Figure 13.5 program using `Lists` and responding to `ItemEvents`.

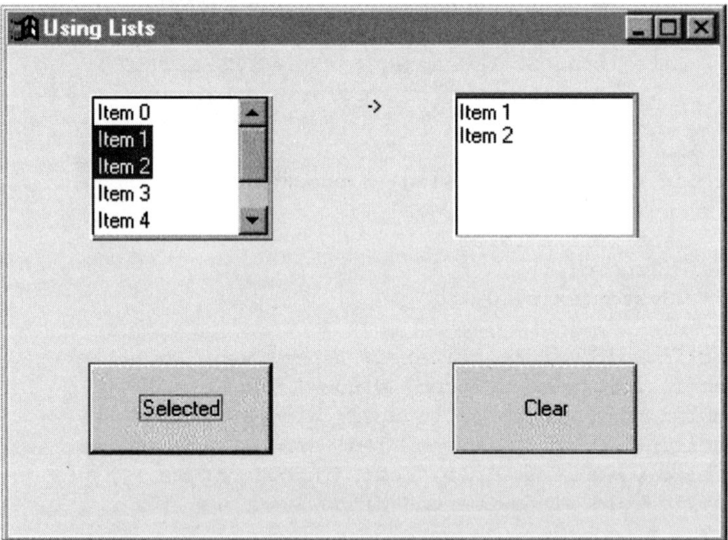

FIGURE 13.6B
Output of program using `Lists` and responding to `ItemEvents`.

When an item in the list is selected the frame background turns red and when an item is deselected the frame background turns blue.

When the "Selected" button is clicked the selected items in the list are transferred to the text area. List items are retrieved as a `String` array `buf` with the `getSelectedItems` method.

```
String buf[] = lSource.getSelectedItems();
```

String array `buf`'s elements are appended to the text area

```
for(int i = 0; i < buf.length; i++)
{
    tTarget.append(buf[i]);
    tTarget.append("\n");
}
```

Note the appending of a new line character, "`\n`", after each array element is appended. Clicking the Clear button clears the text area.

> Scrollbars are automatically added when all items are not visible.

SCROLLBAR COMPONENTS

A **scrollbar** enables user selection from a range of values. A scrollbar with horizontal orientation is shown in Figure 13.7. A scrollbar has a minimum value and a maximum value. The current value of the scrollbar is represented by the position of the **bubble** or **thumb**. The user changes the value of the scrollbar by moving the thumb along the scrollbar. The thumb is moved by dragging with the mouse. Alternately, the thumb can be moved a unit by clicking the **unit increment area** and **unit decrement area.** In addition the thumb can be moved by clicking the **block increment area** and **block decrement area.** Major increments of movement, denoted block increment, are made by a single click in the block increment and decrement areas. Scrollbars can also be displayed with vertical orientation.

Scrollbars are modeled by the **`Scrollbar`** **class.** Constructors of the `Scrollbar` class are summarized in Table 13.4. Selected methods of the `Scrollbar` class are summarized in Table 13.5.

ADJUSTMENT EVENTS

When a scrollbar thumb is moved an **`AdjustmentEvent`** is created. The `AdjustmentEvent` hierarchy is shown in Figure 13.8. To respond to an item event an application must implement an **`AdjustmentListener`** and register

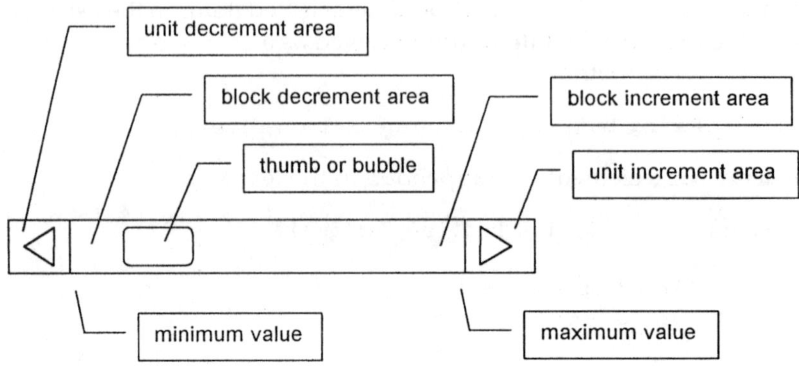

FIGURE 13.7
Typical scrollbar.

TABLE 13.4

Constructors of the `Scrollbar` class

Method	Function
`Scrollbar()`	Creates a scrollbar object with vertical orientation.
`Scrollbar(int` *`orientation`*`)`	Creates a scrollbar object with horizontal or vertical orientation with *`orientation`* values `Scrollbar.HORIZONTAL` and `Scrollbar.VERTICAL`.

TABLE 13.5

Selected methods of the `Scrollbar` class

Method	Function
`int getBlockIncrrement()`	Returns the increment that the scrollbar value increases or decreases when the user clicks in the block increment and decrement areas.
`int getValue()`	Returns the value of the scrollbar.
`void setBlockIncrement(int` *`blockIncrement`*`)`	Sets the increment the scrollbar value is changed when the user clicks in the block increment and decrement areas.
`void setMaximum(int` *`max`*`)`	Sets the maximum value of the scrollbar to the value *`max`*.
`void setMinimum(int` *`min`*`)`	Sets the minimum value of the scrollbar to the value *`min`*.
`void setOrientation{int` *`orientation`*`}`	Orients the scrollbar: *`orientation`* values are `Scrollbar.HORIZONTAL` and `Scrollbar.VERTICAL`.
`void setUnitIncrement(int` *`unitIncrement`*`)`	Sets the increment the scrollbar value is changed when the user clicks in the unit increment and decrement areas.

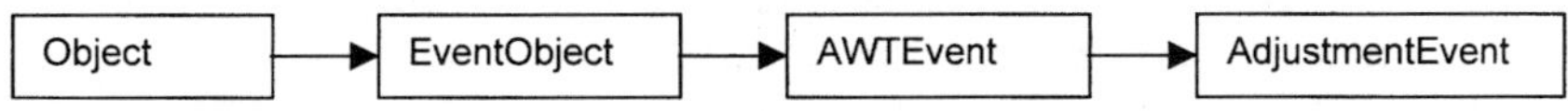

FIGURE 13.8
Adjustment event hierarchy.

the listener to receive item events by calling the list's `addAdjustmentListener` method. The application's response is coded in the `adjustmentValueChanged` method, which is called when an `AdjustmentEvent` is emitted. The `adjustmentValueChanged` method is a member of the `AdjustmentListener` interface.

Five types of action events are defined in the `ActionEvent` class. These events are summarized in Table 13.6.

TABLE 13.6
Action events defined in class `ActionEvent`

Type	Description
`AdjustmentEvent.BLOCK_DECREMENT`	Block decrement area clicked.
`AdjustmentEvent.BLOCK_INCREMENT`	Block increment area clicked.
`AdjustmentEvent.TRACK`	Thumb dragged.
`AdjustmentEvent.UNIT_DECREMENT`	Unit decrement area clicked.
`AdjustmentEvent.UNIT_INCREMENT`	Unit increment area clicked.

The program listing in Figure 13.9a illustrates the creation and manipulation of scrollbars. A scrollbar object `sValue` is created with the statements

```
Scrollbar sValue = new Scrollbar();
```

The scrollbar object is added to the Frame component with the add method in the applications constructor.

```
add(sValue);
```

Scrollbar position and size are set with the `setBounds` method inherited from the `Component` class. A horizontal orientation is set by invoking the `setOrientation` method with argument `Scrollbar.HORIZONTAL`.

```
sValue.setBounds(50, 120, 200, 25);
sValue.setOrientation(Scrollbar.HORIZONTAL);
```

A listener is implemented with the statement

```
sValue.addAdjustmentListener(this);
```

```
import java.awt.*;
import java.awt.event.*;
public class Scrollbars extends Frame
      implements WindowListener,
                     AdjustmentListener
{
```

```
Scrollbar sValue = new Scrollbar( );
TextField tValue = new TextField( );
TextField tType = new TextField( );

public Scrollbars( )
{
     addWindowListener(this);
     setTitle("Using Text Scrollbars");
     setLayout( null );
     setSize(300,185);

     add( sValue );
     add( tValue );
     add( tType );

     tValue.setBounds( 50, 40, 25, 20 );
     tType.setBounds( 50, 80, 150, 20 );
     sValue.setBounds( 50, 120, 200, 25 );
     sValue.setOrientation(Scrollbar.HORIZONTAL);
     sValue.addAdjustmentListener(this);

     setVisible(true);
}

public static void main(String[] args)
{
     Scrollbars list = new Scrollbars();
}

public void adjustmentValueChanged( AdjustmentEvent ae )
{
     int ibuf = ae.getValue( );
     String sbuf = " ";
     tValue.setText(  sbuf.valueOf( ae.getValue() ) );
     if( ae.getAdjustmentType( ) == ae.UNIT_INCREMENT )
          tType.setText("UNIT_INCREMENT");
     if( ae.getAdjustmentType( ) == ae.UNIT_DECREMENT )
          tType.setText("UNIT_DECREMENT");
     if( ae.getAdjustmentType( ) == ae.BLOCK_INCREMENT )
          tType.setText("BLOCK_INCREMENT");
     if( ae.getAdjustmentType( ) == ae.BLOCK_DECREMENT )
          tType.setText("BLOCK_DECREMENT");
     if( ae.getAdjustmentType( ) == ae.TRACK )
          tType.setText("TRACK");
}

public void windowClosing( WindowEvent we )
{
     setVisible(false );
     dispose( );
```

```
            System.exit( 0 );
        }
        public void windowActivated( WindowEvent we ){}
        public void windowClosed( WindowEvent we ){}
        public void windowDeactivated( WindowEvent we ){}
        public void windowDeiconified( WindowEvent we ){}
        public void windowIconified( WindowEvent we ){}
        public void windowOpened( WindowEvent we ){}
}
```

FIGURE 13.9A
Application illustrating creation and use of scrollbars.

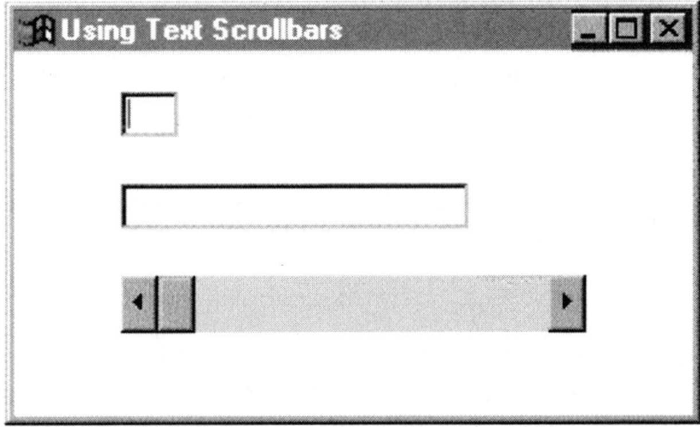

FIGURE 13.9B
Output of application illustrating creation and use of scrollbars.

The Figure 13.9a application's output is illustrated in Figure 13.9b. A horizontal scrollbar is located at the bottom of the frame window. At the top of the frame is a text field that displays the value of the scrollbar. A second text field displays the type of adjustment event created when the thumb of the scrollbar is moved.

The result of dragging the bubble approximately 30% of the length of the scrollbar is illustrated in Figure 13.9c. Note the scrollbar value of 28 and the adjustment type of TRACK. An `AdjustmentEvent` is created whenever the thumb is moved. Subsequently, the `adjustmentValueChanged` method is called with the `AdjustmentEvent` as argument. The value of the scrollbar is displayed in the upper text field and the adjustment type is displayed in the lower text field. The adjustment event's properties include the scrollbar value and can be retrieved with the `getValue` method.

```
int ibuf = ae.getValue();
```

The returned value is assigned to an `int` type variable `ibuff`. The integral value of the buffer is converted to a `String` data type by invoking a `String` object's `valueOf` method with the buffer as argument.

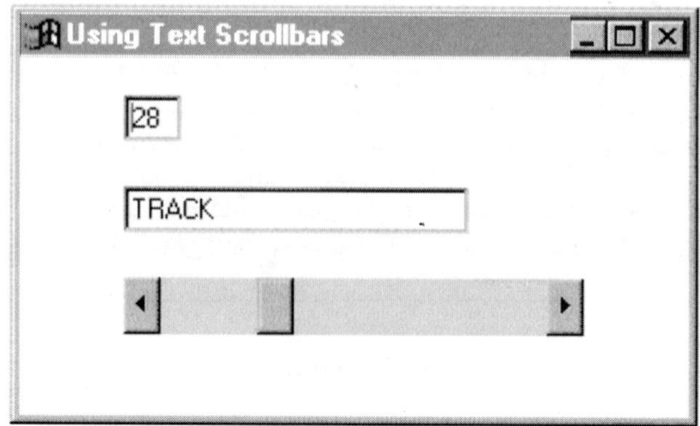

FIGURE 13.9C
Output of application illustrating creation and use of scrollbars.

```
sbuf.valueOf(ae.getValue())
```

The value of the expression is used as argument of the text field, displaying the scrollbar's value, `setText` method.

```
tValue.setText(sbuf.valueOf(ae.getValue()));
```

The type of the adjustment is returned by the `getAdjustmentType` method. The adjustment type is compared with the values defined in the `AdjustmentEvent` class. A positive comparison results in a display of the type in the lower text field. For example,

```
if(ae.getAdjustmentType() == ae.UNIT_INCREMENT)
    tType.setText("UNIT_INCREMENT");
```

Similar comparisons are made regarding the other adjustment types.

> Scrollbars are used for user input of continuously variable values.

CHECKBOX COMPONENTS

Checkboxes provide for entry of either on or off values. When a checkbox is clicked, a check is displayed in the checkbox indicating the checkbox is on. Clicking an on checkbox will clear the check and turn the checkbox off. Checked and unchecked checkboxes are shown in Figure 13.10.

Check boxes have a `true` value when they are checked on and a `false` value when they are off. The on/off property is called the checkbox's **state.**

Checkboxes' components are modeled by the **`Checkbox` class.** Constructors of the `Checkbox` class are shown in Table 13.7. Selected methods of the `Check-`

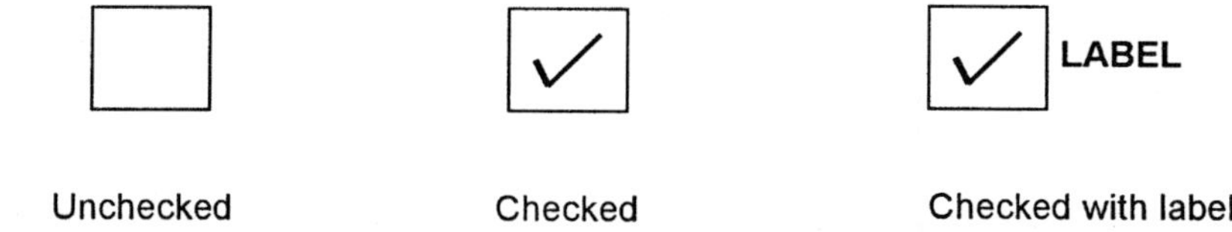

FIGURE 13.10
Checked and unchecked checkboxes.

TABLE 13.7
Constructors of the Checkbox class

Method	Function
`Checkbox()`	Creates a checkbox with no label in the off state.
`Checkbox(String label)`	Creates a checkbox with *label* in the off state.
`Checkbox(String label, boolean state)`	Creates a checkbox with *label* in the off state if *state* is `false` and on if *state* is `true`.

TABLE 13.8
Selected key methods of the Checkbox class

Method	Function
`int getState()`	Returns the state of the checkbox.
`void setLabel(String label)`	Sets the label to the string *label*.
`void setState(boolean state)`	Sets the state to on if *state* is *true* and off if *state* is `false`.

box class are shown in Table 13.8. An application illustrating the use of checkboxes is shown in Figure 13.11a.

```
import java.awt.*;
import java.awt.event.*;
public class Checkboxes extends Frame
      implements WindowListener,
                    ItemListener
{
      Checkbox cValue = new Checkbox();
      TextField tValue = new TextField();

      public Checkboxes()
      {
           addWindowListener(this);
           setTitle("Using Checkboxes");
           setLayout(null);
```

```
            setSize(300,150);

            add( cValue );
            add( tValue );

            tValue.setBounds( 50, 40, 60, 20 );
            cValue.setBounds( 50, 80, 50, 25 );
            cValue.setState(true);
            cValue.addItemListener(this);

            setVisible(true);
        }

        public static void main(String[] args)
        {
            Checkboxes list = new Checkboxes( );
        }

        public void itemStateChanged( ItemEvent ie )
        {
            String sbuf = " ";
            if( ie.getSource( ) == cValue )
                tValue.setText(
sbuf.valueOf( cValue.getState( ) ) );
        }

        public void windowClosing( WindowEvent we )
        {
            setVisible(false );
            dispose( );
            System.exit( 0 );
        }
        public void windowActivated( WindowEvent we ){}
        public void windowClosed( WindowEvent we ){}
        public void windowDeactivated( WindowEvent we ){}
        public void windowDeiconified( WindowEvent we ){}
        public void windowIconified( WindowEvent we ){}
        public void windowOpened( WindowEvent we ){}
}
```

FIGURE 13.11A
Application illustrating use of checkboxes.

As shown in Figure 13.11a, checkbox object `cValue` is created with the statement

```
Checkbox cValue = new Checkbox( );
```

and added to the frame window with the statement

```
add( cValue );
```

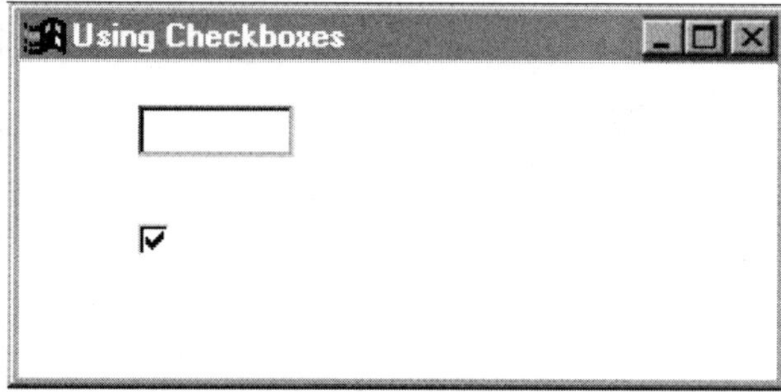

FIGURE 13.11B
Output of application illustrating checkboxes

The checkbox is sized with the setBounds method, set to an on state with the statements

```
cValue.setBounds( 50, 80, 50, 25 );
cValue.setState(true);
```

The application initial window is shown in Figure 13.11b.

Although the checkbox is created to standard size, the bounding rectangle is determined by the `setBounds` method. Clicking anywhere within the bounding rectangle will toggle the checkbox on or off.

> Care should be taken to ensure that the checkbox's bounding rectangle approximates the actual checkbox to minimize user confusion.

Whenever the checkbox is clicked an `ItemEvent` is created. An item listener is added with the statement

```
cValue.addItemListener(this);
```

and implemented with the statement

```
public class Checkboxes extends Frame
    implements WindowListener,
               ItemListener
```

The `itemStateChanged` method is automatically called with the `ItemEvent` as argument. The source of the event is retrieved with the `getSource` method. The state of the `ItemEvent` is retrieved with the `getState` method. These methods are used in the following statements to determine when the checkbox `cValue` emits an `ItemEvent` and to set the text field `tValue` equal to the value of the checkbox.

```
String sbuf = " ";
if( ie.getSource( ) == cValue )
    tValue.setText(
sbuf.valueOf( cValue.getState( ) ) );
```

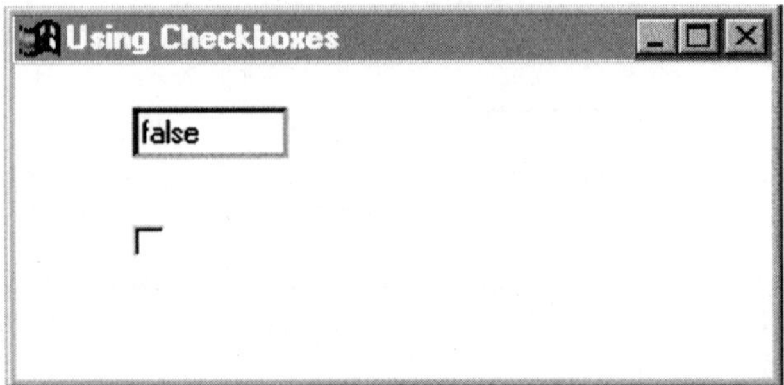

FIGURE 13.11C
Output of application illustrating text fields.

Note the use of the `String` object to convert the `boolean` state value to a `String` value that can be set in the text field.

Figure 13.11c illustrates the value of the checkbox when it is clicked to off.

CHECKBOX GROUP COMPONENTS

In a checkbox group only one checkbox can be checked—that is, in the "on" state—at a time. Checking a checkbox in the group unchecks other checkboxes in the group. A checkbox group is represented by the **`CheckboxGroup`** class. The `CheckboxGroup` class hierarchy is shown in Figure 13.12.

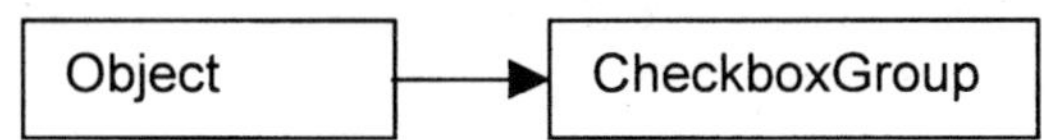

FIGURE 13.12
`CheckboxGroup` class hierarchy.

The `CheckboxGroup` class has one constructor that has no parameters. Selected methods of the `CheckboxGroup` class are shown in Table 13.9. An application illustrating the use of the checkbox group is shown in Figure 13.13a.

TABLE 13.9

Selected methods of the `checkboxGroup` class

Method	Function
`Checkbox getSelectedCheckbox( )`	Returns the checkbox in the group that is in the on state.
`void set getSelectedCheckbox( Checkbox` *`checkbox`*`)`	Sets the checkbox specified, *`checkbox`*, in the group to the on stat

```
import java.awt.*;
import java.awt.event.*;
public class CheckboxGroups extends Frame
      implements WindowListener,
                     ActionListener
{
      CheckboxGroup cbSourceGroup = new CheckboxGroup();
      CheckboxGroup cbTargetGroup = new CheckboxGroup();
      Checkbox cSourceX = new Checkbox("X",false,cbSourceGroup);
      Checkbox cSourceY = new Checkbox("Y",true,cbSourceGroup);
      Checkbox cSourceZ = new Checkbox("Z",false,cbSourceGroup);
      Checkbox cTargetX = new Checkbox("X",false,cbTargetGroup);
      Checkbox cTargetY = new Checkbox("Y",false,cbTargetGroup);
      Checkbox cTargetZ = new Checkbox("Z",false,cbTargetGroup);
      Button bSet = new Button("-->");

      public CheckboxGroups( )
      {
           addWindowListener(this);
           setTitle("Using Checkbox Groups");
           setLayout( null );
           setSize(300,150);

           add( cSourceX );
           add( cSourceY );
           add( cSourceZ );
           add( cTargetX );
           add( cTargetY );
           add( cTargetZ );
           add( bSet );

           cSourceX.setBounds( 50, 40, 50, 25 );
           cSourceY.setBounds( 50, 70, 50, 25 );
           cSourceZ.setBounds( 50, 100, 50, 25 );
           cTargetX.setBounds( 220, 40, 50, 25 );
           cTargetY.setBounds( 220, 70, 50, 25 );
           cTargetZ.setBounds( 220, 100, 50, 25 );
           bSet.setBounds( 130, 70, 40, 30 );

           bSet.addActionListener(this);

           setVisible(true);
      }

      public static void main(String[] args)
      {
           CheckboxGroups list = new CheckboxGroups( );
      }

      public void actionPerformed( ActionEvent ae )
      {
           if( ae.getSource( ) == bSet )
```

```
        {
            Checkbox cTemp;
            cTemp = cbSourceGroup.getSelectedCheckbox();
            if (cTemp == cSourceX)
                cbTargetGroup.setSelectedCheckbox(cTargetX);
            if (cTemp == cSourceY)
                cbTargetGroup.setSelectedCheckbox(cTargetY);
            if (cTemp == cSourceZ)
                cbTargetGroup.setSelectedCheckbox(cTargetZ);
        }
    }

    public void windowClosing( WindowEvent we )
    {
        setVisible(false );
        dispose( );
        System.exit( 0 );
    }
    public void windowActivated( WindowEvent we ){}
    public void windowClosed( WindowEvent we ){}
    public void windowDeactivated( WindowEvent we ){}
    public void windowDeiconified( WindowEvent we ){}
    public void windowIconified( WindowEvent we ){}
    public void windowOpened( WindowEvent we ){}
}
```

FIGURE 13.13A
Application illustrating use of checkbox group.

Checkbox groups, reference by cbSourceGroup and cbTargetGroup, are created with the statements

```
CheckboxGroup cbSourceGroup = new Checkbox-
Group();
CheckboxGroup cbTargetGroup = new Checkbox-
Group();
```

Checkboxes are created and assigned to the group cbSourceGroup with the statements

```
Checkbox cSourceX = new Check-
box("X",false,cbGroup);
Checkbox cSourceY = new Check-
box("Y",true,cbGroup);
Checkbox cSourceZ = new Check-
box("Z",false,cbGroup);
```

Similarly, checkboxes are created and assigned to the group cbTarget with the statements

```
Checkbox cTargetX = new Checkbox("X",false,cbTargetGroup);
```

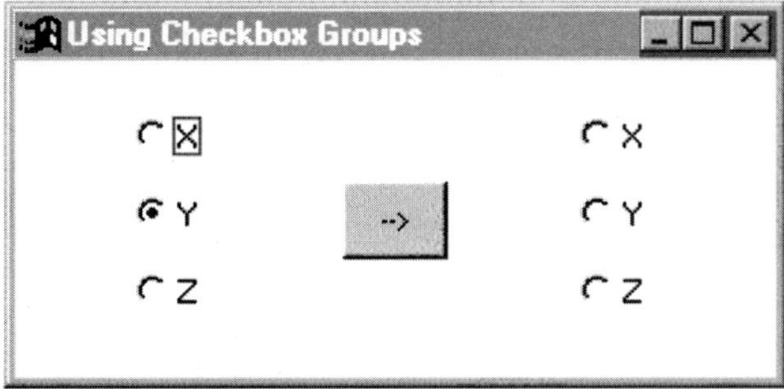

FIGURE 13.13B
Output of an application illustrating use of checkbox groups.

```
Checkbox cTargetY = new Checkbox("Y",false,cbTargetGroup);
Checkbox cTargetZ = new Checkbox("Z",false,cbTargetGroup);
```

Note that the checkbox `cSourceY` is set "on and the other checkboxes `cSourceX` and `cSourceZ` are set "off" initially. Note also that none of the checkboxes in the `cbTargetGroup` are set to the on state initially. This is reflected in the application's output, shown in Figure 13.13b.

The checkboxes in the `cbSourceGroup` are located on the left side of the window. The checkboxes in the `cbTargetGroup` are located on the right side of the window. When the button is clicked, the checkbox in the target group corresponding to the selected checkbox in the source group is selected. For example, if the Y checkbox in the source group is checked and the button is clicked, the Y checkbox in the target group is selected. Consider the implementation of the `ActionPerformed` method shown below:

```
public void actionPerformed( ActionEvent ae )
{
    if( ae.getSource( ) == bSet )
    {
        Checkbox cTemp;
        cTemp = cbSourceGroup.getSelectedCheckbox();
        if (cTemp == cSourceX)
            cbTargetGroup.setSelectedCheckbox(cTargetX);
        if (cTemp == cSourceY)
            cbTargetGroup.setSelectedCheckbox(cTargetY);
        if (cTemp == cSourceZ)
            cbTargetGroup.setSelectedCheckbox(cTargetZ);
    }
}
```

The selected checkbox in the group `cbSourceGroup` is returned with the statement

```
cTemp = cbSourceGroup.getSelectedCheckbox();
```

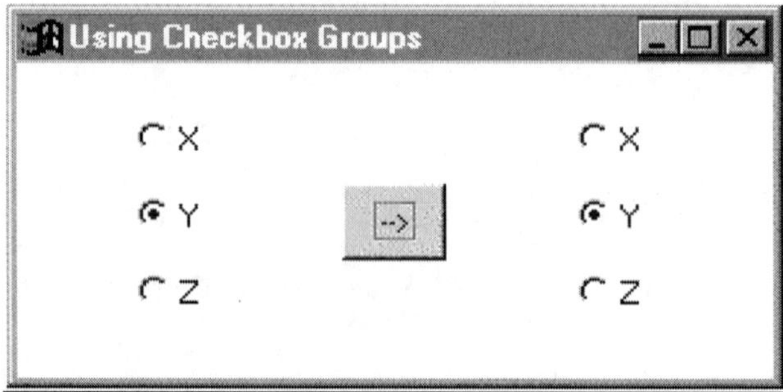

FIGURE 13.13C
Output of an application illustrating the use of checkbox groups.

and assigned to a checkbox object reference `cTemp`. Note the `getSelectedCheckbox` returns null if all of the checkboxes are off.

If the `cTemp` is the object `cSourceX` then the statement

```
cbTargetGroup.setSelectedCheckbox(cTargetX);
```

is executed and the checkbox `cTargetX` in the group `cbTargetGroup` is set "on." If the checkbox argument is `null` or belongs to a different checkbox group, then the `setSelectedCheckbox` method does nothing.
Similarly, `cTargetY` or `cTargetZ` are set on when `cSourceY` or `cSourceZ` is selected when the button is clicked. The program's output when `cSourceY` is selected and the button is clicked is shown in Figure 13.13c.

Note that item events are also created when the checkboxes are selected and deselected.

CHOICE COMPONENTS

Choice components present a pull-down menu of items. A choice component in its normal mode is shown in Figure 13.14a. The choice component appears as a text field with a down arrow located at its left side. When the choice component is clicked, a pull-down menu is presented as a list below the text portion of the choice component. A pull-down menu consisting of the four items is illustrated in Figure 13.14b. Clicking an item in the list selects the item. The selected items appear in the text portion of the choice component.

The choice component is often called a "combobox" as it combines the elements of a text box and a list box. Choice components are modeled by the `Choice` **class.** The `Choice` class has one constructor that has no parameters. The constructor creates a `Choice` component that has no items in its menu. Selected methods of the `Choice` class are summarized in Table 13.10.

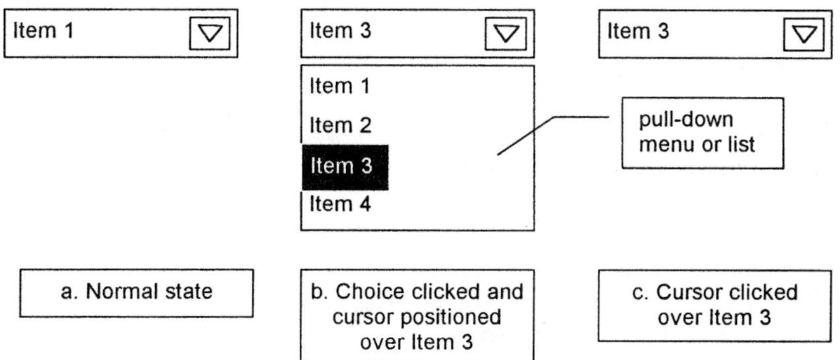

FIGURE 13.14
Choice component in normal, menu pulled down, and item selected modes.

TABLE 13.10
Selected methods of the `Choice` class

Name	Description
`void add( String item )`	Adds the text *item* to the list or menu.
`void addItem( String item )`	Adds the text *item* to the list or menu.
`void addItemListener( ItemListener il )`	Adds the item listener *item* to receive item events form the choice component.
`String getItem( int index )`	Returns the string at the position specified by *index* in the list of items.
`int getItemCount( )`	Returns the number of items in the list or menu.
`int getSelectedIndex( )`	Returns the index of the selected item.
`void insert( String item int index )`	Adds the text *item* to the list or menu at the position specified by *index*.
`void remove( String item )`	Removes the first occurrence of the *item* from the list.
`void remove( int postion )`	Removes the item in the list at *position*.
`void removeAll( )`	Removes all items in the list.
`void select( int postion )`	Sets the selected item in the choice, both text and list portions, to the item at *position*.
`void select( String text )`	Sets the first occurrence of an item that matches *text* to the selected item.

The first item added to the menu becomes the selected item until another item is selected. The index value associated with the first item in the list is zero. If a position is not specified, items are added to the end of the list.

A program illustrating use of the `Choice` component is shown in Figure 13.15a. A `Choice` component, `cChoiceSource`, is created with the statement

```
Choice cChoiceSource = new Choice();
```

The newly created `Choice` component has zero items in its list. Item "Item 1" is added to the choice's menu or list with the statement

```
cChoiceSource.add(new String("Item 1"));
```

Similarly, "Item 2" is added with the statement

```
cChoiceSource.addItem(new String("Item 2"));
```

Additional items are similarly added.

```
import java.awt.*;
import java.awt.event.*;
public class UsingChoices extends Frame
      implements WindowListener,
                    ActionListener
{
      Choice cChoiceSource = new Choice();
      Button bSet = new Button("-->");
      Button bInsert = new Button("<--");
      TextField tTargetField = new TextField();

      public UsingChoices( )
      {
           addWindowListener(this);
           setTitle("Using Choices");
           setLayout( null );
           setSize(300,150);

           add( cChoiceSource );
           add( bSet );
           add( bInsert );
           add( tTargetField );

           cChoiceSource.setBounds( 40, 40, 80, 20 );
           bSet.setBounds( 140, 40, 40, 30 );
           bInsert.setBounds(140, 80, 40, 30 );
           tTargetField.setBounds( 200, 40, 60, 20 );

           bSet.addActionListener(this);
           bInsert.addActionListener(this);

           cChoiceSource.add(new String("Item1"));
           cChoiceSource.addItem(new String("Item2"));
           cChoiceSource.add(new String("Item3"));
           cChoiceSource.add(new String("Item3"));

           setVisible(true);
      }

      public static void main(String[] args)
      {
           UsingChoices choiceObject = new UsingChoices( );
```

```
    }

    public void actionPerformed( ActionEvent ae )
    {
        if( ae.getSource( ) == bSet )
        {
            String sTemp;
            sTemp = cChoiceSource.getSelectedItem();
            tTargetField.setText(sTemp);
        }
        if( ae.getSource( ) == bInsert )
        {
            cChoiceSource.insert(tTargetField.getText(),0);
        }
    }

    public void windowClosing( WindowEvent we )
    {
        setVisible(false );
        dispose( );
        System.exit( 0 );
    }
    public void windowActivated( WindowEvent we ){}
    public void windowClosed( WindowEvent we ){}
    public void windowDeactivated( WindowEvent we ){}
    public void windowDeiconified( WindowEvent we ){}
    public void windowIconified( WindowEvent we ){}
    public void windowOpened( WindowEvent we ){}
}
```

FIGURE 13.15A
Program illustrating choice components.

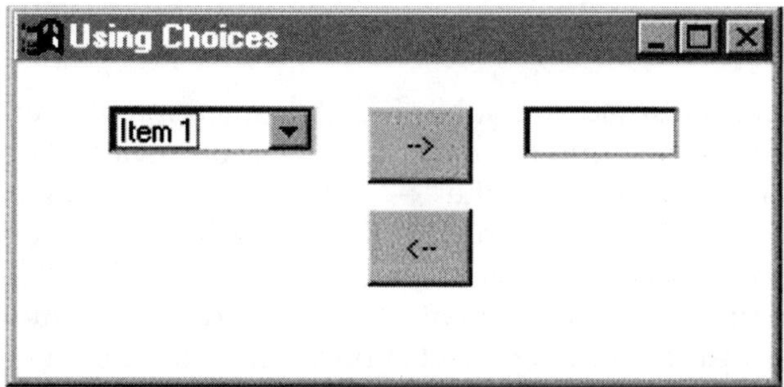

FIGURE 13.15B
Initial output of program illustrating `Choice` components.

The `Choice` component with initial item selection is illustrated in Figure 13.15b. Note that the first item added is initially selected.

When the "→" button is clicked, the selected item in the application's choice box is copied into the text field located in the left side of the applica-

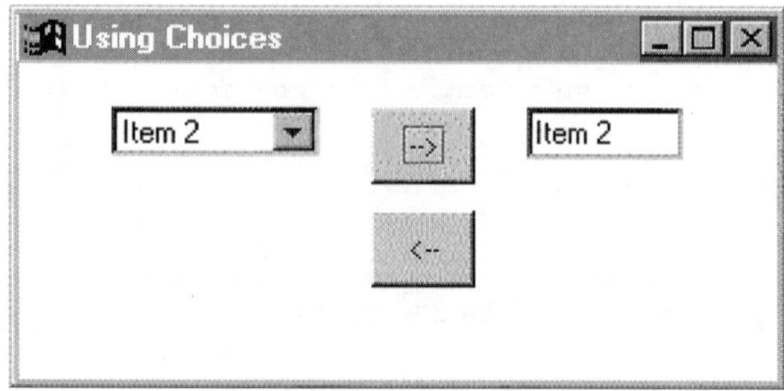

FIGURE 13.15C
Output of program illustrating `Choice` components after item selection.

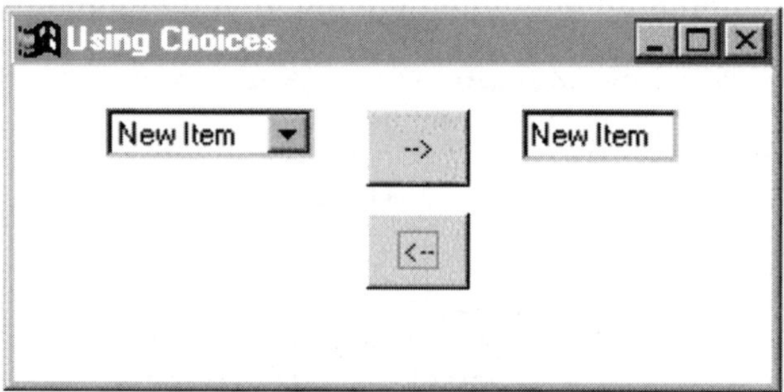

FIGURE 13.15D
Output of program illustrating `Choice` components after item insertion.

tion's window. Figure 13.15c illustrates the circumstance in which the choice "Item 2" was selected and then copied into the text field by clicking the "→" button. The selected item was returned by invoking the `Choice` objects `getSelectedItem` method:

```
cChoiceSource.getSelectedItem()
```

The String object returned was subsequently copied into the text field.

Items can be inserted into the list of a `Choice` component. For example, in the application shown, clicking the "←" button inserts the content of the text field into the `Choice` component list. Figure 13.15d illustrates the case in which the contents of the text field, that is the string "New Item," has been transferred to the `Choice` component by clicking the "←" button.

The string is inserted into the first list position with the `Choice` class `insert` method:

```
cChoiceSource.insert(tTargetField.getText(),0);
```

Note that the first list position has index value zero.

> `Choice` components combine the text and list component attributes into a single component.

Remember, in the Java implementation of the `Choice` component, users cannot input into the text portion.

HANDS-ON EXERCISE 13: LIMITING USERS TO ONE OF MANY OPTIONS

Restrict the user of the Customer Graphical User Interface to picking a selection from platinum, gold and regular.

Radio buttons are used to limit users to a single selection from a group of many. For example, consider the assignment of customer status: Regular, Gold, or Platinum (Figure HO-13.1).

HO-13.1
Customer window.

The checkboxes are in a group and thus only one can be selected at a time. This window background changes to yellow when Gold is selected, gray when platinum is selected, and white when regular is selected.

Step 1: Create a New Project

1. Start Visual J++.
2. Select the New Project… command in the File menu. The New Project dialog box is displayed.
3. Enter **`Customer Type Selection`** in the Name text field.
4. Use the Browse button to enable navigation to the Location `C:\Hands-On\Customer Type Selection`.
5. Click the Open button to create the new project.

Step 2: Create `Customer.java` File

1. Press the CTRL+D keys. The Add Item dialog box is displayed.
2. Select the Class folder.
3. Select the ClassMain icon.
4. Enter **`Customer.java`** in the Name text field.
5. Click the Open button to create the new class.
6. The Code Editor is opened with the file `Customer.java`.

Step 3: Code the Application

Enter code to create a frame window, checkboxes, checkbox group, and handlers for the window events and item selection events.

```
import java.awt.*;
import java.awt.event.*;

public class Customer extends Frame
      implements WindowListener, ItemListener
{
      CheckboxGroup typeGroup = new CheckboxGroup();
      Checkbox regularCheckbox = new
               Checkbox("Regular",true,typeGroup);
      Checkbox goldCheckbox = new
               Checkbox("Gold",false,typeGroup);
      Checkbox platinumCheckbox = new
               Checkbox("Platinum",false,typeGroup);

      public Customer( ) {

           setLayout( null );
           add(regularCheckbox);
           add(goldCheckbox);
           add(platinumCheckbox);
           regularCheckbox.setBounds(300,100,80,20);
           goldCheckbox.setBounds(300,120,80,20);
           platinumCheckbox.setBounds(300,140,80,20);
           regularCheckbox.addItemListener(this);
           goldCheckbox.addItemListener(this);
           platinumCheckbox.addItemListener(this);

           addWindowListener(this);
           setTitle("Customer");
           setSize(400,200);
           setVisible(true);
      }

      public static void main(String[] args)
```

```
        {
             new Customer();
        }

        public void itemStateChanged(ItemEvent ie) {
             if(ie.getSource()==regularCheckbox)
             {
                  setBackground(Color.white);
             }
             if(ie.getSource()==goldCheckbox)
             {
                  setBackground(Color.yellow);
             }
             if(ie.getSource()==platinumCheckbox)
             {
                  setBackground(Color.gray);
             }
        }

        public void windowClosing( WindowEvent we ) {
             setVisible(false );
             dispose( );
             System.exit( 0 );
        }
        public void windowActivated( WindowEvent we ){}
        public void windowClosed( WindowEvent we ){}
        public void windowDeactivated( WindowEvent we ){}
        public void windowDeiconified( WindowEvent we ){}
        public void windowIconified( WindowEvent we ){}
        public void windowOpened( WindowEvent we ){}
}
```

Step 4: Build the Basic Application

1. Press the CTRL+SHIFT+B keys.
2. Debug the code as required.

Step 5: Build the Basic Application

1. Press the CTRL+SHIFT+B keys.
2. Debug the code as required.

Step 6: Set the Project Properties

1. Select the Customer Input Output Properties… option in the Project menu. The properties dialog box is displayed.
2. Click the OK button to accept the default properties.

Step 7: Execute the Application

1. Press the F5 key. The application starts and the application window is displayed
2. Check Gold. The background turns yellow.
3. Check Platinum. The background turns platinum.
4. Check White. The background turns white.
5. Close the application.

Step 8: Save All the Project Files

Press the CTRL+SHIFT+F5 keys.

Step 9: Close the Project

Select the Close All option in the File menu.

Step 10: Exit Visual J++ the Project

Select the Exit option in the File menu.

Challenge 1: Modify the application to color the checkbox to the background color.

SUMMARY

1. List boxes provide for user selection of one or multiple items from a list.
2. Scrollbars enable users to select values from a continuum.
3. Checkboxes allow users to select multiple conditions.
4. Checkbox groups restrict user selection to only one of the grouped check boxes.
5. Choice controls display a list selection in a text box.
6. Item events are created by lists, checkboxes, and choice component selection and deselection.
7. Moving the thumb or bubble of a scrollbar creates an adjustment event.
8. Lists, scrollbars, choice, checkboxes and checkbox groups are useful components in the development of Graphical User Interfaces.

KEY WORDS AND CONCEPTS

`Adjustment` class
`AdjustmentEvent`
block decrement area
block increment
block increment area
bubble
checkbox
`Checkbox` class
checkbox group
`CheckboxGroup` class
choice
`Choice` class
combobox
`DESELECTED`
`ItemEvent` class
`ItemListener`
list
`List` class
pull-down menu
scrollbar
`Scrollbar` class
`SELECTED`
state
thumb
unit decrement area
unit increment area

REVIEW QUESTIONS AND EXERCISES

1. Mark the following questions true or false.

   (T/F) Selecting a list item does not create an action event.

   (T/F) Items cannot be entered by typing in the textbox portion of a choice component.

   (T/F) Several checkboxes can be set "on" in a checkbox group.

   (T/F) Scrollbar adjustments are incremental values set during creation.

   (T/F) Lists can be used where multiple item selection is required.
2. How is a list box enabled for multiple selections?
3. Create and test an application that retrieves the index and textual content of selected items from a multiple selection list.
4. What interface is required to implement item events?
5. How are minimum and maximum values of a scrollbar set?
6. What interface is required to implement adjustment events?
7. Can scrollbars be set horizontally?
8. How is the initial state of a checkbox determined?
9. Compare list and choice components.

10. Create and test a checkbox group that enables user indication of martial status.

PRACTICE WITH VISUAL J++

Create a user interface that provides for the following:

a. Setting of a temperature value with a scrollbar.
b. Selection of the scale as either Centigrade or Fahrenheit using a checkbox group.
c. Selection of the month, day, and year using choice boxes.
d. Selection of the ambient weather summary with a list selection.

Part 4

Communications, Networking and Advanced Topics

14

Stream I/O

OBJECTIVES

After reading this chapter, you will be able to:

1. Understand the stream and file concept.
2. Use the `java.io` package to create, write to, and read from streams.
3. Demonstrate the ability to create and use persistent data.

OVERVIEW

Java views the flow of data to and from files as a stream. The `java.io` package provides the classes required to read and write to streams. The package includes classes that model the flow of data from a file. Classes that model the flow of data to a file are also included. In addition, the classes also contain a class that models files including file attributes and methods for examining file properties. Although not examined in the following, Java also includes methods for modeling random access storage.

STREAMS

All input and output are treated as character operations on a stream of characters, which is reflected in use of the term **stream.** The streams can contain non-text or binary values as well as the printable characters. All input and output are to streams, including input and output associated with keyboards and display terminals, communication networks, and files. Streams

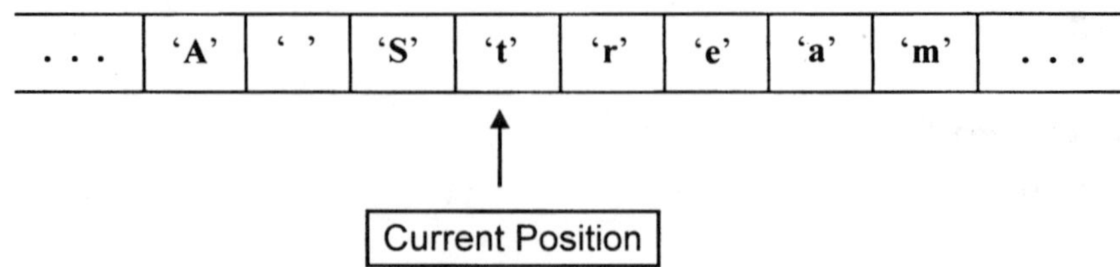

FIGURE 14.1
Visualization of a stream and the position of input and output.

are modeled by stream objects. Typically, the standard input stream object, **System.in,** is the keyboard and usually the standard output stream object, **System.out,** is the display monitor.

A stream can be viewed as an endless ribbon of cells. Each cell contains a byte of data. Input and output are performed at a specific stream location. This point of access is referred to as the current position of the stream. Figure 14.1 presents a stream visualization.

The stream can be visualized as moving, thereby enabling the contents of successive cells to be accessible at the current position. Contents of the cell at the current position can be read or extracted. Similarly, new information can be written or inserted into the cell. This view of a stream parallels the operation of reading data sequentially from a disk or tape drive. An alternate visualization, in which the current position moves to another cell, is equally valid and is representative of the operation of direct access devices.

Alternately, a stream can be viewed as a pipeline connecting two objects through which data is conveyed. In this visualization, the stream is said to have a `source` and `sink`. Data inserted into the stream at the source is said to be written into the stream. Data extracted at the sink is said to be read from the stream. This visualization of a stream resembles our conceptual view of a communication channel.

In both views, data is read from the stream a byte at a time. Similarly, new data is written to the stream a byte at a time. This mode of access is referred to as sequential file access.

SEQUENTIAL FILE PROCESSING

Classes in the `java.io` package provide the functionality required to read and write to streams. The `java.io` package class hierarchy is summarized in Figure 14.2. The input and output stream classes are summarized in Table 14.1.

The **`File` class** models the name of a file or directory. Selected constructors of the `File` class are summarized in Table 14.2.

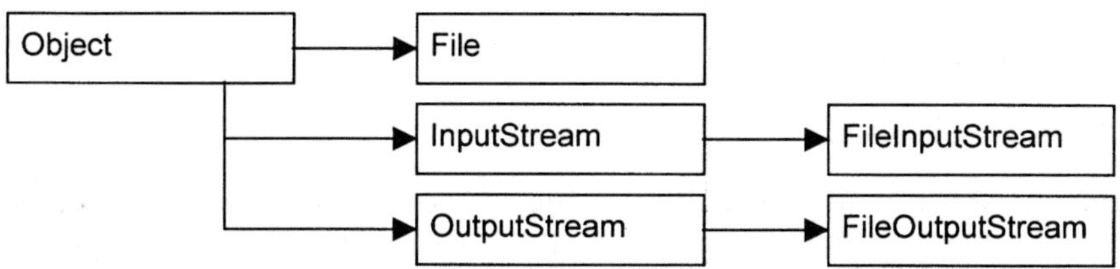

FIGURE 14.2.
Input/output stream class hierarchy.

TABLE 14.1
Summary of selected input/output stream hierarchy classes

Class	Description
`File`	Name of a file or directory
`FileInputStream`	An input stream for reading from an input file.
`FileOutputStream`	A output stream for writing to an output file.
`InputStream`	An abstract class representing an input stream of bytes.
`OutputStream`	An abstract class representing an output stream of bytes.

TABLE 14.2
Selected constructors of the `File` class

Name	Description
`File(String` *`path`*`)`	Creates a file object with path specified by *`path`*.
`File(String` *`path`*`, String` *`name`*`)`	Creates a file object with path specified by *`path`* and file name specified by *`name`*.

An instance of the class `File` representing the file with path `C:\stream.dat` is created with the statements

```
String fileName = "c:\\stream.dat";
File stream;

stream = new File(fileName);
```

Note that the backslash character in the path is replaced with a double backslash. Further note that the pathname must follow the conventions of the host operating system. In addition, note that the path name may be the absolute path name or a path relative to the current directory.

The `File` class also contains methods for ascertaining file properties.

STREAM CREATION

Output streams are modeled by the **FileOutputStream** class objects. Selected constructors of the FileOutputStream class are shown in Table 14.3.

TABLE 14.3

Selected constructors of the FileOutputStream class

Name	Description
FileOutputStream(File *file*)	Creates a file output stream object for the File object *file*.
FileOutputStream(String *name*)	Creates a file output stream object with path *name* that can be written to.

An object of the FileOutputStream class is created with the statement

```
FileOutputStream outStream =
    new FileOutputStream(fileName);
```

One form of the constructor provides for a string variable in which the file name can be specified. In the example, the file name is specified by the string variable fileName, which has been assigned the value "c:\\stream.dat".

Selected methods of the FileOutputStream class are shown in Table 14.4. A program illustrating creation of a stream is illustrated in Figure 14.3a. The statement

```
import java.io.*;
```

provides for the access to the class FileOutputStream declared in the java.io package.

TABLE 14.4

Selected methods of the FileOutputStream class

Name	Description
void close()	Closes the file output stream object and releases associated resources.
void write(byte *dataArray*)	Writes *dataArray.Length* bytes from the array *dataArray* to the file output stream.
void write(int *byteOfData*)	Writes the byte *byteOfData* to the file output stream.

The object outStream is instantiated as an object of type FileOutputStream. The object is instantiated in a try block. When the FileOutputStream object is created and the file is open a test is performed regarding the success of the object creation and the file opening. If the creation is unsuccessful or the file is not opened, an IOException object is created and thrown. For example, if

```
import java.io.*;

class CreateStream
{
      public static void main(String[] args)
      {
            String fileName = "C:\\stream.dat";

            try {
                  FileOutputStream outStream =
                        new FileOutputStream(fileName);

                  outStream.close();
            }
            catch (IOException ioe) {
                  System.out.println("Create error");
            }
      }
}
```

stream opened for output

stream closed

FIGURE 14.3A
Creating a file using the `FileOutputStream`.

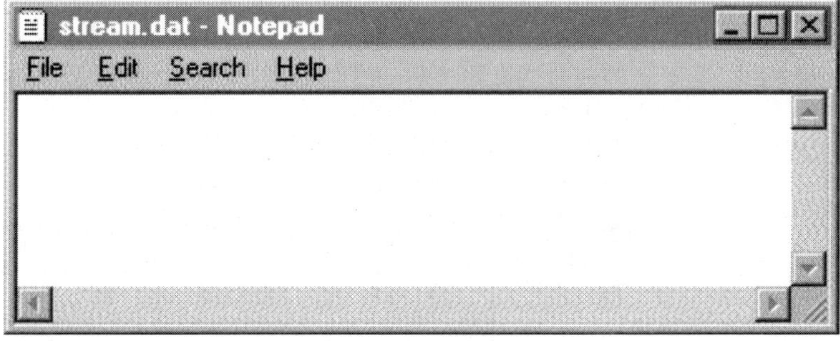

FIGURE 14.3B
Product of the program creating a file using the `FileOutputStream` class.

the path is invalid, the file cannot be opened and an exception will be thrown. In the program, the thrown `IOException` is caught in the `catch` block. The message "`Create error`" is displayed when an `IOException` is thrown.

The file is closed by the statement

```
outStream.close();
```

The file is opened in Notepad (Figure 14.3b). Note that the Notepad window is empty, as printable characters have not been written to the file.

WRITING TO A STREAM

The `FileOutputStream` method `write(int b)` writes the specified byte to the output stream. For example,

```
outStream.write('H');
```

inserts the character 'H" in the input stream. The write method throws an exception when the write operation is unsuccessful. A program illustrating writing to an output stream is shown in Figure 14.4a. The `FileOutputStream` object is created and a file is opened for output. The `write()` method is used to sequentially insert bytes into the stream.

```
import java.io.*;

class WriteStream
{
      public static void main(String[] args)
      {
            String fileName = "c:\\stream.dat";
            FileOutputStream outStream;

            try {
                  outStream =
                        new FileOutputStream(fileName);

                  outStream.write( 'H' );
                  outStream.write( 'e' );
                  outStream.write( 'l' );
                  outStream.write( 'l' );
                  outStream.write( 'o' );
                  outStream.write( '!' );
                  outStream.close( );
            }
            catch (IOException ioe) {
                  System.out.println(ioe.toString());
            }
      }
}
```


FIGURE 14.4A
Writing to a file using the `FileOutputStream` class method **`write()`**.

The `write()` method throws an `IOException` and thus is enclosed in a `try` block which is followed by a `catch` block to catch any thrown exceptions. Note that the write method is called successively, writing a character to the file output stream each time.

The contents of the file are shown in Figure 14.4b. The file contains the characters "Hello!" written by the successive invocations of the write method.

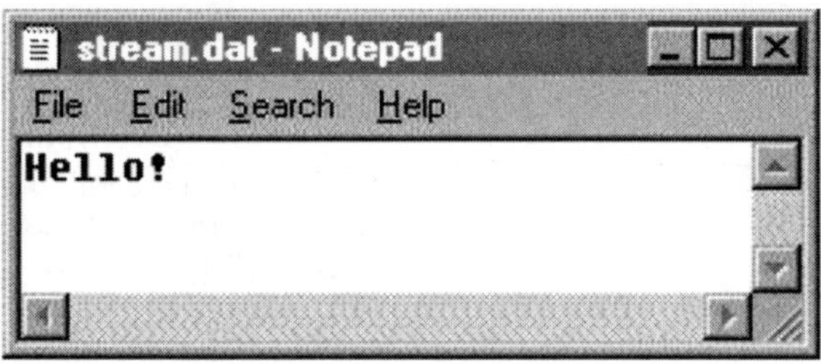

FIGURE 14.4B
Content of file produced by program writing to a file using the `FileOutputStream` class method `write()`.

READING FROM A STREAM

Input streams—that is, stream objects enabling reading of a file's contents—are modeled by the **`FileInputStream`** class. Selected constructors of the `FileInputStream` are summarized in Table 14.5.

TABLE 14.5

Selected constructors of the `FileInputStream` class

Name	Description
`FileInputStream(File` *`file`*`)`	Creates a file input stream object for the `File` object *`file`*.
`FileInputStream(String` *`name`*`)`	Creates a file input stream object with path *`name`* that can be read from.

Input streams, to read from a specified file, are created with `FileInputStream` class constructors. For example, the constructor

```
FileInputStream(fileName)
```

creates an input stream object and opens the file named for input. Data input from a file are said to be **extracted** from the stream. The following statements create an input stream object and open the file `stream.dat` for input.

```
String fileName = "stream.dat";
FileInputStream inStream;
try {
    inStream =
        new FileInputStream(fileName);
```

`FileInputStream` constructors throw a `FileNotFoundException` if the file is not found. Consequently, the `FileInputStream` constructor is enclosed in a `try` block. Note that the `FileNotFoundException` is derived from the `IOException` and thus can be handled in the `catch` block catching `IOExcep-`

TABLE 14.6

Selected methods of the `FileInputStream` class

Name	Description
`int available()`	Returns the number of bytes that can be read from the file without blocking.
`void close()`	Closes the file output stream object and releases associated resources.
`int read()`	Reads the next byte of data from the file input stream.
`int read(byte` *`dataArray []`*`)`	Reads `dataArray.length` bytes from the file input stream into the array *`dataArray`*.
`long skip(long` *`noBytes`*`)`	Skips over *`noBytes`* bytes in the file input stream and return the actual number of bytes skipped.

`tions`. Selected methods of the `FileInputStream` class are shown in Table 14.6.

The `FileInputStream` method `available()` returns the number of the bytes that can be read from the stream without blocking. Blocking is a method of handling large numbers of stream input or output bytes in smaller groups called **blocks.** The method is invoked by the stream's `FileInputStream` object.

The `FileInputStream` method `read()` reads the byte at the current position in the stream. Most operating systems provide for indicating when the **end of a file,** often denoted **EOF** or end-of-file, is encountered—that is, the end of the stream. The read method returns –1 when the end of file is encountered. A program illustrating use of the `FileInputStream` class and its methods to read a file are shown in Figure 14.5a.

When the file is opened, the current position is the first byte in the stream or file. Each read statement returns the byte at the current position and moves the stream one position to the left. Consequently, successive reads input successive bytes from the stream. The program's output is shown in Figure 14.5b.

The number of bytes that can be read from the stream without blocking is returned by the `available()` method. In this case, the number of bytes is small, and the number represents the number of data bytes in the file. The `read()` method is invoked by the input stream object.

```
iBuf = inStream.read();
```

The method returns an integer value which is assigned to an integer variable. Each value read from the stream is displayed. Note the returned values: 72, 101, 108, 108, 111, 33. These values correspond to the ASCII values of the characters "H", "e," "l," "l," "o," "!." The `-1` values represent the value returned by the `read()` method when the end-of-file is encountered. The `read()` method throws an `IOException` when an error occurs.

```
import java.io.*;

class readStream
{
      public static void main(String[] args)
      {
            String fileName = "stream.dat";
            FileInputStream inStream;

            try {
                  inStream =
                        new FileInputStream(fileName);

                  iBuf = inStream.available( );      —— bytes in stream

                  int iBuf;
                  System.out.println("No of bytes
                              : " + iBuf );

                  iBuf = inStream.read(  );          ⎫
                  System.out.println( iBuf );        ⎪
                  iBuf = inStream.read(  );          ⎪
                  System.out.println( iBuf );        ⎪
                  iBuf = inStream.read(  );          ⎪
                  System.out.println( iBuf );        ⎪
                  iBuf = inStream.read(  );          ⎪
                  System.out.println( iBuf );        ⎬ —— extracting
                  iBuf = inStream.read(  );          ⎪    bytes from
                  System.out.println( iBuf );        ⎪    stream
                  iBuf = inStream.read(  );          ⎪
                  System.out.println( iBuf );        ⎪
                  iBuf = inStream.read(  );          ⎪
                  System.out.println( iBuf );        ⎪
                  iBuf = inStream.read(  );          ⎪
                  System.out.println( iBuf );        ⎭
                  iBuf = inStream.read(  );
                  System.out.println( iBuf );
            }
            catch (IOException ioe) {
                  System.out.println(ioe.toString());
            }

      }
}
```

FIGURE 14.5A
Reading from a file using the `FileInputStream` class method `read`.

Consider the revised program in Figure 14.6a that uses the end-of-file value to control extraction of data from the stream and casts the value returned by the `read()` method. The statement

```
int iBuf = inStream.read();
```

reads the first byte in the stream. The integral value returned is then tested in the expression of a `while` statement. As long as the value of the byte just read is not –1, indicating an end-of-file condition, the `while` loop continues.

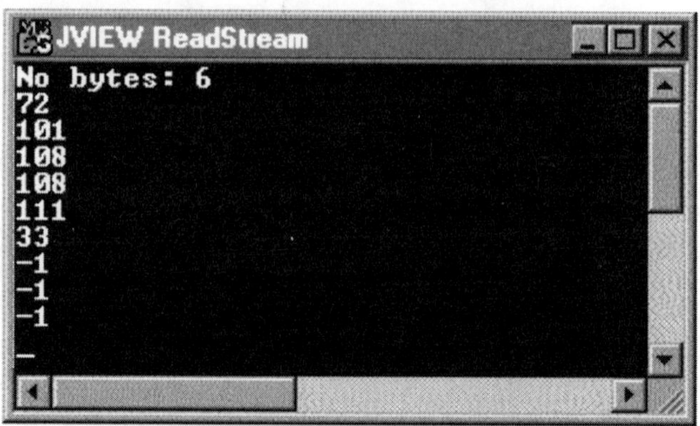

FIGURE 14.5B
Output of program reading from a file using the `FileInputStream` class method `read`.

```
import java.io.*;

public class ReadStream
{
      public static void main(String[] args)
      {
            String fileName = "c:\\stream.dat";
            FileInputStream inStream;

            try {
                  inStream =
                        new FileInputStream(fileName);

                  System.out.println( "No bytes: " +
                                    inStream.available(  ) );

                  int iBuf = inStream.read(  );
                                                         end-of-file test
                  while ( iBuf != -1 ) {
                        System.out.println( (char)iBuf );
                        iBuf = inStream.read(  );
                  }
            }
            catch (IOException ioe) {
                  System.out.println(ioe.toString());
            }
      }
}
```

FIGURE 14.6A
Testing for end-of-file condition during reading from a stream.

```
while (iBuf!= -1) {
    System.out.println((char)iBuf );
    iBuf = inStream.read();
}
```

In the `while` block, the value returned by the read is cast to `char` and the next byte is read from the stream. The program output is shown in Figure 14.6b.

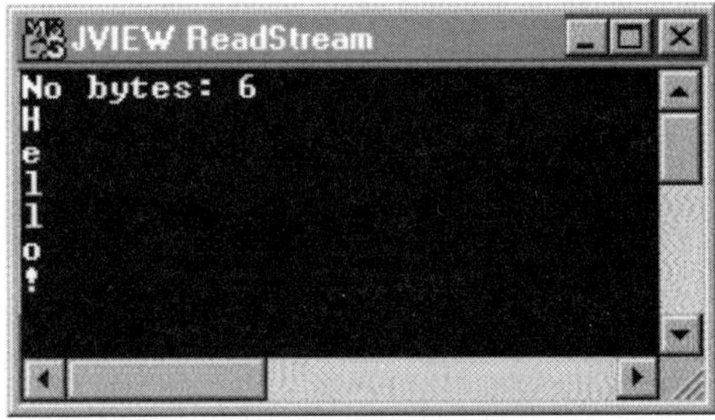

FIGURE 14.6B
Output of program testing for end-of-file condition during reading from a stream.

Note that the cast of the integral value returned to `char` type results in the display of the character representation of the `byte` value. Note also that the end-of-file condition is used to terminating attempts to extract bytes from the stream.

> Test for end-of-file conditions in reading from an input stream.

FILE PROPERTIES

Properties of a file are represented by instances of the `File` class. Properties of the file are retrieved by invoking the class' methods. Selected File class methods are summarized in Table 14.7. The program in Figure 14.7a illustrates the use of `File` class methods to ascertain a file's properties.

In Figure 14.7a, an instance of the class `File` representing the file with path `C:\stream.dat` is created with the statements

TABLE 14.7
Selected methods of the File class

Method	Function
`File(`*`fileName`*`)`	Creates and instance of the File class that represents the specified by the argument *`fileName`*.
`delete()`	Deletes the file represented by the object invoking the method.
`length()`	Returns the number of bytes in the file represented by the invoking object.
`getPath()`	Returns the path of the file represented by the invoking object.

```
import java.io.*;

public class FileProperties
{
      public static void main(String[] args)
      {
            String fileName = "c:\\stream.dat";
            File stream;

            stream = new File(fileName);

            System.out.println( "No bytes: " + stream.length(  ) );
            System.out.println( "Path:     " + stream.getPath(  ) );
      }
}
```

File class methods

FIGURE 14.7A
Retrieving file properties using `File` class methods.

```
String fileName = "c:\\stream.dat";
File stream;

stream = new File(fileName);
```

Note that the backslash character in the path is replaced with a double backslash.

The length of the file and path are retrieved and displayed with the following statements:

```
System.out.println("No bytes: " + stream.length());
System.out.println("Path:     " + stream.getPath());
```

The program's output is shown in Figure 14.7b.

The File class also contains fields representing file attributes. These fields are summarized in Table 14.8.

The separators are operating-system-dependent. The ";" string and ';' character are used to separate file names in a list of files, for example, a path list. Similarly, the "\" string and '\' character are used to separate the directory and file components of a file name, often referred to as the file's path. These fields are initialized to the system's parameters. Note that the backslash character is represented by the double backslash character in Java.

FIGURE 14.7B
Output of program retrieving file properties using `File` class methods.

TABLE 14.8

`File` class fields

Name	Type	Meaning
`pathSeparator`	`String`	Character used to separate file names in a sequences of files in a list.
`pathseparatorChar`	`char`	Character used to separate file names in a sequences of files in a list.
`separator`	`String`	Character used to separate directory and file components in a file name.
`separatorChar`	`char`	Character used to separate directory and file components in a file name.

A double backslash must be used as the separator in file paths.

HANDS-ON EXERCISE 14: READING AND WRITING CUSTOMER INFORMATION TO A FILE

An application illustrating storage and retrieval of customer account balances.

Most information used in business must be persistent. That is, it must persist from one execution of an application to another. This is generally accomplished by saving the information in a persistent type of storage, a disk file for example. Consider the Customer application window shown in Figure HO-14.1. The window consists of two text fields. The first contains the customer's identification code and the second the customer's name. The application contains a single menu with two options: Save and Exit. This application writes the customer information displayed to a disk file `C:\Customer.dat`.

Step 1: Create a New Project

1. Start Visual J++.
2. Select the New Project… command in the File menu. The New Project dialog box is displayed.
3. Enter **`Customer Input Output`** in the Name text field.
4. Use the Browse button to enable navigation to the Location `C:\Hands-On\Customer Input Output`.
5. Click the Open button to create the new project.

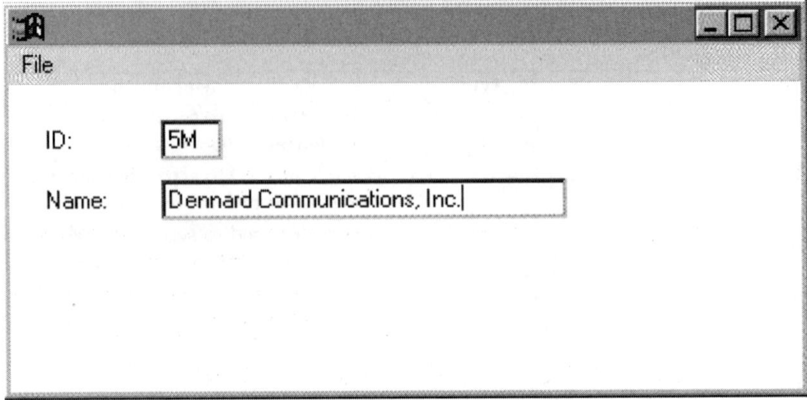

HO14.1
Customer window.

Step 2: Create `Customer.java` File

1. Press the CTRL+D keys. The Add Item dialog box is displayed.
2. Select the Class folder.
3. Select the ClassMain icon.
4. Enter **`Customer.java`** in the Name text field.
5. Click the Open button to create the new class.
6. The Code Editor is opened with the file `Customer.java`.

Step 3: Code the Application

Enter code to create a frame window, text fields, and handlers for the window events and menu action events.

```
import java.awt.*;
import java.awt.event.*;
public class Customer extends Frame
      implements WindowListener, ActionListener
{
      MenuBar menuBar = new MenuBar();

      Menu menuFile = new Menu("File");
      MenuItem itemSave = new MenuItem("Save");
      MenuItem itemExit = new MenuItem("Exit");

      Label labelID = new Label("ID:");
      TextField textID = new TextField(10);
      Label labelName = new Label("Name:");
```

```
    TextField textName = new TextField(50);

    public Customer( ) {
        setMenuBar(menuBar);
        menuBar.add(menuFile);
        menuFile.add(itemSave);
        menuFile.add(itemExit);

        itemSave.addActionListener(this);

        setLayout( null );
        add(labelID);
        add(textID);
        add(labelName);
        add(textName);
        labelID.setBounds(20,60,40,20);
        textID.setBounds(80,60,30,20);
        labelName.setBounds(20,90,40,20);
        textName.setBounds(80,90,200,20);

        addWindowListener(this);
        setSize(400,200);
        setVisible(true);
    }

    public static void main(String[] args)
    {
        new Customer();
    }

    public void actionPerformed(ActionEvent ae) {
    }

    public void windowClosing( WindowEvent we ) {
        setVisible(false );
        dispose( );
        System.exit( 0 );
    }
    public void windowActivated( WindowEvent we ){}
    public void windowClosed( WindowEvent we ){}
    public void windowDeactivated( WindowEvent we ){}
    public void windowDeiconified( WindowEvent we ){}
    public void windowIconified( WindowEvent we ){}
    public void windowOpened( WindowEvent we ){}
}
```

Step 4: Build the Basic Application

1. Press the CTRL+SHIFT+B keys.
2. Debug the code as required.

Step 5: Add Code to Write to a Disk File

Modify the `actionPerformed` method to save the customer information in the file `Customer.dat`.

```
public void actionPerformed(ActionEvent ae) {
          if(ae.getSource()==itemSave)
          {
               String fileName = "C:\\Customer.dat";
               FileOutputStream outStream;
               DataOutputStream dataOutStream;
               try {
                    outStream =
                           new FileOutputStream(fileName);
                    dataOutStream =
                         new DataOutputStream(outStream);
                    dataOutStream.writeChars(textID.getText());
                    dataOutStream.writeChars(textName.getText());
               }
               catch (IOException ioe) {
                    System.out.println(ioe.toString());
               }
          }
}
```

Note that the `FileOutputStream` is wrapped in the `DataOutputstream`!

Step 6: Build the Basic Application

1. Press the CTRL+SHIFT+B keys.
2. Debug the code as required.

Step 7: Set the Project Properties

1. Select the Customer Input Output Properties... option in the Project menu. The properties dialog box is displayed.
2. Click the OK button to accept the default properties.

Step 8: Execute the Application

1. Press the F5 key. The application starts and the application window is displayed.
2. Enter the following information:

ID:	5M
Name:	Dennard Communications, Inc.

3. Select the Save menu. The menu closes.
4. Close the application.

Step 9: Open the File Customer.dat

Open the file `C:\Customer.dat`.

FIGURE HO-14.2
File `Customer.dat`

2. Close the file.

Step 10: Save All the Project Files

Press the CTRL+SHIFT+ F5 keys.

Step 11: Close the Project

Select the Close All option in the File menu.

Step 12: Exit Visual J++ the Project

Select the Exit option in the File menu.
Challenge: Modify the application to retrieve and display Customer data from the `Customer.dat` file an Open option is selected in the File menu.

SUMMARY

1. Java file access uses a stream concept.
2. Files and interfaces for input and output are contained in the package `java.io`.
3. The `File` class contains fields representing file properties and methods for retrieving file attributes.
4. `FileIntputStream` and `FileOutputStream` classes model input and output streams, respectively.
5. The `write` method of the `FileOutputStream` class inserts bytes into a stream.
6. The `write` read of the `FileInputStream` class extracts bytes from a stream.

KEY WORDS AND CONCEPTS

blocks
current position
end-of-file (EOF)
extract
`File` class
`FileInputStream`
`FileOutputStream`
insert
`Java.io`
sequential file access
stream

REVIEW QUESTIONS AND EXERCISES

1. What is a stream?
2. List the standard streams.
3. What class contains fields and methods regarding file properties?
4. What type objects are required to implement the following:

   Reading from a stream______________________________

   Writing to a stream________________________________

   Creating a stream ________________________________

   Determining a file's properties______________________

5. What has occurred when the `read` method returns `–1`?
6. Create and test programs to use the `write(byte b[])` and `read(byte b[])` methods.
7. Review the following Java classes: `BufferedInputStream` and `BufferedOutputStream`.
8. Review Java classes that facilitate reading and writing 16-bit Unicode characters.

PRACTICE WITH VISUAL J++

Review the Java class `RandomAccessFile`. Develop an application illustrating use of the `RandomAccessFile` class to store a customer record containing the customer's identification code and account balance.

15

Network Programming

OBJECTIVES

After reading this chapter, you will be able to:

1. Understand the major concepts that underlie the World Wide Web.
2. Access network resources with the `URL` and `URLConnection` classes in network programming.
3. Understand layouts as used in network browser environments.
4. Implement Java layouts.
5. Create applets using the Java Applet class.
6. Understand Web pages and the use of Hypertext Markup Language.
7. Demonstrate the development and implementation of a Welcome applet.

OVERVIEW

The U. S. Department of Defense network project started in the 1960s has evolved into the World Wide Web, which connects millions of computers. Java includes classes enabling network programming. These classes provide for modeling the location of network resources as well as establishing communication links between network computers. Hypertext Markup Language facilitates the viewing of text files, stored on the network and referred to as Web pages, in network browsers. The language also provides for encoding links to other network resources which are retrieved when clicker by the viewer. Java provides a class for modeling programs that can be included in Web pages; these programs are called applets.

WORLD WIDE WEB

The World Wide Web started with a 1960s U.S. Department of Defense project, ARPANET, to provide a reliable, secure, and robust communications network that could still operate even if a part failed. Throughout the 1970s and 1980s, the network grew as universities not involved in the original project linked onto the network. To overcome system disparities the network adopted communication protocols. A **protocol** specifies how systems or computers will treat information resources. By adhering to the protocol, diverse computer systems can communicate. These protocols have evolved over time and are now widely used. Several widely used protocols are summarized in Table 15.1.

TABLE 15.1

Widely used protocols

Protocol	Use
HTTP	Requesting and sending Web documents.
FTP	File transfer from one system to another
Telnet	Remote login and use of server resources
File	Local system file access

Other protocols provide for services such as email transmission and group access.

Hypertext Transfer Protocol

In 1989, Tim Berners-Lee at the European Particle Physics Laboratory (CERN) proposed a new protocol for hypertext documents. **Hypertext** documents are text documents that contain links to other documents. By simply clicking on the hypertext link, the reader retrieves the related document. This capability marks the beginning of the World Wide Web we know today.

The basis of Berners-Lee's concept is a **client program** called a **browser** and a **server program.** All servers, running on an Internet host, respond to client requests. The requests specify a document using a unique address called a **uniform resource location.**

The first Web browser with a graphical interface, called Mosaic, was developed by Marc Andreessen and an associate, Eric Bina, in 1993. The rest is history!

Internet Protocol Number

Every machine connected to the Internet has a unique address called the **Internet Protocol (IP)** number. This address is not user-friendly, in that it is composed of four 8-bit numbers separated by periods. A typical IP address is 111.222.001.255. Additionally, a TCP/IP (Transfer Control Protocol/Internet Protocol) port number can be added to the address.

The IP address has an alias that can be used in its place, the **host name.** Technically, the host name consists of two parts: a host name and a domain name. The domain name is an extension identifying country or type of organization. For example, government organizations belong to **.gov,** educational institutions to **.edu,** and business organizations to **.com.** Note that the use of the term "host name" is used to identify the name part of the IP address as well as the name part and domain part combination.

Uniform Resource Locators

Network resources, such as files or directories, are uniquely named and located on the World Wide Web with a **Uniform Resource Locator (URL).** A typical URL is shown in Figure 15.1.

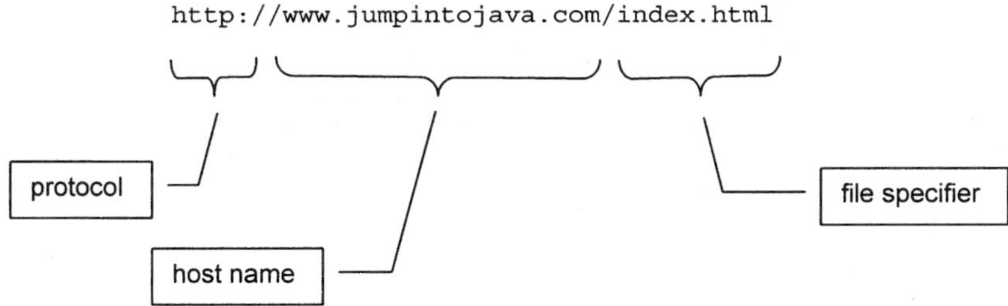

FIGURE 15.1
A typical URL.

The URL indicates that the document is to be handled with the **HTTP protocol.** Thus both the client system and the server system are able to communicate and transfer this file by using the HTTP protocol. The URL also specifies the machine on which the file is located with the host **name.** The host name begins with double slashes (//). Finally, the specific file on the machine is specified with the URL **file specifier.** The file specifier follows the host name separated by a slash (/).

> URLs specify the machine on which the file is located, how the file is handled, and the file's name.

Hypertext Markup Language

Web pages are created with a computer language, **Hypertext Markup Language or HTML.** A simple text file, not requiring compilation, the language prescribes tags that are placed within the text. These tags specify how a block of text is to be displayed. For example, a tag can specify the text be displayed as a title or as a hyperlink. Web pages are stored on computers continuously connected to the internet, commonly called servers. These files are retrieved by Web clients and displayed using a Web browser. The browser uses these tags to create the image displayed.

NETWORKING CLASSES

The package `java.net` includes classes and interfaces for implementing network-based applications. Included in the package are the `URL` and `URLConnection` classes, Figure 15.2.

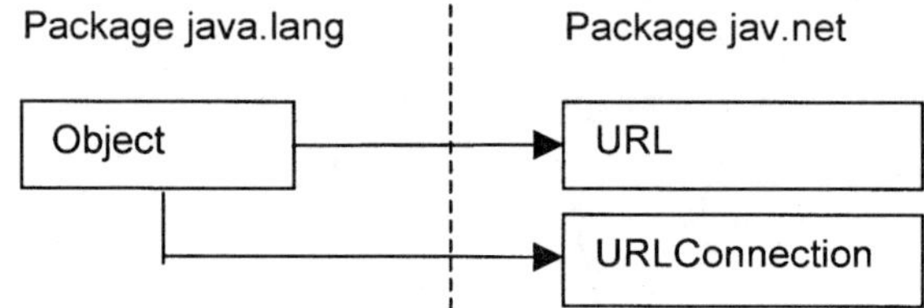

FIGURE 15.2
Selected members of the `java.net` package.

`URL` Class

The **`URL` class** models the World Wide Web Uniform Resource Locator. Constructors for the URL class are summarized in Table 15.2. Selected methods of the `URL` class are summarized in Table 15.3.

The `getContent( )` method performs the combined operation of opening a connection and getting the contents, for example

```
OpenConnection( ).getContent( )
```

TABLE 15.2

Selected `URL` class constructors

Name	Description
`URL(String` *`url`*`)`	Creates a URL object representing the string *`url`*.
`URL(String` *`protocol`*`, String` *`host`*`, String` *`file`*`)`	Creates a URL object representing the *`protocol`*, *`host`*, and *`file`* names.

TABLE 15.3

Selected URL class methods

Name	Description
Object getContent( )	Returns the contents of the URL.
String getFile( )	Returns the URL file name.
String getHost( )	Returns the URL host name.
int getPort( )	Returns the URL port number or –1 if the port is not set.
String getProtocol( )	Returns the URL protocol name.
String host, String file)	*protocol*, *host*, and *file* names.
String getRef( )	Returns the URL anchor or reference.
URLConnection openConnection( )	Returns a URLConnection object representing a connection to the resource represented by the URL.
InputStream openStream( )	Opens a connection and returns an input stream object for reading from the resource represented by the URL.

Similarly, the openStream( ) method performs the combined operation of opening a connection and creating an input stream form reading from the URL, for example

```
OpenConnection( ).getInputStream( )
```

URLConnection Class

A URLConnection object is created and returned by the URL class method openConnection( ). The connection is to the resource referenced by the URL. The characteristics of the connection are specified by the **URLConnection** class fields, Table 15.4.

Selected fields, allowUserInteraction, doInput, doOutput, ifModifiedSince, and useCaches represent those available to the connection and must be set or modified after the connection is created but before the connection is established with the connect method. These URLConnection class fields are set and retrieved with set and get methods summarized in Table 15.5.

The field url represents the location of the resource for which the connection is established. The value or the url field is returned by the getURL method.

Selected URLConnection class methods are summarized in Table 15.6.

The connect method opens a communication link to the resource referenced by the url. The connect method is ignored if the connected field is true (which means the connection is already established). The connection is implicitly performed when methods like getContentType or getContent are called. After the connection is opened, trying to set the various options, such as doInput or allowUserInteraction, is an error.

TABLE 15.4

Selected `URLConnection` class fields

Name	Description
`boolean allowUserInteraction`	Allows user interaction if `true` and no user interaction if `false`. The default value is set by the last invocation of the `setDefaultAllowUserInteraction ( )` method.
`boolean connected`	`True` if a communication link is established to the resource referenced by the `URL` and `false` otherwise.
`boolean doInput`	`True` if the application intends to read data from the resource referenced by the `URL` using the connection and `false` otherwise. The default value is `true`.
`boolean doOutput`	`True` if the application intends to write data to the resource referenced by the `URL` using the connection and `false` otherwise. The default value is `false`.
`long ifModifiedSince`	The time of modification expressed as the number of seconds since January 1, 1970 if the resource has been modified. The resource will not be fetched by protocols that skip fetching unless the resource has been modified since the `ifModifiedDate`. The default value is 0, in which case the resource will always be fetched.
`URL url`	URL of the resource to which the connection is established. The default value of the `url` field is the value of the `URL` argument in the `URLConnection` constructor.
`boolean useCaches`	When `true` the application can use caching when possible. The default value is set by the last invocation of the `setDefaultUseCaches( )` method.

TABLE 15.5

`URLConnection` class methods for setting and getting selected field values

Field	Set and get method
`allowUserInteraction`	`getAllowUserInteraction` `setAllowUserInteraction`
`doInput`	`getDoInput` `setDoInput`
`doOutput`	`getDoOutput` `setDoOutput`
`ifModifiedSince`	`getIfModifiedSince` `setIfModifiedSince`
`useCaches`	`getUseCaches` `setUseCaches`

TABLE 15.6

Selected `URLConnection` class methods

Name	Description
`abstract void connect( )`	Opens a communication to the resource referenced by the URL.
`Object getContent( )`	Retrieves the contents of the location referenced by the `URLConnection`'s `url`.
`String getContentEncoding( )`	Retrieves the content encoding of the location referenced by the `URLConnection`'s `url`. A `null` value is returned if the content encoding is not known.
`int getContentLength( )`	Returns the content length of the location referenced by the `URLConnection`'s `url`. A `-1` value is returned if the content length is not known.
`String getContentType( )`	Returns the content type of the resource referenced by the `URLConnection`'s `url`. A `null` value is returned if the content type is not known.
`long getDate( )`	Returns the date of the resource referenced by the `URLConnection`'s `url` in terms of the number of seconds since January 1, 1970. The value `0` is returned if the date is not known.
`long getExpiration( )`	Returns the expiration date of the resource referenced by the `URLConnection`'s `url` in terms of the number of seconds since January 1, 1970. The value `0` is returned if the expiration date is not known.
`InputStream getInputStream( )`	Returns an input stream object that enables reading from the connection.
`long getLastModified( )`	Date the resource referenced by the `URLConnection`'s `url` was last modified expressed in terms of the number of seconds since January 1, 1970. The value `0` is returned if the last modified date is not known.
`OutputStream getOutputStream( )`	Returns an output stream object that enables writing to this connection.
`URL getURL( )`	Returns the value of the `URLConnection`'s `url` field.

Using the **URLConnection**

The `URLConnection` is created by invoking the `URL` class `openConnection` method. After the `URLConnection` object is created the following option values can be manipulated:

- `allowUserInteraction;`
- `doInput;`
- `doOutput;`
- `ifModifiedSince;` and
- `useChaches.`

Additionally, the general request properties can be modified with the `setRequestProperty` method.

Once a `URLConnection` object is created, the connection is established with the `connect` method. When the connection is established, the `URLConnection` instance can be used to read from and write to the `url` resource referenced.

BUILDING A WEB CONNECTION APPLICATION

A Web connection application is shown in Figure 15.3a. First, the user enters a Uniform Resource Location in the upper text field. Then, the user

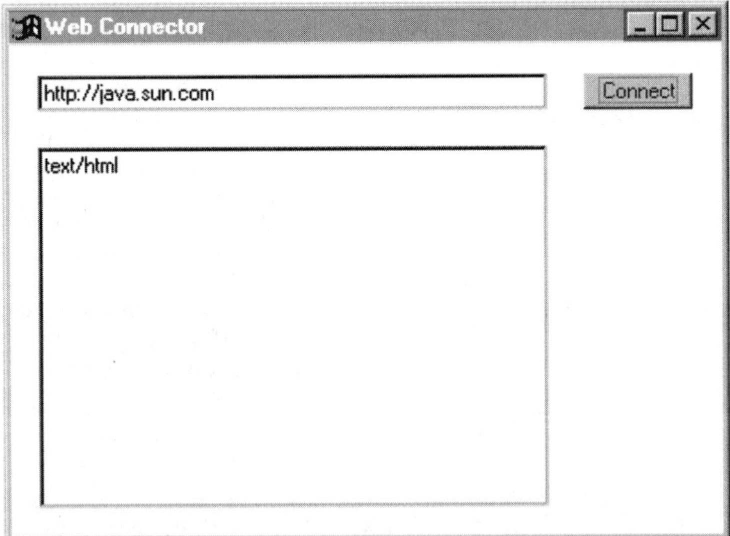

FIGURE 15.3A
Web connection application.

clicks the Connect button. The application creates and opens a connection to the resource referenced in the upper text field. Finally, the application retrieves the type of content comprising the resource. The application will work on any computer that is able to access Web locations with Internet Explorer or Netscape Navigator.

The Web application program is shown in Figure 15.3b.

```
import java.awt.*;
import java.awt.event.*;
import java.net.*;
import java.io.*;
public class WebConnector extends Frame
      implements  WindowListener, ActionListener
{
      TextField docAddress = new TextField();
      TextArea  docProperties = new TextArea();
      Button    bConnect = new Button("Connect");

      public static void main(String[] args)
      {
           WebConnector webConnection = new WebConnector();
      }

      public WebConnector()
      {
           addWindowListener(this);
           setTitle("Web Connector");
           setSize(400,300);
           setVisible(true);
           add(docAddress);
           add(docProperties);
           add(bConnect);
           docAddress.setBounds(20,40,280,20);
           docProperties.setBounds(20,80,280,200);
           bConnect.setBounds(320,40,60,20);
           bConnect.addActionListener(this);
      }

      public void windowClosing( WindowEvent we )
      {
           setVisible(false );
           dispose( );
           System.exit( 0 ) ;
      }
      public void windowActivated( WindowEvent we ){}
      public void windowClosed( WindowEvent we ){}
      public void windowDeactivated( WindowEvent we ){}
      public void windowDeiconified( WindowEvent we ){}
      public void windowIconified( WindowEvent we ){}
      public void windowOpened( WindowEvent we ){}
```

```
        public void actionPerformed( ActionEvent ae )
        {
              if( ae.getSource( ) == bConnect )
              {
                    try
                    {
                          URL url = new URL(docAddress.getText());
                          URLConnection urlCon =
                                  url.openConnection( );
                          urlCon.connect();
                          docProperties.append(
                             urlCon.getContentType() + "\n");
                    }
                    catch (MalformedURLException mfURLe)
                    {
                          docProperties.append( "Invalid URL \n");
                          docProperties.append( mfURLe.toString()
                                + "\n");
                    }
                    catch (IOException ioe)
                    {
                          docProperties.append( "IO Exception \n");
                          docProperties.append( ioe.toString()
                                + "\n");
                    }
              }
        }
}
```

FIGURE 15.3B
Web connection application.

Assume the user has entered a URL in the text field, `docAddress`, and clicked the button `Connect`. A `URL` object is created using the contents of the `docAddress` text field as the constructor argument with the statement

```
URL url = new URL(docAddress.getText());
```

If the content of the text field is not a valid URl value, a `MalformedURLException` is thrown. The exception is caught in the `catch` block and an error message is displayed in the lower text area.

The `URL` class `openConnection` method is invoked to create a `URLConnection` object with the statement

```
URLConnection urlCon = url.openConnection( );
```

If the connection object cannot be thrown, an `IOException` is thrown. The exception is caught is a catch block and an error message is displayed.

A connection link is established by invoking the `URLConnection` class method `connect` with the statement

```
urlCon.connect();
```

Again, if the connection link is not established, an `IOException` error is thrown and caught in the following `catch` block.

Finally, the `getContentType` method is invoked to return the content type of the resource referenced by the URL.

```
docProperties.append(urlCon.getContentType() + "\n");
```

The content type is returned as a string, which is appended to the `docProperties` text area.

Note the import statement

```
import java.net.*;
```

enabling access to the package `java.net` classes, interfaces, and exceptions. Similarly, note the statement

```
import java.io.*;
```

that enables access to the package `java.io`, specifically the class `IOException`.

> Java classes, interfaces, and exceptions in the package java.net provide for network programming.

LAYOUT MANAGERS

Layout managers arrange components in a container. These layout managers automatically resize the components and arrange them relative to one another for all display dimensions or resolutions. This is a major advantage in applications to be deployed or differing platforms such as experienced in the Internet. The disadvantage is that they do not provide the flexibility required for many businesses enterprise applications in which Graphical User Interface space is at a premium.

Java Layouts

Java provided layouts are summarized in Table 15.7. The layout managers are inherited directly from the `Object` class. The layout managers implement the **`LayoutManager` interface** that contains the methods used by the layout managers to arrange the components in a container.

TABLE 15.7

Layout manager classes summary

Class Name	Description
`BorderLayout`	Components are resized to fill one of the five regions: North, East, South, West, and Center.
`CardLayout`	Each component is treated as a card in a stack in which only one component is visible at a time.
`FlowLayout`	Components are arranged left-to-right in centered lines until the lines are filled.
`GridBagLayout`	Components are arranged in one or more rectangular grid cells.
`GridLayout`	Components are arranged singularly in equal-sized rectangular cells.

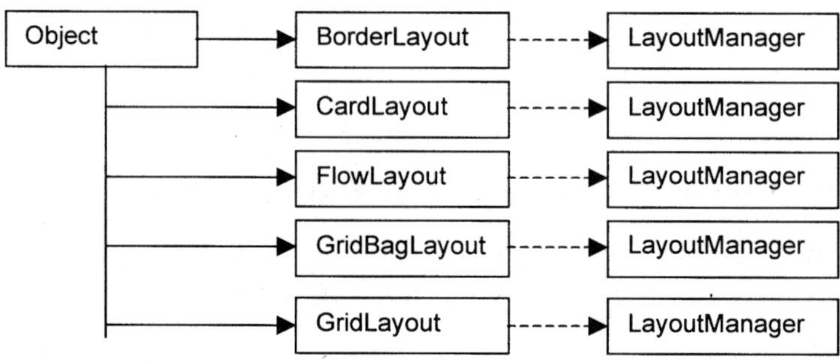

FIGURE 15.4
`Layout` class and interface hierarchy.

Classes encapsulating the provided arrangements are summarized in Figure 15.4.

Using the Borderlayout

The `BorderLayout` arranges the components in five regions, as illustrated in Figure 15.5. Components are added with the following version of the `add( )` method accepting two parameters. The first parameter is a reference to the component to be added and the second parameter is a string specifying the region of the container in which the component is to be placed. For example, the statement

```
Add( cmp, "East");
```

specifies that the component `cmp` is to be added in the East region of the container.

The components are sized to fit the specified region. The North and South regions are stretched horizontally to fit the container. The East and West

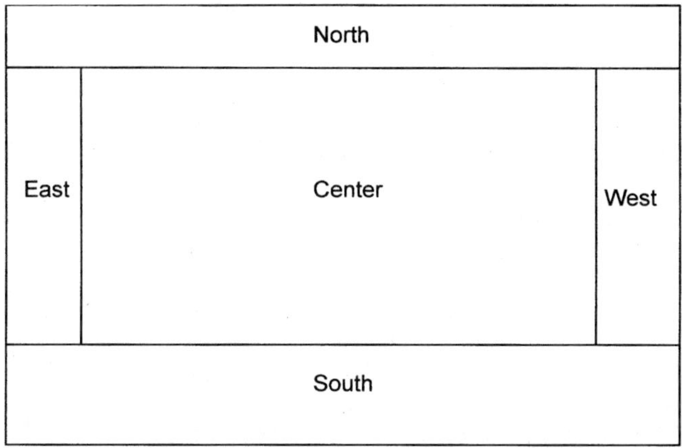

FIGURE 15.5
Five regions of the `BorderLayout`.

regions are stretched vertically to fit the container. The Center region stretches both horizontally and vertically to fit the remaining size.

The `BorderLayout` is the default `Frame` layout. Components added to a container with border layout without a region specified are positioned in the Center region.

A listing of an application illustrating the use of `BorderLayout` is shown in Figure 15.6a. Output of the application utilizing `BorderLayout` is shown in Figure 15.6b.

```
import java.awt.*;
import java.awt.event.*;
public class GUI extends Frame
      implements WindowListener
{
      public GUI( )
      {
           addWindowListener(this);

           setTitle("Graphic User Interface");

           setSize(400,200);

           add("North", new Button("North") );
           add("East", new Button("East") );
           add("South", new Button("South") );
           add("West", new Button("West") );
           add(new Button("Center") );

           setVisible(true);
      }
```

```
    public static void main(String[] args)
    {
        GUI gui = new GUI();
    }

    public void windowClosing( WindowEvent we )
    {
        setVisible(false );
        dispose( );
        System.exit( 0 );
    }
    public void windowActivated( WindowEvent we ){}
    public void windowClosed( WindowEvent we ){}
    public void windowDeactivated( WindowEvent we ){}
    public void windowDeiconified( WindowEvent we ){}
    public void windowIconified( WindowEvent we ){}
    public void windowOpened( WindowEvent we ){}
}
```

FIGURE 15.6A
`BorderLayout` application.

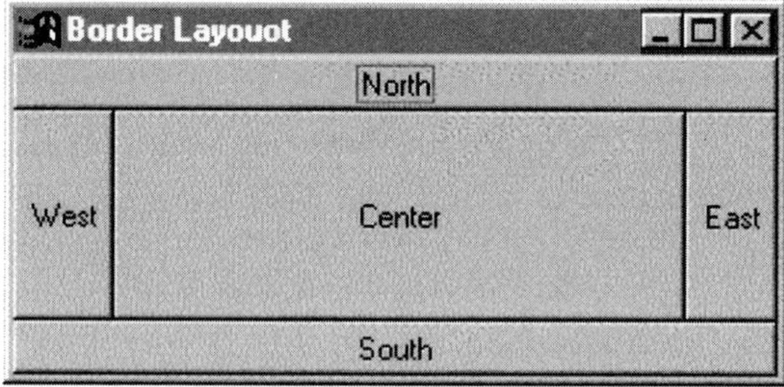

FIGURE 15.6B
`BorderLayout` application output.

Built-in layout managers are an advantage for user interfaces that will be used on displays with differing sizes or resolutions, for example, applications developed for the Internet.

APPLETS

Applets are programs intended to be imbedded inside other applications. For example, applets are typically included inside Web pages that are displayed on Browsers.

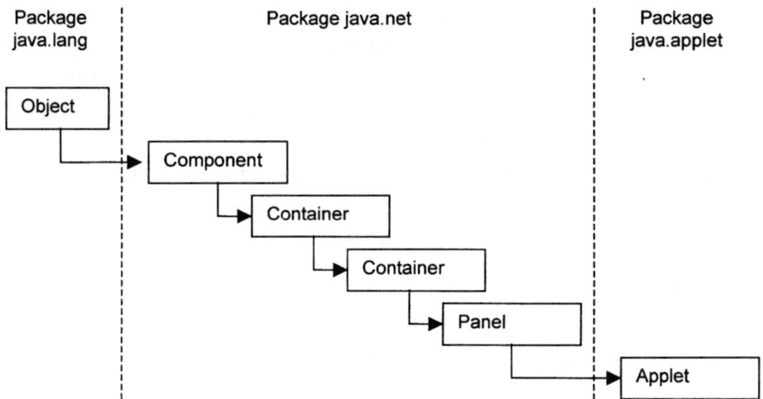

FIGURE 15.7
`Applet` class hierarchy.

TABLE 15.8
Selected `Applet` class methods

Class Name	Description
`void destroy( )`	Called by the browser after the applet is loaded and prior to calling the `start` method.
`String getAppletInfo( )`	Returns a string containing information about the applet.
`void init( )`	Called by the browser after the applet is loaded and prior to calling the `start` method.
`void resize(int width, int height)`	Requests that the applet be resized to width and height specified by *`width`* and *`height`* arguments.
`void showStatus(String text)`	Requests that the string *`text`* be displayed in the applet viewer's status window.
`void start( )`	Called by the browser to inform the applet to start its execution.
`void stop( )`	Called by the browser to inform the applet to stop its execution.

`Applet` Class

Appelets are modeled by the **`Applet` class,** which is declared in the package `java.applet` (Figure 15.7). The `Applet` class has a single constructor that takes no arguments. Selected methods of the `Applet` class are summarized in Table 15.8.

Applet initialization is implemented by overriding the `init` method. The `start` method is called by the browser every time the Web page containing the applet is visited. Actions to be performed every time the Web page is visited should be placed in an override of this method. The `stop` method is called whenever the Web page is replaced by another page and just before the applet is destroyed. Actions to be performed under these conditions should be placed in an override of this method. The `stop` method is always called prior to the `destroy` method. Actions, such as thread manipulation and resource reclamation, should be placed in an override of the `destroy` method. Similarly, the method `getAppletInfo` method must be overridden to return a `String` with applet information such as author and version.

Additionally, the `Applet` class inherits the `add`, `repaint`, `update`, and `paint` method.

Using the `Applet` Class

An applet that will display a Welcome window is shown in Figure 15.8. Initially, the applet is shown with a top button captioned "Display Greeting" and a blank window. When the button is clicked by the user, the greeting "Welcome to Java!" is displayed.

```
import java.applet.*;
import java.awt.*;
import java.awt.event.*;
public class Greeting extends Applet
      implements ActionListener
{
      boolean displayFlag = false;
      Button bDisplayGreeting = new Button("Display Greeting");

      public void init( )
      {
           setLayout(new BorderLayout());
           setBackground(new Color(0,255,255));
           add(bDisplayGreeting,"North");
           bDisplayGreeting.addActionListener(this);
           resize(400,150);
      }
      public void actionPerformed( ActionEvent ae)
      {
           if(ae.getSource() == bDisplayGreeting)
           {
                displayFlag = true;
                repaint();
           }
      }

      public void paint(Graphics g)
      {
           if(displayFlag == true)
           {
```

```
                g.setFont(new Font("Seirf",Font.BOLD +
                                                Font.ITALIC,36));
                g.setColor(new Color(255,0,0));
                g.drawString("Welcome to Java!", 50, 100);
            }
        }
}
```

FIGURE 15.8
Welcome applet.

The applet `Greeting` is declared with the statement

```
public class Greeting extends Applet
    implements ActionListener
```

The `Applet` class method `init` is called when the applet is started. The `init` method statement

```
setLayout(new BorderLayout());
```

implements the `BorderLayout` in the applet. As a result, the button, `bDisplayGreeting`, is added in the North position with the statement

```
add(bDisplayGreeting,"North");
```

The applet color and initial size are set with the statements

```
setBackground(new Color(0,255,255));
resize(400,150);
```

Note that an `ActionListener` is implemented in the `Greeting` applet declaration and that a listener is implemented for the `bDisplayGreeting` button in the `init` method.

When the `bDisplayButton` is clicked by the user, an `ActionEvent` object is created and the `actionPerformed` method called. Whenever the source of the action event is the `bDisplayGreeting` button a variable `displayFlag` is set to value `true` and the applet `repaint` method is called with the statements

```
displayFlag = true;
repaint();
```

The `repaint` method brings about a call to the applet's `update` method. In turn, the `update` method calls the applet's `paint` method.

If the `displayFlag` is set to the value true, the following `paint` method statements are executed:

```
g.setFont(new Font("Seirf",Font.BOLD +
                            Font.ITALIC,36));
g.setColor(new Color(255,0,0));
g.drawString("Welcome to Java!", 50, 100);
```

These statements display the text "Welcome to Java!" in bold, italicized, bright red letters in the Applet window. Note that the initial value of the `displayFlag` is `false` and consequently the text "Welcome to Java!" is not displayed when the applet is painted initially. Moreover, subsequent painting of the applet, resulting from user minimizing, maximizing, or moving to other pages, will not result in the message's display. However, once the Display Greeting button is clicked, the message will be displayed the next time the applet is painted.

> Applets are programs intended to be imbedded in other applications such as Web pages.

HYPERTEXT MARKUP LANGUAGE

Hypertext Markup Language (HTML) is a system for marking and tagging text in documents that are published on the World Wide Web. In addition to specifying how document text is displayed, the language also includes specifying link to other documents and inclusion of Applets.

The HTML file Welcome.htm including the Welcome applet is shown in Figure 15.9. The applet, defined in `Greeting.class` file, is included in the applet with the statement

```
<APPLET CODE=Greeting.class

          HEIGHT=200 WIDTH = 450 VIEWASTEXT></APPLET>
```

The applet is specified by

```
CODE=Greeting.class
```

The applet is initially shown with the height and width parameters set in the statement

```
HEIGHT=200 WIDTH = 450
```

The applet is resized in the `init` method.

The `VIEWASTEXT` entity is added to instruct the text editor to show the applet as text in the Source view.

```
<HTML>
<HEAD>
<META NAME="GENERATOR" Content="Microsoft Visual Studio 6.0">
<TITLE>Welcome Page</TITLE>
</HEAD>
<BODY>
<P>
<APPLET CODE=Greeting.class
        HEIGHT=200 WIDTH = 450 VIEWASTEXT></APPLET>
</P>
</BODY>
</HTML>
```

Applet tags and Greeting.class inclusion

FIGURE 15.9
`welcome.htm` file.

HANDS-ON EXERCISE 15: IMPLEMENTING A WELCOME APPLET

The Welcome applet as it would appear in a Web browser is shown in Figure HO-15.1. The applet window is initially displayed without the greeting. The greeting is displayed when the user clicks the Display Greeting button. The Display button is located at the top of the applet window, the North position of the border layout.

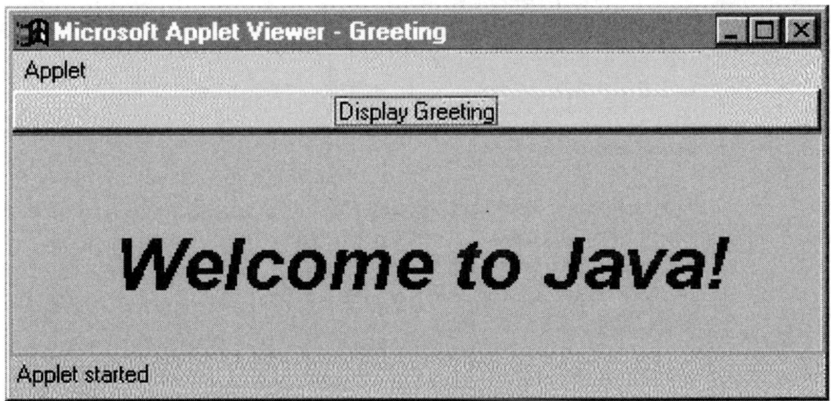

FIGURE HO-15.1
The Welcome applet as displayed in a viewer or Web browser.

Step 1: Create Visual J++ Project Welcome

1. Launch Visual J++.
2. Click the New Project option in the File Menu.

3. Enter **`Welcome`** in the Name text field and **`C:\Hands-On\Welcome`** in the Location text field.
4. Click the Open button to create the Welcome project.

Step 2: Create the `Greeting.java` File

1. Click the Add Class... option in the Project menu.
2. Enter **`Greeting.java`** in the Name text field.
3. Click the open button to create the `Greeting.java` file. The Text Editor opens with the `Greeting.java` file.

Step 3: Declare the Greeting class

1. Open the `Greeting.java` file in the Text Editor (if not proceeding directly from the previous step.)
2. Enter the code shown in Figure HO-15.2.

```
import java.applet.*;
import java.awt.*;
import java.awt.event.*;
public class Greeting extends Applet
      implements ActionListener
{
      boolean displayFlag = false;
      Button bDisplayGreeting = new Button("Display Greeting");

      public void init( )
      {
           setLayout(new BorderLayout());
           setBackground(new Color(0,255,255));
           add(bDisplayGreeting,"North");
           bDisplayGreeting.addActionListener(this);
           resize(400,150);
      }
      public void actionPerformed( ActionEvent ae)
      {
           if(ae.getSource() == bDisplayGreeting)
           {
                displayFlag = true;
                repaint();
           }
      }

      public void paint(Graphics g)
      {
```

```
            if(displayFlag == true)
            {
                g.setFont(new Font("Seirf",Font.BOLD +
                                        Font.ITALIC,36));
                g.setColor(new Color(255,0,0));
                g.drawString("Welcome to Java!", 50, 100);
            }
        }
}
```

FIGURE HO-15.2
Greeting class declaration.

Step 4: Create the HTML Page

1. Click the Add Web Page... in the Project menu. The Add Item dialog box is displayed with the Web Page folder opened in the New Tab window pane (Figure HO-15.3).

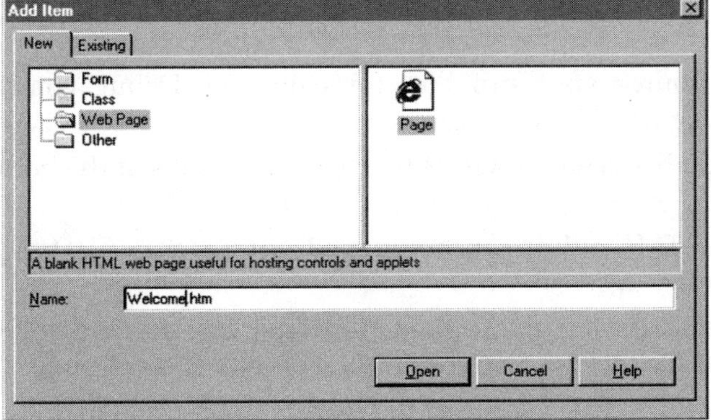

FIGURE HO-15.3
Add Item dialog box opened with Web Page folder selected.

2. Enter **`Welcome.htm`** in the Name text field.
3. Click the Open button. The file `Welcome.htm` is opened in the HTML Editor.

Step 5: Edit the `Welcome.htm` File

1. The HTML Editor provides for editing HTML file. The Editor provides three views of the file (Table HO-15.1).
2. In the Source view, the Text is colored according to its function. The default color scheme is summarized in Table HO-15.2.

TABLE HO-15.1

HTML Editor views

View	Description
Design	Only the body of the document, HTML script between the <BODY> tags, is displayed.
Source	The `.htm file` is displayed as text.
Quick	The `.htm file` is displayed as it would appear in a browser including interactivity—that is, the links in the code are active.

TABLE HO-15.2

Default colors used in Source view

Color	Function
Purple	Tag Text
Brown	Element Name
Red	Entity/Tag Attribute
Blue	Operator String
Gray	Comment

3. The Source view provides for using the Debug commands for debugging HTML code.
4. Toggle between the views by clicking the tabs at the bottom of the editor.
5. The HTML editor can be opened with a `.htm` file for editing in several ways:
   - Select the file in the Open File dialog box that is displayed when the Open File… option in the File menu is selected;
   - Double click the file in the Project Explorer; and
   - Select the HTML Editor option in the Open With dialog box that is displayed when the file is selected in the Project Explorer and the Open With… option is selected in the View menu.
6. Click the Source tab. The HTML Editor Source view is displayed (Figure HO-15.4).
7. Enter the script

```
<APPLET CODE=Greeting.class
        HEIGHT=200 WIDTH=450 VIEWASTEXT></APPLET>
    in the BODY.
```

8. Click the Design tab. The HTML Editor Design view is displayed (Figure HO-15.5).
9. Note that the applet is shown in its default size: 200 by 450.

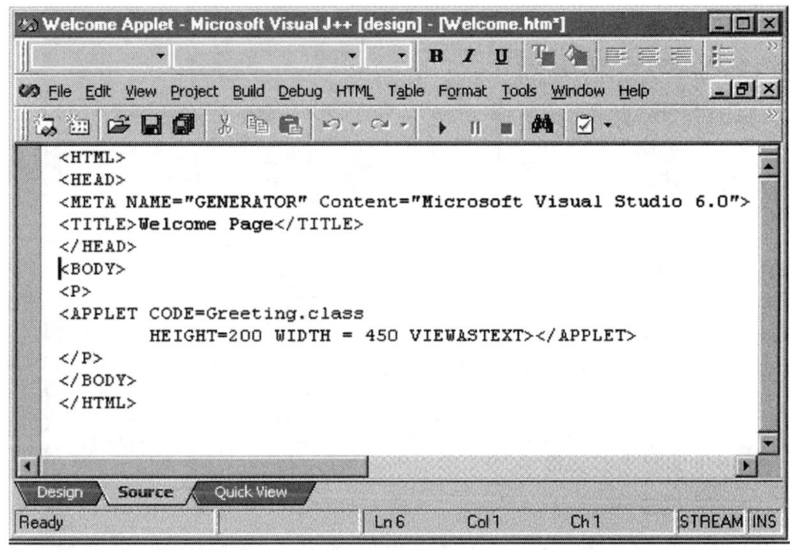

FIGURE HO15.4
Welcome.htm file opened in the HTML Editor Source view.

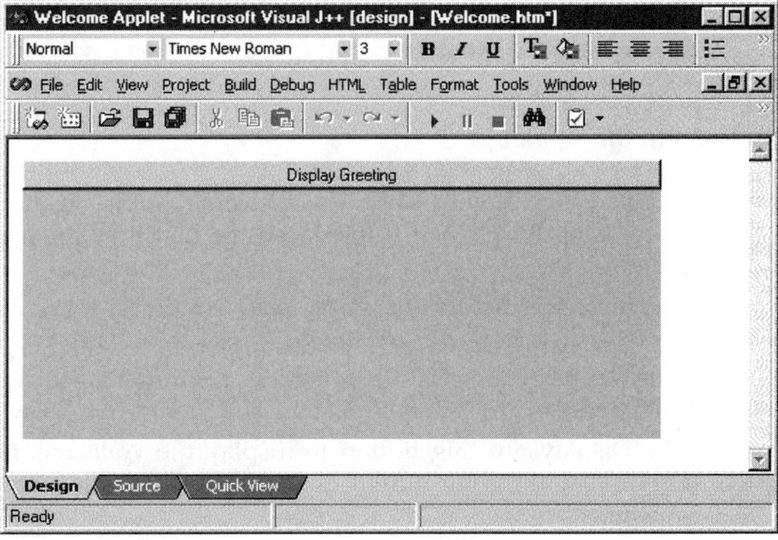

FIGURE HO-15.5
`Welcome.htm` file opened in the HTML Editor Design view.

Step 5: Build the Project

1. Click the Build option in the Build menu.
2. Debug the source code files as required.

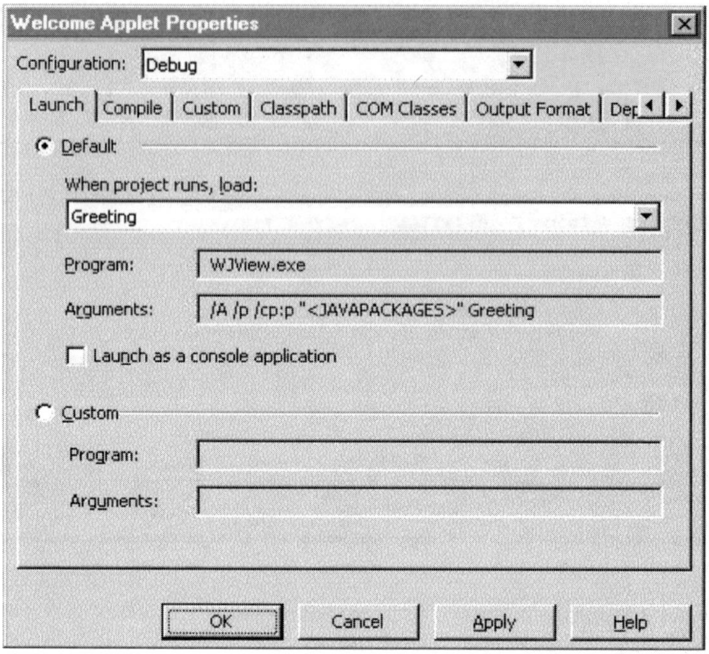

FIGURE HO-15.6
Welcome Applet Properties dialog box.

Step 6: Execute the Applet

1. Click Start in the Debug menu. The Welcome Applet Properties dialog box (Figure HO-15.6) is displayed the first time the applet is executed.
2. Click OK to accept the default properties.
3. The Applet Viewer is displayed with the applet started (Figure HO-15.7). The Applet Viewer is a browserless environment in which the applet appears and behaves exactly as it would in a browser.
4. Click the Display greeting button to display the Welcome. Note that the Display Greeting button is resized to fit as the window size is changed.
5. Click the Exit button in the right corner of the Applet Viewer title bar to stop the applet and close the viewer. Alternately, click the Quit option in the Applet menu.

Step 7: View the Web Page Welcome in a Browser

1. Launch Internet Explorer offline.
2. Click the Open option in the File menu (Figure HO-15.8).

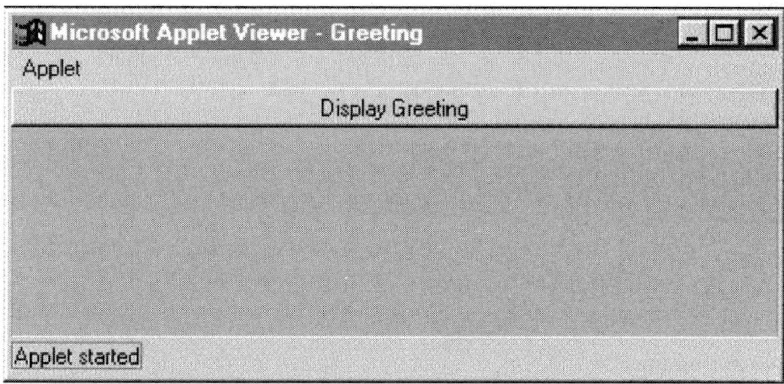

FIGURE HO-15.7
Welcome Applet displayed in applet viewer.

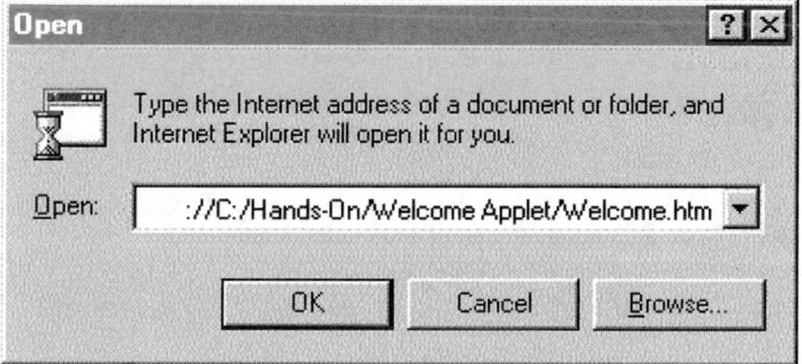

FIGURE HO-15.8
Open option in the File menu.

3. Enter `file://C:/Hands-On/Welcome Applet/Welcome.htm` in the Open text field.
4. Click OK.
5. The Web page `Welcome.htm` is displayed in the browser.
6. Close the browser.

SUMMARY

1. The World Wide Web is a reliable and robust communication network.
2. Package `java.net` includes classes modeling network entities and processes.
3. The class `URL` models a network Uniform Resource Locator.

4. Network connection links are implemented with the `URLConnection` class.

   Understand concept of connected and connectionless network communication

   Gain familiarity with the package java.net

   Demonstrate understanding and facility of accessing network files

   Demonstrate understanding and facility client/server techniques

KEY WORDS AND CONCEPTS

`.com`
`.edu`
`.gov`
applets
`Applet` class
`BorderLayout` class
`CardLayout` class
client program
file specifier
`FlowLayout` class
`GridBagLayout` class
`GridLayout` class
host name
HTTP protocol
Hypertext
Hypertext Markup Language (HTML)
Internet Protocol (IP)
`java.applet` package
layout managers
`LayoutManager` interface
protocol
server program
Uniform Resource Locator (URL)
`URL` class
`URLConnection` class

REVIEW QUESTIONS AND EXERCISES

1. Mark the following true or false.
   a. Failure of an Internet link will cause failure of the entire network.
   b. Protocols enable diverse hardware and systems to communicate.
   c. A network connection link is established automatically when a `URLConnection` object is created.
   d. `URLConnection` methods can be used to ascertain the properties of a network resource.
   e. Java layout resizes automatically to fit the window.

f. Applets run inside other applications and as stand-alone applications.

g. Border layouts can be implemented in applets as well as applications.

h. `ActionEvent` objects are created when users click over applet controls.

i `.htm` files can be viewed on network browsers.

j. Method `main` is the applet entry point.

2. Is 111.222.333.444 a valid IP address? *Hint:* Can 333 be represented as a 8-bit number?
3. Modify the Web Connector application to retrieve the following:
   a. Date
   b. Expiration date
   c. Content length
4. ***Challenge:*** Modify the Web Connector to read and display the content of a URL with type text content. *Note that the next step is a browser!*
5. Develop and test applications using the following layouts:
   a. `CardLayout`
   b. `FlowLayout`
   c. `GridLayout`
   d. `GridBagLayout`
6. Develop an applet that contains buttons for the colors red, white, and blue and will change background color to reflect the button clicked.
7. Develop an applet that will trace mouse movement over the applet.

PRACTICE WITH VISUAL J++

1. Modify the Welcome applet to include a text field for entry of the user name. Include the name entered as part of the greeting, for example

   Username, Welcome to Java!

2. View the applet in the Quick View of the HTML Editor.

3. Revise the colors in the HTML Editor. Select Options in the Tools menu to option the Options Dialog box. HTML colors are selected in the Fonts and Color option under Text Editor.

16

Threads

OBJECTIVES

After reading this chapter, you will be able to:

1. Understand concept of threads.
2. Use threads to implement application processes that run concurrently.
3. Demonstrate the use of threads in applets.

OVERVIEW

Threads enable multitasking in an application. Threads also enable the division of a process' tasks among multiple processors when they are available. Multiprocessing and multitasking both work to improve processing throughput. Java implements threads, which can be used to improve a throughput of a process.

MULTITASKING, MULTIPROCESSING, AND THREADS

Statements comprising a program are executed sequentially beginning with the first statement and ending with the last. From time to time, statements may be repeated in a loop or statements may be skipped in a decision, but generally speaking, the flow is from the first statement to the last statement. Following the execution of any specific statement, an outsider looking in would see a computer element pointing to the next statement to be executed, memory elements containing the values of the programs variables and objects, memory elements containing the remaining program statements to be executed, as well as memory elements containing the prior program statements. These elements, along with other resources, such as open files, are a **process.**

These elements can be stored, releasing the computer's processor to execute other programs. At some point in time, the stored elements can be retrieved and reinstated in the computer. The program can then resume execution as if nothing had happened. Note that the values of all the variables and objects have been remembered. Similarly, the steps of the program have been remembered. Moreover, the very next statement to be executed has been remembered. In this way, the computer's central processing unit can be alternately used by several process.

The control of flow through a process is called a **thread.** Practically, a thread represents one of many possible tasks required to complete and application. For example, an order entry application may consist of many tasks: user data entry, database access, and printing or faxing. Each of these tasks can be viewed as a thread (Figure 16.1).

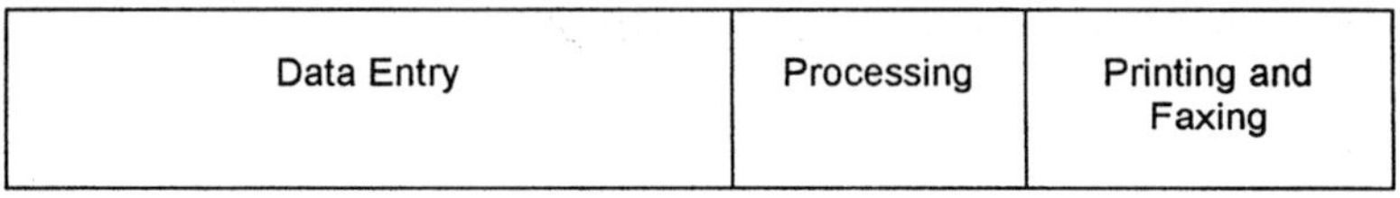

Data Entry	Processing	Printing and Faxing

FIGURE 16.1
Order entry threads executed sequentially.

In a **multitasking** environment, slow and time-consuming input and output tasks can be assigned to separate threads, while the data lookup and retrieval task can be assigned to a different thread. The execution of the tasks will appear to the user to overlap. This is in contrast to a single threaded operation, in which each task must be performed sequentially, each task being fully completed before the next task can begin. Figure 16.2 illustrates the impact of multitasking. Most of the time during data entry is spent waiting for the operator to enter more data. During this period the central processor is not being used and is available to execute other threads. In a multitasking environment, the processor executes other threads during this waiting period; for example, processing orders already entered.

An alternative is the assignment of the various tasks to a different computer, which can also be used to improve throughput, referred to as **multiprocessing.**

The ability to alternately execute several processes is called time slicing. Actually, *the computer's processor can work on only one thread at a time.* A multitasking environment is obtained by enabling the processor to switch back and forth between events. Various schemes exist to determine when the processor will switch from one thread to another. An example of time slicing is illustrated in Figure 16.3. In this case, execution of a lengthy Process 1 is halted; the elements are stored and the execution of Process 2 initiated. After

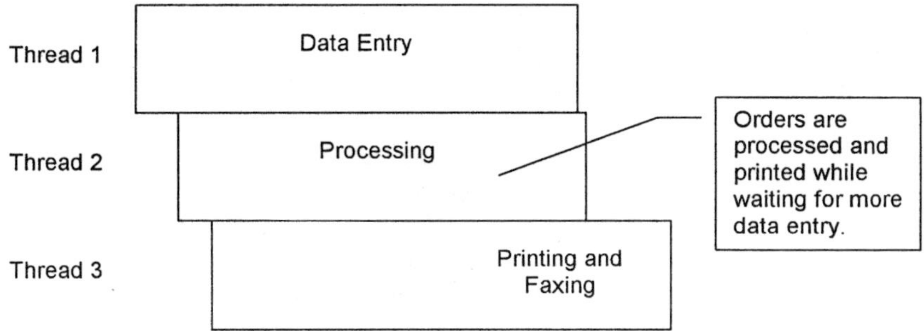

FIGURE 16.2
Multitasking environment.

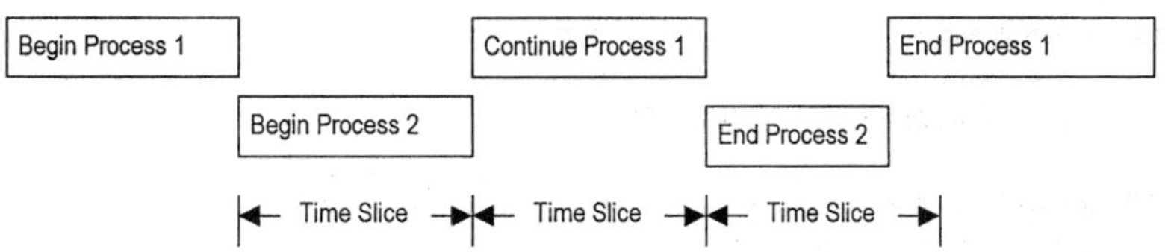

FIGURE 16.3
Time slicing.

a period, execution of Process 2 is halted, its elements stored, the elements of Process 1 retrieved and reinstated in memory, and execution of Process 2 resumed. After another period, execution of Process 1 is again halted, the elements stored, the elements of Process 2 retrieved and reinstated in memory, and its execution resumed. Finally Process 2 ends and the execution of Process 1 resumes.

Typically, the time allocated to a typical process is dependent upon the process's priority. In the illustration, the processes are assumed to have equal priority and thus are allocated equal time slices. Initially Process 1 is executing. When Process 2 appears, execution of Process 1 is suspended and execution of Process 2 is initiated. Thereafter, the processes are allocated equal time. Process 1 resumes execution immediately following the end of Process 2 even though Process 2 did not use all of its allotted time.

> Improved throughput and concurrent operation are the advantages enabled with using threads.

Java provides the `Thread` class, `ThreadGroup` class, and `Runnable` interface for implementing threads. These Java classes and interface are shown hierarchally in Figure 16.4.

The Java Virtual Machine can support multiple threads at the same time. These threads may be processed by time-slicing on a single processor, processing on multiple processors, or time-slicing on multiple processors.

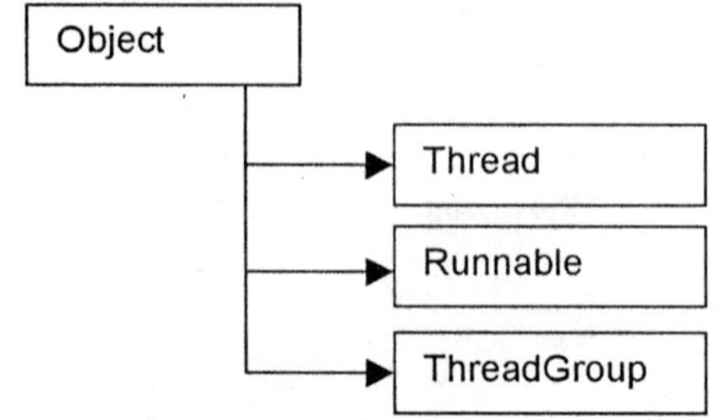

FIGURE 16.4
`Thread` class, `ThreadGroup` class, and `Runnable` interface hierarchy.

CREATING THREAD OBJECTS

The **`Thread`** class models a program thread. The class has several constructors, summarized in Table 16.1.

When a name is not specified, newly created threads are automatically named "`Thread-n`" where `n` is an integer value. Selected methods of the `Thread` class are summarized in Table 16.2

An object of the `Thread` type class is a thread. Similarly, an object of a class inheriting the `Thread` class is a thread. The class `ThreadType` (Figure 16.5) extends the `Thread` class. The entry point of a thread type is the `run` method. The `run` method is called by the Java Virtual Machine when the object's `start` method is invoked. The thread's `run` method contains the code that is executed when the `run` method is called. In this case the `run` method is a loop in which the number of iterations conceptually represents the length of time needed to complete the thread's program steps.

TABLE 16.1
Selected `Thread` class constructors

Name	Description
`Thread()`	Creates a new thread.
`Thread(String` *`name`*`)`	Creates a new thread named *`name`*.
`Thread(ThreadGroup` *`threadGroup`*`)`	Creates a new thread named belonging to the thread group *`threadGroup`*.
`Thread(ThreadGroup` *`threadGroup`*`, String` *`name`*	Creates a new thread named belonging to the thread group *`threadGroup`* and named *`name`*.

TABLE 16.2

Selected `Thread` class methods

Name	Description
`Thread currentThread( )`	Returns a reference to the currently executing thread.
`void destroy( )`	Destroys the thread.
`String getName( )`	Returns the thread's name.
`int getPriority( )`	Returnsthe thread's priority.
`ThreadGroup getThreadGroup( )`	Returns the thread's thread group.
`void interrupt( )`	Interrupts the currently executing thread.
`boolean isAlive( )`	Returns `true` if the thread has been started and has not died, `false` otherwise.
`boolean isDaemon( )`	Returns `true` if the thread is daemon and `false` otherwise.
`void join( )`	Waits for the thread to die.
`void join(long milsec)`	Waits at most *milsec* milliseconds for the thread to die.
`void join(long milsec, int` *nansec*`)`	Waits at most *milsec* milliseconds plus *nansec* nanoseconds for the thread to die.
`void resume( )`	Resumes a suspended thread.
`void run( )`	Calls the run method of a Runnable object if existing, otherwise nothing.
`void setDaemon(boolean` *setOn* `)`	Marks the thread as daemon when *setOn* is `true`.
`void setName(String name)`	Changes the thread name to *name*.
`void setPriority(int` *newPri* `)`	Changes the thread priority to the value *newPri*.
`void sleep(long` *milsec*`)`	Causes the thread to suspend execution for *milsec* milliseconds.
`void sleep(long` *milsec*`, int` *nansec*`)`	Causes the thread to suspend execution for *milsec* milliseconds plus *nansec* nanoseconds.
`void start( )`	Causes the thread to begin execution.
`void stop( )`	Causes the thread to end execution.
`void suspend( )`	Causes the thread to suspend execution until it is resumed.
`void yield( )`	Causes the thread to pause enabling other threads to execute.

In this case, the thread is a looping process that performs `noOperation` iterations. During each loop the percentage of completion of the process is computed with the statement

```
percentComplete = 100*i/noOperations;
```

The value is evaluated as the percentage of the loop completed, specified by the variable i, relative to the number of loop required to complete the process, specified by the variable `noOperations`. The computed value is stored in the variable `percentComplete`. Additionally, in each iteration the thread pauses:

```
sleep(100);
```

```
import java.awt.*;
import java.awt.event.*;
public class ThreadType extends Thread
{
      private int percentComplete;
      private int noOperations;
      public ThreadType(int no){
           noOperations = no;
      }
      int getPercentComplete(){
           return percentComplete;
      }

      public void run(){
           try {
                for (int i = 1; i < noOperations; i++)
                {
                     percentComplete = 100*i/noOperations;
                     sleep(100);
                }
           }
           catch(InterruptedException ie){
                return;
           }
      }
}
```

FIGURE 16.5
Class `ThreadType` extends class `Thread`.

In this case, the thread pauses running for 100 milliseconds. Other threads can run during this pause in operation. The `sleep` method throws an `InterruptedException` if the pause in running is interrupted. The exception is caught in the following `catch` in which control is simply returned.

The number of loop iterations performed by the thread is passed as an argument in the thread's constructor and stored in the variable `noOperations`.

The thread's percentage of completion can be retrieved while the thread is active by calling the method `getPercentComplete`.

> Objects of the ThreadType class are threads.

USING THREADS

A thread is launched or spawned by creating a thread object. For example, the statements

```
ThreadType redThread = new ThreadType(200);
ThreadType blueThread = new ThreadType(100);
```

create separate threads, `redThread` and `blueThread`, of the `ThreadType` class, which was declared in Figure 16.5. Note that the `redThread` type represents a thread with 200 operations whereas the `blueThread` represents a thread with only 100 operations. The `redThread` represents an operation that will take twice as long to complete as the `blueThread`. The threads are started by invoking the thread's `start` method; for example

```
redThread.start();
```

and

```
blueThread.start();
```

The Java Virtual Machine will call the thread's `run` method. The thread then runs until its state is changed.

An application illustrating use of threads is shown in Figure 16.6a. In this application, two `ThreadType` objects are launched representing a long red operation and a shorter blue.operation. Although the `redThread` is started first, the `blueThread` finishes first as the `redThread` sleeps periodically, enabling the `blueThread` to run. Similarly, the `blueThread` also sleeps periodically enabling the `redThread` to run for a period of time.

```
import java.awt.*;
import java.awt.event.*;
public class UsingThreads extends Frame
      implements ActionListener, WindowListener
{
      public int percentCompleteRed, percentCompleteBlue;
      Button bStart = new Button("Start");
      boolean startFlag = false;
      ThreadType redThread = new ThreadType(200);
      ThreadType blueThread = new ThreadType(100);
      public static void main(String[] args){
           UsingThreads ut = new UsingThreads();
           while(true){
                     ut.repaint( );
           }
      }
      UsingThreads( ){
           setTitle("Using Threads");
           setSize(300,200);
           setLayout(null);
           addWindowListener(this);
           add(bStart);
           bStart.setBounds(20,40,40,30);
           bStart.addActionListener(this);
           setVisible(true);
```

```
	}

	public void paint(Graphics g){
		if(startFlag){
			percentCompleteRed =
				redThread.getPercentComplete() ;
			g.setColor(Color.red);
			g.drawLine(100,100,
100 + percentCompleteRed,100);
			percentCompleteBlue =
				blueThread.getPercentComplete() ;
			g.setColor(Color.blue);
			g.drawLine(100,120,
100 + percentCompleteBlue,120);
		}
		formatFrame(g);
	}

	private void formatFrame(Graphics g) {
		g.setColor(Color.black);
		g.drawString("Red Thread",20,100);
		g.drawString("Blue Thread",20,120);
		g.drawLine(100,125,100,130);
		g.drawLine(200,125,200,130);
		g.drawString("0%",100,145);
		g.drawString("100%",200,145);
	}

	public void actionPerformed(ActionEvent ae){
		if(ae.getSource()==bStart){
			redThread.start();
			blueThread.start();
			startFlag = true;
		}
	}
	public void windowClosing(WindowEvent we){
		setVisible(false);
		dispose();
		System.exit(0);
	}
	public void windowActivated(WindowEvent we){}
	public void windowClosed(WindowEvent we){}
	public void windowDeactivated(WindowEvent we){}
	public void windowDeiconified(WindowEvent we){}
	public void windowIconified(WindowEvent we){}
	public void windowOpened(WindowEvent we){}
}
```

FIGURE 16.6A
Spawning `ThreadType` threads.

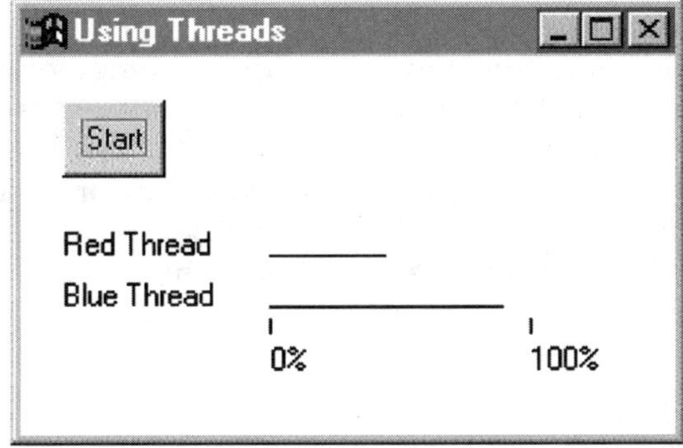

FIGURE 16.6B
Output of application spawning `ThreadType` threads.

The two threads are started by clicking a Start button. The thread's `run` methods are called by the Java Virtual Machine. The value of each thread's `percentComplete` variables is continually displayed as a graphical line in the frame window (Figure 16.6b).

Although the relatively larger `redThread` was started first, the `blueThread` is completing faster. Therein lies the value of threads: multiple threads can be executed concurrently, enabling fast throughput on threads representing short jobs.

> The first thread started need not be the first thread to end.

THREAD CHARACTERISTICS

Threads have types, states, priorities and can be synchronized.

Types

There are two types of threads, regular and daemon. Regular threads are user-type threads. An application ends when all user threads have stopped executing. Daemon threads are threads that continue running even after all other threads have stopped. The garbage collector is a daemon thread. A thread inherits the daemon property. In addition, a thread can be set to daemon status by invoking the `setDeamon` method before the thread has been started. An `IllegalThreadStateException` is thrown if the `setDeamon` method is invoked after the thread has been started. The `isDeamon` method can be used to ascertain if a thread is daemon.

Priorities

All blocked threads compete for access to the computer's processor. The processor can only execute one thread at a time. The processor shows preference to threads based on thread priority. In general, threads with a higher priority are executed before threads with a lower priority. Threads inherit priority. In addition, priority can be set with the `setPriority` method. The `Thread` class includes member fields defining the range of priority values. The members are summarized in Table 16.3.

TABLE 16.3
Thread fields declaring priority values

Name	Description
`MIN_PRIORITY`	Lowest priority
`NORM_PRIORITY`	Default priority
`MAX_PRIORITY`	Highest priority

Assigning a priority higher or lower than `MAX_PRIORITY` and `MIN_PRIORITY` causes an exception. The priority of a thread is determined by invoking the `getPriority` method.

The `yield` and `join` methods also determine how a thread gains access to the processor. A thread gives up access to the processor to threads of equal or higher priority when the `yield` method is invoked. A thread invoking the `join` method is suspended until the joined thread dies.

> The processor executes threads with higher priority first.

Synchronizing Threads

Consider the situation in which two different threads have access to the same resource. The result can often depend on which thread accesses the resource first. In cases where access must be ordered and controlled, Java uses the **`synchronize`** keyword. Using the `synchronized` keywords effectively locks an object so that a thread cannot enter a synchronized method if another thread is paused in any synchronized method with the same object. The syntax and operation of the `synchronize` statement is shown in Figure 16.7.

A block of code containing the `synchronize` keyword is executed in three steps. First the expression following the keyword `synchronize` is evaluated. The resulting object referenced by the expression is locked, thereby refusing access to other threads. Then the block of code is executed. Finally, the lock is released.

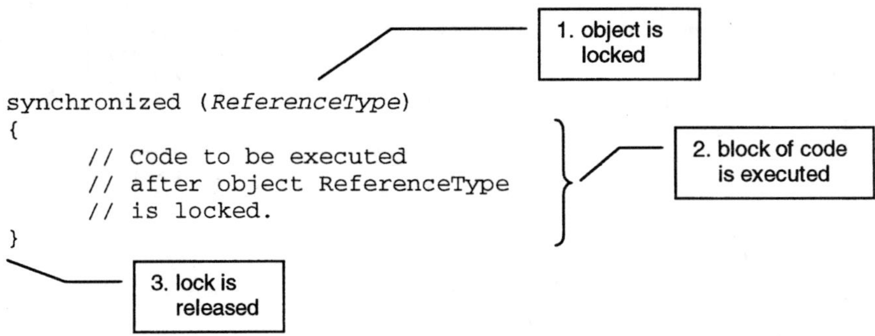

FIGURE 16.7
Syntax and operation of the `synchronize` statement.

TABLE 16.4
`Object` class methods facilitating thread management

Name	Description
`void notify()`	Wakes up a single thread waiting on the object's monitor.
`void notifyAll()`	Wakes up all threads waiting on the object's monitor.
`void wait()`	Waits to be notified to assume ownership of the object's monitor and resume execution.
`void wait(long` *`maxTime`*`)`	Waits at most *`maxTime`* in milliseconds to be notified to assume ownership of the object's monitor and resume execution.
`void wait(long` *`maxTime`*`, int` *`addTime`*`)`	Waits at most *`maxTime`* in milliseconds and *`addTime`* in microseconds to be notified to assume ownership of the object's monitor and resume execution.

Serializing Threads

In some processes, it may be necessary to ensure that a thread waits for another thread to finish with a resource before it accesses the resource. The `Thread` class inherits `Object` class methods that can be used to manage threads including serializing thread access to a resource. These methods are summarized in Table 16.4.

Thread States

Threads are started, suspended, resumed, stopped, and destroyed. The impact of these methods on a thread object are described by the thread state diagram illustrated in Figure 16.8. A thread is said to be in the **New state** when it has been created but not started by invoking the `start` method.

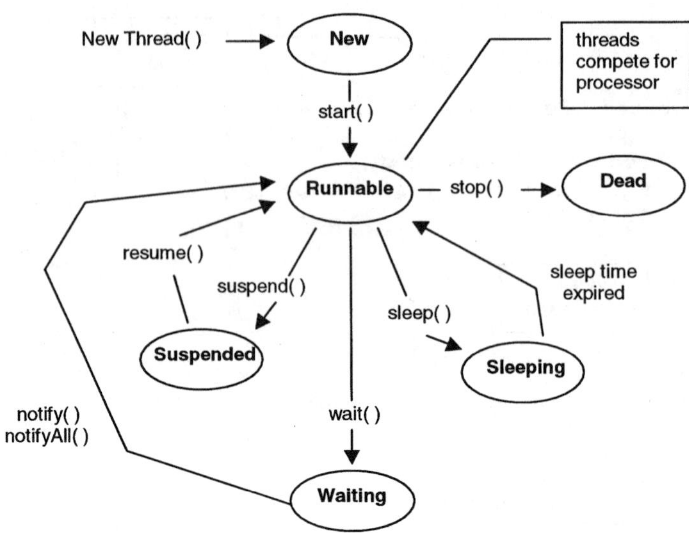

FIGURE 16.8
Thread state diagram.

When the thread's start method is invoked the thread moves to the **Runnable state.** A thread in the **Runnable state** competes with other threads for access to the computer's processor. A thread is moved to the **Sleeping state** by invoking the `sleep` method. Threads remain in the Sleeping state until the sleep time expires. Threads are moved to the Suspended state with the `suspend` method. Threads remain in the **Suspended state** until they are moved to the Runnable state by the `resume` method. Additionally, threads can be placed in the **Waiting state** with the `wait` method. Threads remain in the Waiting state until returned to the Runnable state by the `notify` or `notifyAll` methods. When the `run` method ends or the `stop` method is invoked the thread moves to the Dead state. A thread in the **Dead state** is no longer executing. Threads in the Dead state cannot be resurrected and returned to the Runnable state. A **ThreadDeath** exception is thrown when the stop method is invoked for a dead thread.

A thread that has been started and has not been stopped is said to be **alive.** The `isAlive` method can be used to ascertain the state of a thread. The method returns a `true` value if the thread is alive and `false` otherwise. In time, dead threads are destroyed by the Java Virtual Machine.

THREAD GROUPS

A thread group is a set of threads. Thread groups can include other thread groups in a tree-like structure. A thread group can access information regard-

TABLE 16.5

Selected `ThreadGroup` constructors

Name	Description
`ThreadGroup(String name)`	Creates a thread group whose parent is the currently running thread and with name *name*.
`ThreadGroup(ThreadGroup parent, String name)`	Creates a thread group whose parent is the thread group *parent* and whose name is *name*.

ing itself but cannot access information about other thread groups or its parent thread group. Thread groups are modeled by the ThreadGroup class. Constructors of the ThreadGroup class are summarized in Table 16.5.

`ThreadGroup` class methods enable thread group access including the retrieval of information pertaining to the group, setting the group's maximum priority, as well as resuming, suspending, and stopping threads in the group. Methods of the `ThreadGroup` class are summarized in Table 16.6.

Thread groups can be given Daemon status. A thread group with Daemon status is automatically destroyed when all of its threads have stopped or its last thread group is destroyed.

RUNNABLE INTERFACE

Any class whose instances are to be intended to be executed as threads should implement the **Runnable interface.** The class `Thread` implements the interface `Runnable`. Every class implementing the interface `Runnable` must declare a method `run`. The declared `run` method should not have any arguments and is called by the thread's `start` method. An application illustrating use of the `Runnable` interface to implement threads is shown in Figure 16.9a. The class `ImplementsRunnable` implements the `Runnable` interface

```
public class ImplementsRunnable
    implements Runnable
```

A thread reference `irThread` is declared as a member of the class

```
Thread irThread;
```

An instance of the thread is created, named, and started in the class's constructor with the statements

```
irThread = new Thread(this, "irThread");
irThread.start( );
```

The `run` method is called automatically by the Java Virtual Machine in response to the `start` method called in the constructor. The `run` method, in

TABLE 16.6

Selected ThreadGroup methods

Name	Description
int activeCount( )	Returns the number of active threads in the thread group.
int activeGroupCount( )	Returns the number of active thread groups parented by the thread group.
void destroy( )	Destroys the thread group and all subgroups.
int enumerate(Thread[] *list*)	Copies all active threads in the group and its subgroups into the array *list*. Returns the number of threads copied into the array list.
int enumerate(Thread[] *list*, boolean *include*)	Copies all active threads in the group into the array *list*. Active threads in subgroups are included if the argument *include* is true. Returns the number of threads copied into the array list.
int enumerate(ThreadGroup[] *list*, boolean *include*)	Copies references to all active thread subgroups into the array *list*. References to active thread subgroups of the subgroups are included if the argument *include* is true. Returns the number of thread subgroup references copied into the array list.
int enumerate(ThreadGroup[] *list*)	Copies references to all active thread subgroups into the array *list*. Returns the number of thread subgroup references copied into the array list.
int getMaxPriority( )	Returns the maximum priority that a thread in the group can have.
String getName( )	Returns the name of the invoking thread group.
ThreadGroup getParent( )	Returns the parent thread group.
boolean isDaemon( )	Returns true if the thread group is Daemon and false otherwise.
boolean isDestroyed( )	Returns true if the thread group has been destroyed and false otherwise.
void list( )	Lists information about the thread group to the Standard output device.
boolean parentOf(ThreadGroup *trdgrp*)	Returns true if the thread group is the parent of the thread group *trdgrp* and false otherwise.
void resume( )	Resumes all processes in the thread group.
void setDaemon(boolean *state*)	Marks the thread group as Daemon if *state* is true and not Daemon or normal if *state* is false.
void setMaxPriority(int *priority*)	Sets the thread group priority to the value *priority*. Threads in the group with a higher authority are not effected by the change.
void stop( )	Stops all process in the thread group by calling the stop method for all threads in the group and subgroups.
void suspend( )	Suspends all process in the thread group by calling the suspend method for all threads in the group and subgroups.

this simplified case, is a nine-iteration loop. During each iteration of the loop, the message "irThread running" is displayed at the Standard output and the thread sleeps for one second. Note the `sleep` method is a checked exception and thus is enclosed in a `try` block and accompanying `catch` block.

```
public class ImplementsRunnable
      implements Runnable
{
      Thread irThread;

      public static void main(String[] args)
      {
           System.out.println("Main starting");
           ImplementsRunnable ir = new ImplementsRunnable( );
           Thread.currentThread().getThreadGroup().list();
           System.out.println("Main ending");
      }

      ImplementsRunnable( ){
           irThread = new Thread(this, "irThread");
           irThread.start( );
      }

      public void run(){

           for(int i = 1; i < 10; i++){
                System.out.println("irThread running");
                try {
                     Thread.sleep(1000);
                }
                catch (InterruptedException ie){
                     return;
                }
           }
      }
}
```

FIGURE 16.9A
Using the `Runnable` interface to implement a thread.

The Figure 16.9a application's output is shown in Figure 16.9b. The actual output may vary somewhat depending on the time required by the various components of the processes to complete. The relationship between the application's output and the spawned threads is illustrated in Figure 16.9c.

Note that when thread `irThread` sleeps, the other thread, `main`, is provided the opportunity to execute. This application clearly illustrates the use of threads to enable multiple processes to executed "concurrently."

Improved throughput is provided by the use of threads.

```
JVIEW ImplementsRunnable
Main starting
irThread running
java.lang.ThreadGroup[name=main,maxpri=10]
    Thread[main,5,main]
    Thread[irThread,5,main]
irThread running
Main ending
irThread running
irThread running
irThread running
irThread running
irThread running
irThread running
irThread running
```

FIGURE 16.9B
Output of application using the `Runnable` interface to implement a thread.

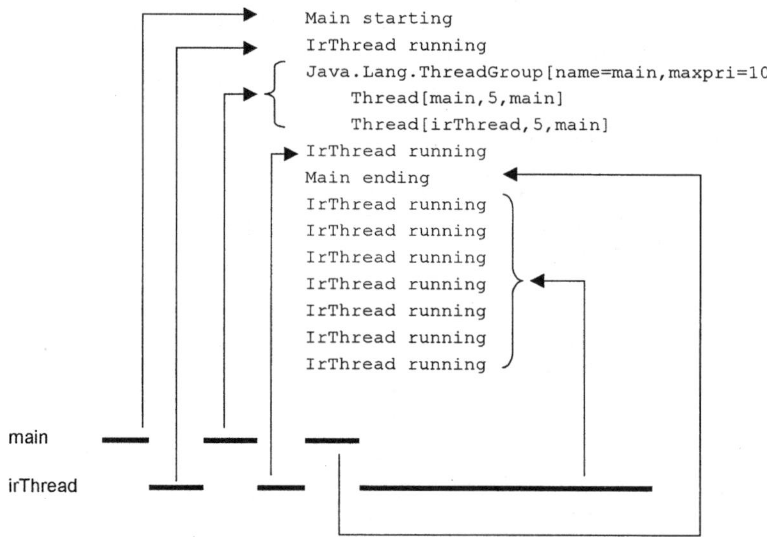

FIGURE 16.9C
Relating output of application using the `Runnable` interface to implement a thread to running threads.

HANDS-ON EXERCISE 16: IMPLEMENTING A PANNING WELCOME

The Web page, shown in Figure HO-16.1, contains a panning welcome marquee—a current must in all Web site home pages. The message, "Welcome to Java!" moves from left to right across the top of the page.

HO-16.1
Web page with panning Welcome marquee.

Step 1: Create Visual J++ Project Panning Welcome

1. Click the New Project option in the File menu.
2. Enter **`Panning Welcome`** in the Name text field and **`C:\Hands-On\Panning Welcome`** in the Location text field.
3. Click the Open button to create the Panning Welcome project.

Step 2: Create the Welcome.java File

1. Click the Add Class... option in the Project menu.
2. Enter **`Welcome.java`** in the Name text field.
3. Click the Open button to create the `Welcome.java` file. The Text Editor opens with the `Welcome.java` file.

Step 3: Declare the Welcome Class

1. Open the Welcome.java file in the Text Editor (if not proceeding directly from the previous step).

2. Enter the code shown in Figure HO-16.2.

```
import java.applet.*;
import java.awt.*;
public class Welcome extends Applet
      implements Runnable
{
      Thread marqueeThread = null;
      String sText = "Welcome to Java!";
      int iOffset = 0;
      public void paint(Graphics g) {
           g.setFont(new Font("Seirf",Font.BOLD +

Font.ITALIC,36));
           g.setColor(new Color(255,0,0));
           g.drawString(sText, iOffset, 40 );
           iOffset++;
           }
      public void init() {
           if (marqueeThread == null )
           {
                marqueeThread = new Thread(this);
                marqueeThread.start();
                resize(400,150);
           }
      }
      public void stop() {
           if (marqueeThread != null ) {
                marqueeThread.stop();
                marqueeThread = null;
           }
      }
      public void run() {
           int nSleepTime = 20;
           while(true) {
                try {
                     repaint();
                     marqueeThread.sleep(nSleepTime);
                }
                catch (InterruptedException e) {
                     stop( );
                }
           }
      }
}
```

FIGURE HO-16.2
Welcome class declaration.

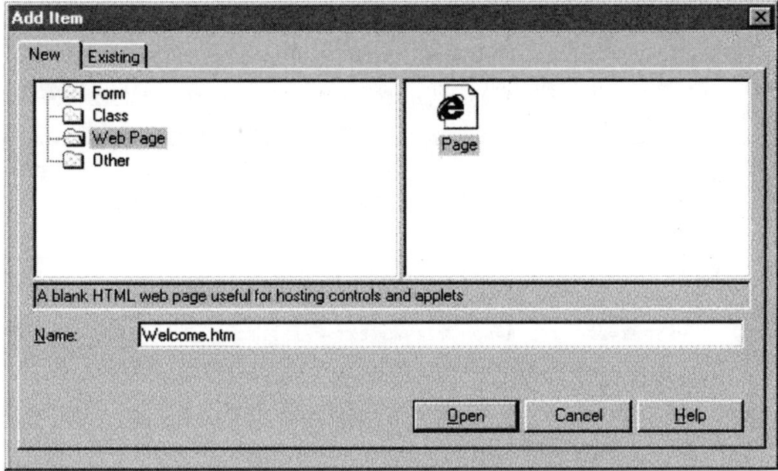

FIGURE HO-16.3
Add Item dialog box opened with Web Page folder selected.

Step 4: Create the Welcome HTML Page

1. Click the Add Web Page… in the Project menu. The Add Item Dialog Box is displayed with the Web Page folder opened in the New Tab window pane (Figure HO-16.3).
2. Enter **`Welcome.htm`** in the Name text field.
3. Click the Open button. The file Welcome.htm is opened in the HTML Editor.

Step 5: Edit the `Welcome.htm` File

1. Click the Source tab in the HTML Editor opened with the `Welcome.htm` file.
2. Enter the script

```
<APPLET CODE=Welcome.class
   HEIGHT=200 WIDTH=450 VI WASTEXT></APPLET>
```

in the BODY. The edited `Welcome.htm` code is shown in Figure HO-16.4.

Step 6: Set Project Properties

1. Select Panning Welcome Properties… in the Project menu.
2. The Planning Welcome Properties dialog box is displayed (Figure HO-16.5).

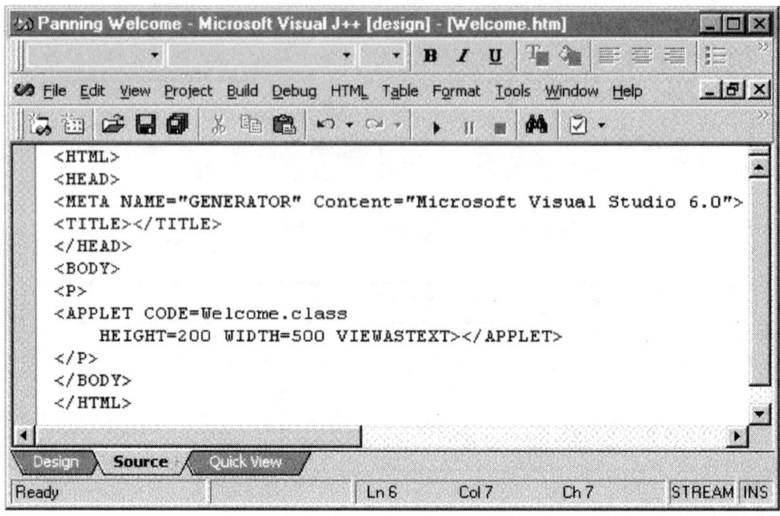

FIGURE HO-16.4
`Welcome.htm` file opened in the HTML Editor source view.

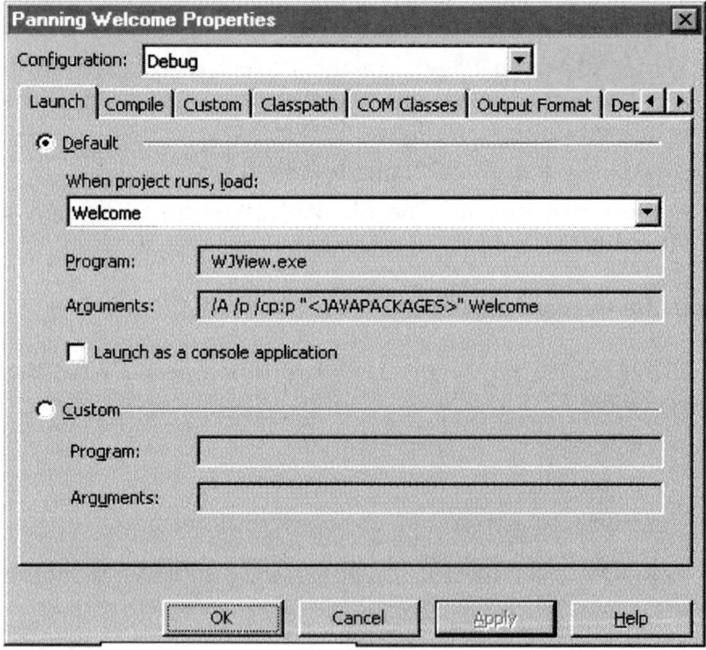

FIGURE HO-16.5
Project Properties dialog box.

3. Select Welcome in the When the project runs, load: choice box.
4. Click the Apply button.
5. Click the OK button.

Step 6: Execute the Application

1. Click the Start option in the Debug menu. The Welcome applet is displayed in the Applet Viewer.
2. Note that the welcome message pans across the screen from left to right.

Step 7: Observe Panning Message Properties

Does the panning stop when:

a. The window is resized?
b. The window is maximized or minimized?
c. Another window is given the focus?

Step 8: Close Project

1. Click the Exit button in the right corner of the Applet Viewer title bar to stop the applet and close the viewer. Alternately, click the Quit option in the Applet menu.
2. Select the Close All option in the File menu.
3. Select the Exit option in the File menu.

SUMMARY

1. Java uses threads to implement multitasking.
2. Threads are implemented with the `Thread` class and the `Runnable` interface.
3. Threads have priorities that can be used to control processor access.
4. Threads can be synchronized.
5. Threads can be serialized using the `wait`, `notify`, and `notifyAll` methods.
6. Throughput can be improved by using threads.

KEY WORDS AND CONCEPTS

alive
Daemon thread
Dead state
multiprocessing
multitasking
New state
process
`Runnable` interface
Runnable state
Sleeping state
Suspend state
synchronize
thread
`Thread` class
thread group
`ThreadDeath` exception
`ThreadGroup` class
time slicing
Waiting state

REVEIW QUESTIONS AND EXERCISES

1. Mark the following true or false:
   a. Multiple threads can be executed by the processor simultaneously.
   b. Threads can improve throughput.
   c. Threads cannot be used where processes must be synchronized.
   d. Newly created threads start automatically.
   e. Thread functionally is expressed in the run method.
   f. Threads in the Dead state can be resurrected.
2. Modify the application Using Threads to spawn additional threads with various priorities. Answer the following questions:
   a. Will a higher priority thread with a large number of operations finish first?
   b. Will a low priority thread with a low number of operations finish first?
3. Create an application to demonstrate synchronization.

PRACTICE WITH VISUAL J++

1. Use Thread Window.
   - Open the Implementing Runnable application.
   - Position the cursor at the first statement in the `main` method.
   - Select the Run To Cursor option in the Debug menu.
   - Select the Debug Windows option in the View menu. A submenu with window options is displayed.
   - Select the Threads option in the Debug Windows submenu. The Threads window is displayed (Figure 16.10).
   - Step through the application and observe the Threads window.

2. Modify the Welcome to include another marquee panning at a different rate and vertical offset.
3. ***Challenge:*** Modify the applet to pan the message from right to left. When the message slides to the left side, restart the panning message on the right side.

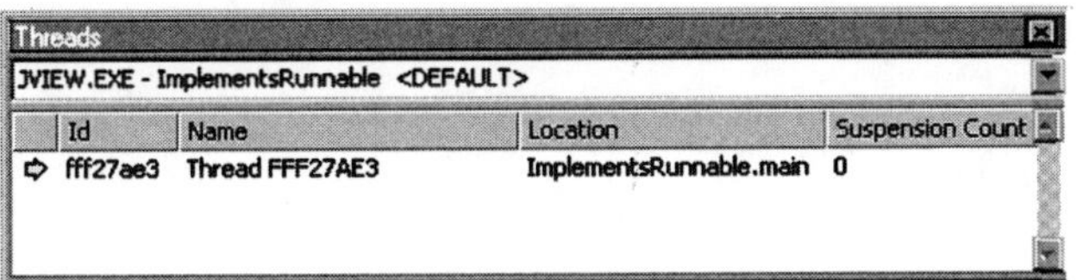

FIGURE 16.10
Threads window.

Appendixes

Appendix A

ASCII Character Codes

The American Standard Code for Information Interchange (ASCII) is the character code adopted by the American Standards Association. The 8-bit character set is capable of representing the characters required for normal discourse, communication control, and error checking. Originally, seven bits were used to represent printable and control characters and the 8th bit was used for parity checking.

The first thirty-two characters are used to control transmission—for example, acknowledgment and negative acknowledgment—and formatting such as line feeding, bell ringing, and backspacing. These characters are not printable. They appear on numerous display devices as periods.

Control and Formatting Characters

Code			Code		
Character	**Decimal**	**Hex**	**Character**	**Decimal**	**Hex**
NULL	00	00	DLE	16	10
SOH	01	01	DCI	17	11
STX	01	02	DC	18	12
ETX	03	03	DC3	19	13
EOT	04	04	DC4	20	14
ENQ	05	05	NAK	21	15
ACK	06	06	SYN	22	16
BEL	07	07	ETB	23	17
BS	08	08	CAN	24	18
HT	09	09	EM	25	19
LF	10	0A	SUB	26	1A
VT	11	0B	ESC	27	1B
FF	12	0C	FS	28	1C
CR	13	0D	GS	29	1D
SO	14	0E	RS	30	1E
SI	15	0F	US	31	1F

Printable characters include the following:

Printable Characters

	Code			Code	
Character	**Decimal**	**Hex**	**Character**	**Decimal**	**Hex**
Space	32	20	P	80	50
!	33	21	Q	81	51
"	34	22	R	82	52
#	35	23	S	83	53
$	36	24	T	84	54
%	37	25	U	85	55
&	38	26	V	86	56
'	39	27	W	87	57
(	40	28	X	88	58
)	41	29	Y	89	59
*	42	2A	Z	90	5A
+	43	2B	[	91	5B
`	44	2C	\	92	5C
-	45	2D	]	93	5D
,	46	2E	^	94	5E
/	47	2F	-	95	5F
0	48	30	"	96	60
1	49	31	a	97	61
2	50	32	b	98	62
3	51	33	c	99	63
4	52	34	d	100	64
5	53	35	e	101	65
6	54	36	f	102	66
7	55	37	g	103	67
8	56	38	h	104	68
9	57	39	i	105	69
:	58	3A	j	106	6A
;	59	3B	k	107	6B
<	60	3C	l	108	6C
=	61	3D	m	109	6D
>	62	3E	n	110	6E
?	63	3F	o	111	6F
@	64	40	p	112	70
A	65	41	q	113	71
B	66	42	r	114	72
C	67	43	s	115	73
D	68	44	t	116	74
E	69	45	u	117	75
F	70	46	v	118	76
G	71	47	w	119	77
H	72	48	x	120	78
I	73	49	y	121	79
J	74	4A	z	122	7A
K	75	4B			
L	76	4C			
M	77	4D			
N	78	4E			
O	79	4F			

With the introduction of the PC, IBM extended the character codes to include 128 additional characters. In the extended character set the parity bit is used for additional character codes rather than error detecting. The extended characters include printable alphanumeric and graphical characters

Java uses the **Unicode** 16-bit character set, which is capable of representing characters worldwide including accented and ideographic characters. Unicode is a superset of the ASCII character set. The first 128 Unicode characters are ASCII characters.

Appendix B

Managing Project Files

A Visual J++ project consists of several files. The files associated with a typical project are shown in the Window Explorer, Figure B.1. Each file and folder in the project corresponds to a file and folder on the computer's hard disk.

FIGURE B.1
Project files displayed in Window Explorer.

FIGURE B.2
Project files displayed in Project Explorer.

Visual J++ has a Project Explorer window that also displays the files comprising a project (Figure B.2). Projects have the following files:

- A **project file,** which keeps track of the project's components, and is stored with the `.vjp` file extension, in this case `Welcome.vjp`.
- One or more **class files,** each of which describes one of the project's classes, stored with the `.java` file extension, in this case `Greeting.java`.

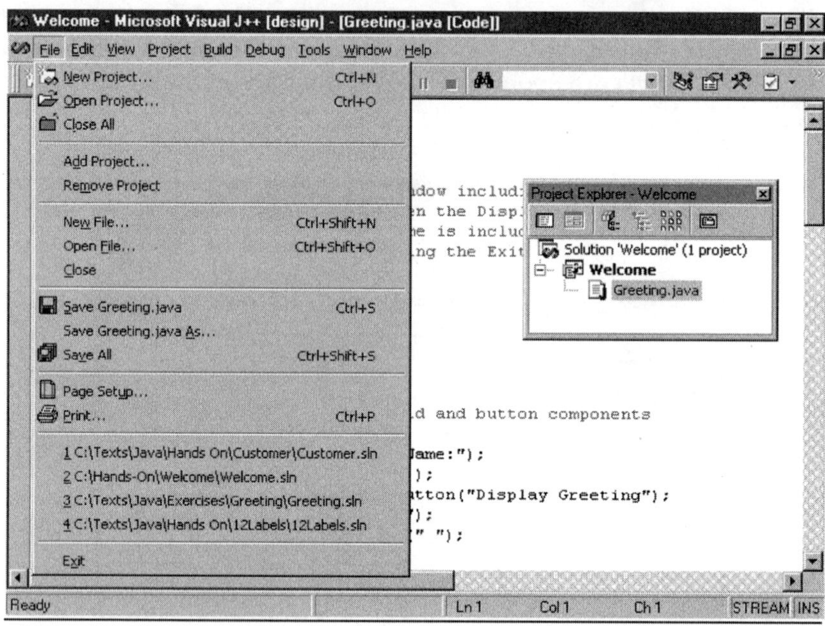

FIGURE B.3
File menu.

VISUAL J++ FILE MENU

Commands for managing the project's files are contained in the File menu (Figure B.3). The menu contains the commands for creating, opening existing, and closing projects; creating, opening existing, closing, and saving the project's files; printing; and exiting the development environment.

New Project

The **New Project... command** is selected to create a new project. When selected, the New Project window, Figure B.4, is displayed with the New tab selected.

The New tab displays window panes enabling the selection of templates for creating Windows applications, applets, controls, database applications, utility applications, and distribution facilities. A template category folder is selected in the left pane and a specific template is selected in the right pane. The folders are expanded and collapsed by clicking the + and – signs. In Figure B.3, the Visual J++ Projects folder is selected in the left pane. The Empty Project template is selected in the right pane. In general, these templates generate skeletal files, forms, and program code typically used in the specified application. The project's name is entered in the Name textbox. The name is added to the full path name specified in the Location textbox. The project's files are stored in a folder at the location specified. The Browse button is clicked to display a window enabling navigating to a specific location.

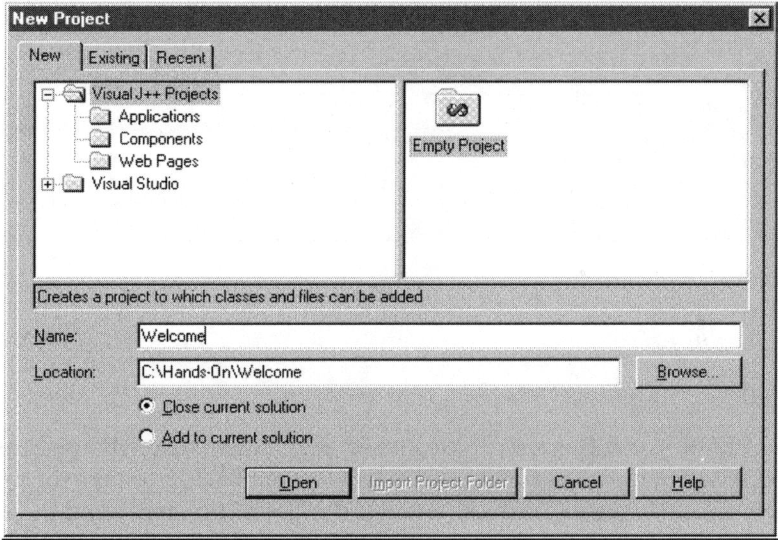

FIGURE B.4
New Project window.

Selecting the Existing tab results in the display of windows that enable the identification and selection of existing projects.

The Recent tab displays a window that lists the projects that have been recently opened.

Open Projects

When the **Open Project... command** is selected, the New Project window is displayed with the Existing tab selected. The Look in dropdown list and Move-Up folder icon provide for navigating the systems directories. The Files of Type dropdown list enables specification of the type of files that will be listed. Files are listed in the large window. A selected file is listed in the File Name textbox and is opened when the Open button is clicked.

In Figure B.5, the Visual J++ project file `Welcome.vjp` is selected and opened when the Open button is clicked.

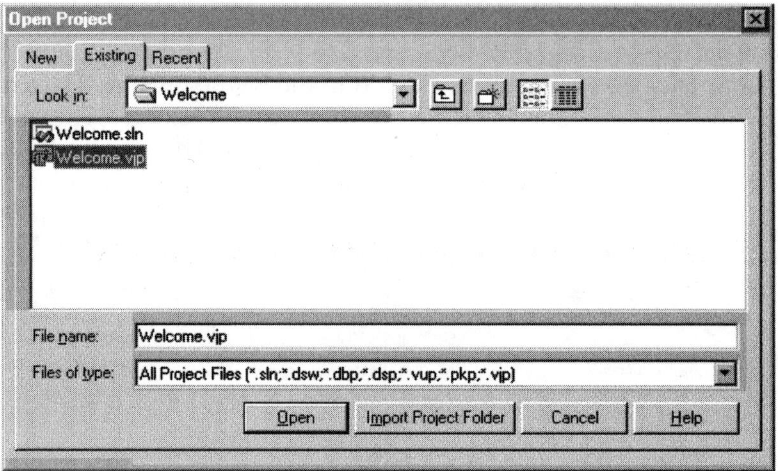

FIGURE B.5
Open Project window.

Close All

All of the files and the project are closed when the **Close All command** is selected. The Microsoft Visual Studio may prompt the User to save the project and form files. The listed files are saved if the Yes button is clicked.

After the files are saved, the project is closed.

Add Project

The **Add Project command** is used to add a project to the existing project.

Delete Project

An existing project component project of a project is removed with the **Delete Project command.**

New File

The New File window with the New tab selected is displayed when the **New File command** is selected. Folders, for various types of files, are shown in the left window pane. When a folder is selected, the specific file types in the folder are listed in the right pane. When a type of file is selected in the right pane and the Open button is clicked the file is created and opened for editing.

The Existing tab and Recent tab enable the user to open existing and recent files and add them to the project.

Open File

The **Open File command** displays the Open File window with the Existing tab selected. The window enables the selection and opening of existing files. New and Recent tabs provide for opening new and recent files.

Close File

The file associated with the currently selected project item is closed when the **File command** is selected. For example, if a form is selected for editing in the Form Designer, the Form Designer will be closed if the Close command is selected.

Save

The **Save command** saves the currently selected project, form, or code component. All programmers are familiar with the exasperation and frustration associated with the loss of hours of work resulting from a system hang-up that requires rebooting or a power loss that unilaterally shuts down the computer. Each occurrence results in the solemn pledge to routinely save all files periodically. After many instances of this type, most programmers learn to save their files periodically by regularly clicking the Save command option.

Take the pledge now—

I will save my files regularly!

The currently selected project item file name is added to the Save command in the menu.

The **Save As command** provides for saving the selected item to a new file.

All of the project's files are saved when the **Save All command** is selected.

Rename

A selected item in the Project Explorer window can be renamed when the **Rename command** is selected.

Print

Selected project items, such as a form's code, are printed with the **Print command.** The **Page Setup command** provides for selecting the printer; print properties such as page size, margins and orientation; and number of copies.

Recent Files

Recently used Visual J++ files are listed at the bottom of the File menu above the Exit command. Clicking a entry will result in the file opening.

Exit

The **Exit command** closes the current project and exits Visual J++. The system prompts the saving of any changed project files prior to closing.

File Menu Shortcuts

Many of the File menu commands can be executed directly by pressing selected keys. These shortcuts are summarized in Table B.1.

TABLE B.1

Selected File Menu Shortcuts

Key sequence	Command
Ctrl + N	Open New Project
Ctrl + Shift + N	Open New File
Ctrl + Shift + O	Open File
Ctrl + S	Save
Ctrl + Shift + S	Save All

The File menu commands are used to manage the project's files and folders.

Index